THE MEGASTRUCTURE OF THE EURASIAN WORLD

through the Prism of Geology, Archaeology, and History

THE MEGASTRUCTURE OF THE EURASIAN WORLD

through the Prism of Geology, Archaeology, and History

Evgenij Chernykh

ACADEMIC STUDIES PRESS

BOSTON

2025

Library of Congress Cataloging-in-Publication Data

Names: Chernykh, E. N. (Evgeniĭ Nikolaevich), 1935- author
Title: The megastructure of the Eurasian world through the prism of
  geology, archaeology, and history / Evgenij Chernykh.
Other titles: Megastruktura Evraziĭskogo mira skvoz' prizmu geologii,
  arkheologii, istorii. English
Description: Boston : Academic Studies Press, 2025. | Includes
  bibliographical references and index.
Identifiers: LCCN 2025016144 (print) | LCCN 2025016145 (ebook) | ISBN
  9798897830022 hardback | ISBN 9798897830039 adobe pdf | ISBN
  9798897830046 epub
Subjects: LCSH: Prehistoric peoples--Eurasia | Social archaeology--Eurasia
  | Geology, Stratigraphic--Pleistocene | Geology, Stratigraphic--Holocene
Classification: LCC GN849 .C5413 2025 (print) | LCC GN849 (ebook) | DDC
  950/.1--dc23/eng/20250616
LC record available at https://lccn.loc.gov/2025016144
LC ebook record available at https://lccn.loc.gov/2025016145

ISBN 9798897830022 (Hardback)
ISBN 9798897830039 (Adobe PDF)
ISBN 9798897830046 (ePub)

Book design by Lapiz Digital Services
Cover design by Ivan Grave

Published by Academic Studies Press
1007 Chestnut Street
Newton, MA 02464, USA
press@academicstudiespress.com
www.academicstudiespress.com

Part Two
To the Heights of Globalism

Part Three
Reflections on the Essence and Architectonics of the Surrounding World

We Are All Children of Planet Earth

About four and a half billion years ago, the mysterious and powerful forces of the cosmos gave birth to the solar system and Earth, with its "encircling" atmosphere and biosphere. This distinguished it from other neighboring planets. Cosmic forces have always influenced our planet, but the effect has never been evenly distributed. Earth was subject to various, arrhythmic powerful "explosions." These cataclysms caused reconstructions of both the inner workings of the planet as well as of its surface—the lithosphere and hydrosphere. It was during that time as well that various supercontinents formed and fragmented the planet's surface leading to the emergence of such giant areas such as Eurasia, whose megastructure is one of my core research interests.

In addition to the supercontinents and continents that we are so accustomed to today, Earth developed a biosphere and all-encompassing biocenoses, fantastic in their diversity. Approximately three or, as commonly thought today, more than three million years ago, the genus *Homo*—the Earth's brainchild—emerged in the Earth's biosphere. This genus soon rose to become the master of the planet or at least aspired and claimed to be one ("soon" in the context of the billions of years of the planet). These masters oversaw an insignificantly small fraction of the Earth's life—only 0.07% of its entire geohistory.

The ascent of humans to their leading position was gradual, as reflected in the anthropological evolution of the genus *Homo* from archaic humans such as *Homo erectus* ("upright man"), *Homo ergaster* ("working man"), and *Homo habilis* ("handy man"). This process culminated in the emergence of *Homo sapiens* ("wise man") or even *Homo sapiens-sapiens*—"very intelligent men," that is, already the modern humans. Anthropological progress was paralleled by the expansion of the *Homo* species across the planet's land and sea. Human cultures were subject to distant cosmic forces, yet humans either did not recognize them or had very diverse and original explanations for them.

Humans have pondered the mysteries of the universe—that is, the emergence of Earth and its development, the emergence of humankind, the effects of geological macrostructures on human cultures—for a very long time. It is possible that the predecessors of *Homo sapiens* also began to concern themselves with such matters, but this would be impossible to prove. Once *Homo sapiens* emerged, we know with certainty that they began to think about such issues. There have been a multitude of hypotheses developed on the subject, but they all lend themselves to a rather simple systematization which I will touch upon in the final sections of this book. Clearly, modern scholarly interpretations of these complex problems disagree and will be briefly summarized below.

Cardinal metamorphoses of the planet's geological macrostructures caused changes in worldviews of *Homo* cultures as we as human beings cannot be indifferent to its early history. We remain completely dependent on the planet and its state of being, regardless of the all-encompassing power that people imagine to have over it. It is worth mentioning, that "the Eurasian world" does not only include the continent of Eurasia as a geocosmic phenomenon, but also the innumerable human groups and their cultures that have inhabited Eurasia for millions of years.

In a certain respect, this book stems from my earlier monograph, *The Cultures of Homo: Challenging Essays about Humankind's Multi-Million Year History*, first published in Russian in 2019 and two years later translated into English (Chernykh 2021). Moreover, it is related to my even earlier works (see, for example: Chernykh 1992; Chernykh 2009; 2013, volumes 1 and 2; Chernykh 2017). The present study, however, focuses on two large groups of foundational issues which determined the megastructure of the entire Eurasian world but which have been narrowly dealt with in my earlier publications. The discussion here will overlap with some of the materials presented in my earlier works and will depend on points made previously.

Nevertheless, this monograph marks a clear shift towards reconstructing large-scale events of a geocosmic nature. Some of them occurred in the distant past (even on the scale of geological time), while others are rather recent and to this day continue to influence and impact the character and dynamics of the geomorphology and geoecology of Eurasia as well as the various human groupings and their cultures on which they depend. Therefore, readers of this book will encounter broad geological reconstructions in most chapters, but largely in those that deal with the first group of problems that outline the main planetary components of the Eurasian of geomorphology-geoecological areas of Eurasia (geoareas). In essence, these geoareas have served as geofoundations for the development and formation of the Eurasian world, as the first part of this book, "On the planetary origins of the formation of the Eurasian world," demonstrates.

In the next section, entitled "In a whirlpool of perceptions of the essence and architectonics of the surrounding world," the focus is on comparing various perceptions of planetary changes developed by human cultures and their responses to them. It offers a brief analysis of the most important ideological (religious) systems or doctrines, developed by various Eurasian groups and cultures including shamanism or animism to monotheism of the Abrahamic religions. In researching these matters, one cannot ignore geology and important geological events. It would be extremely difficult to gain a full understanding of the global Eurasian world without them.

For non-historians, a fundamental difference between history as a scholarly discipline and as a historical narrative needs to be explained. An analysis, based on written historical documents, as well as methods and techniques of history as a scholarly discipline is a scholarly historical reconstruction. A narrative produced by a certain people, culture (i.e. about a monument), or a person is a historical narrative. For our purposes, in the majority of cases, it is not necessary to describe historical research and relevant scholarship is simply cited in the footnotes. This is an interdisciplinary book, combining geological, archaeological, anthropological, ethnological, linguistic studies and aims to be seen as a historical narrative.

Finally, the remaining sections of the book focus on the so-called millennial eve of the Modern period. The reason for this choice is quite simple. This transitionary period was marked by what is now known as the Great Geographical Discoveries. The worldwide discoveries of the fifteenth to seventeenth centuries ultimately led to the sixth and last Great Intercontinental

Migration of people. It was characterized by a dramatic restructuring of the world. Eurasia gained a special new position following the discoveries of the new continents of Australia and America and the re-discovery of the African cradle of the *Homo* species. At that time, the megastructure of Eurasia seemed largely complete. A study of all the countless materials of the Modern Era would require a tremendous effort, which this rather small volume is not equipped to handle. Much attention was paid to them in a number of my recent work (see, for example: Chernykh 2009, 351–507; Chernykh 2013, volume 2, 203–330; Chernykh 2019, 227–366; Chernykh 2017, 471–618). Therefore, the temporal scope of this book does not include the Modern period. Nevertheless, the readers will find here some brief forays into the Modern period whenever they are useful to better understand and illuminate events of the past.

Acknowledgements

This book would hardly have been completed without the support of the staff of the Laboratory of Natural Scientific Methods at the Institute of Archaeology of the Russian Academy of Sciences, in which I have worked practically all my academic life. It is my great pleasure to express my gratitude to everyone who took part in the discussion of various scholarly directions of this book. I would like to thank especially Lyubov Boleslavovna Orlovskaya, who carefully edited the entire text, as she had done for all my previous publications. I also wish to express a deep gratitude to the translator of the book, Irina Vladimirovna Bogatyreva, who worked on a number of my earlier works. And, of course, I am thankful to my wife, Elena Yurievna Lebedeva, for reading with great attention the entire draft of the book and for her insightful comments on the contents and the structure of the work.

Part One

THE PLANETARY ORIGINS OF THE EURASIAN WORLD

Chapter 1

The Eurasian World in the Formation of the Megastructure

1.1. The Formation of Earth

An introductory quotation below on the formation of Earth as part of the solar system is taken from the book of the famous American geologist Robert M. Hazen (born 1948), *The Story of Earth: The First 4.5 Billion Years, from Stardust to a Living Planet*. Due to his original style, Hazen's writings might appear as philosophical reflections not common among scholars of such a well-grounded and strict science as geology:

> Our solar system, with its glowing central star and varied planets and moons, is a relative newcomer to the cosmos—a mere 4.567 billion years old. The stage was set for our planet's birth much, much earlier, at the origin of all things—the Big Bang—about 13.7 billion years ago, by the latest estimates. That moment of creation remains the most elusive, incomprehensible, defining event in the history of the universe. It was a singularity—a transformation from nothing to something that remains beyond the purview of modern science or the logic of mathematics. If you would search for signs of a creator god in the cosmos, the Big Bang is the place to start. In the beginning, all space and energy and matter came into existence from an unknowable void. Nothing. Then something. The concept is beyond our ability to craft metaphors. Our universe did not suddenly appear where there was only vacuum before, for before the Big Bang there was

no volume and no time. Our concept implies emptiness—before the Big Bang there was nothing to be empty in.

Then in an instant there was not just something, but everything that would ever be, all at once. Our universe assumed a volume smaller than an atom's nucleus. That ultra-compressed cosmos began as pure homogeneous energy, with no particles to spoil the perfect uniformity. Swiftly the universe expanded, though not into space or anything else outside it (there is no outside to our universe). Volume itself, still in the form of hot energy, emerged and grew. As existence expanded, it cooled. The first subatomic particles appeared a fraction of a second after the Big Bang—electrons and quarks, the unseen essence of all the solids, liquids, and gases of our world, materialized from pure energy. . . . Things were still ridiculously hot and remained so for about a half-million years, until the ongoing expansion eventually cooled the cosmos to a few thousand degrees-sufficiently cold for electrons to latch on to nuclei and form the first atoms. The overwhelming majority of these first atoms were hydrogen—more than 90 percent of all atoms—with a few percent helium and a trace of lithium thrown in. That mix of elements formed the first stars. (Hazen 2012, 7–8)

Around the same time, the entire solar system, of which Earth is a part, was formed. Earth is nearly 4.6 billion years old.[1] Interestingly, in examining archaeological sites and reading archaeological scholarship, one might get the feeling that archaeological monuments, including ancient settlements and cemeteries, are located on immutable land, as if frozen in time, since it never gets any special attention. But as soon as you move into the domain of geology, land starts to suddenly lose all its stability, its appearance begins to vary, and its huge components appear to be always in motion—rising and plunging into the abyss of the ocean. Yet to have a comprehensive understanding of how these two disciplines—archaeology and geology—relate to each other, one needs to combine their approaches. This is not an easy task.[2]

1.2. Four Eons in the History of Earth

Historians as well as archaeologists are not accustomed to dividing Earth's history into the so-called eons, common in geology, even though the archaeological time scale composed of successive eras is much more similar to the geological rather than the historical view of history. The terminology of geoperiodization is based on the Greek language with eon meaning "century" or "era" in Greek.

Geologists subdivide the history of Earth into four successive eons.

---

[1] By the latest estimates, Earth is 4,567 million years old (see Bouvier, Meenakshi, 2010)—the number cited by Hazen. However, for us this precise number is not as important.

[2] One more important note: an archaeologist reading this book should remember that the geologic time scale (from eons to eras, to periods, to epochs, and finally to centuries) differs greatly from the time scale adopted by archaeology.

The first eon was the Hadean (from Hades, the Greek god of the underworld), lasting 600 million years. The planet's surface at the time was smelting from overheating; no traces of sedimentary rocks can be found.

The next, the Archaean, lasting 1,500 million years (from 4,000 to 2,500 million years ago), means "beginning" in Greek. It was marked by the first indisputable sedimentary rocks—traces of the future lithosphere, which plays such a great role in the Earth's structure.

The Proterozoic, existing about 2,000 million years (from 2,500 to 542 million year ago), comes from *protero* ("former") and *zoic* ("life") and is the longest eon. It was marked by the formation of an ocean, close to the modern one in size, and the emergence of multicellular organisms such as sponges and mushrooms. Soils were formed because of bacteria and unicellular algae that lived not only in oceanic waters, but also on land in water films between mineral particles located in areas of partial flooding near water bodies.

The last eon, the Phanerozoic, lasting from 542 million years ago to the present, means "visible life" in Greek (from *phanerós*, "visible"). It was characterized by a sharp deepening in the differentiation of endogenous and exogenous geological processes; formation of giant blocks of the lithosphere and the movement of lithospheric plates; formation of oceans and continents; emergence of skeletal animals; distribution of plants on land; and development of the biosphere, which gave rise to the genus *Homo* later in the eon.

Let us now have a closer look at the last eon in the history of Earth and compare it with the earlier ones.

1.3. Three Phanerozoic Eras and the Emergence of the Genus *Homo*

Humans emerged at the very end of the Phanerozoic. This eon is subdivided into three years eras: Paleozoic (from 542 to 251 million years ago), Mesoproterozoic (from 251 to 66 million years ago), and the Neoproterozoic (from 66 million years ago to the present).

The last Neproterozoic Era is further subdivided into three periods: the Paleogene period (from 66 to 23 million years ago), the Neogene period (from 23 to 2.58 million years ago), and the Quaternary or Antropogen period (from 2.58 million years ago to the present).

And finally, the Quaternary or Antropogen period is divided into two epochs: the Pleistocene (from 2.58 to 0.0117 million years ago) and the Holocene (from 0.0117 million years ago to the present time).

There is an almost universal consensus that the origins of the genus *Homo* can be traced to the Quaternary period, which is why it is also known as the Antropogen period. However, recent discoveries in South Africa indicate that the human race is even older, placing its origins at the end of the Neogene or the Tertiary period (Fig. 1.1)—a designation, now considered obsolete by many geologists.

The study of the oldest traces of the Paleolithic at Lomekwi-3 near Lake Turkana in Kenya is undoubtedly of great importance. These remains suggest that the genus *Homo* might indeed be 3 million years old, since Lomekwi-3 is considered 3.3 million years old (Harmand et al. 2015). It is likely that the

Figure 1.1. Diagram of the chronological sequence of the most important geological epochs and formations in the history of the Earth.

beginning of the Antropogen will eventually be associated with the end of the Pliocene, but even then the history of *Homo* would appear as an insignificant fraction of the entire Phanerozoic period, amounting to no more than 0.55% of the total length of the final eon (Fig. 1.2). Moreover, Antropogen would only account for the minuscule 0.07 % of the entire geohistory of planet Earth. Indeed, the genus *Homo* seems a mere micro-bio-addition to Earth.

1.4. Geospheres of Planet Earth and Eduard Suess

The Phanerozoic eon was marked by the formation of the main geospheres of Earth, which served as peculiar membranes of the entire planet. It was the famous Austrian geologist and public figure Eduard Suess (1831–1914), who first understood and conceptualized their importance for the formation, development and sustenance of life on Earth. In 1875 he introduced the notions of geosphere and biosphere (the worldwide sum of ecosystems):

Just as one can identify concentric spheres on the Sun, we can probably identify spheres on Earth, each of which is in extensive communication with its neighboring sphere. There is the atmosphere, the hydrosphere, and finally the lithosphere.

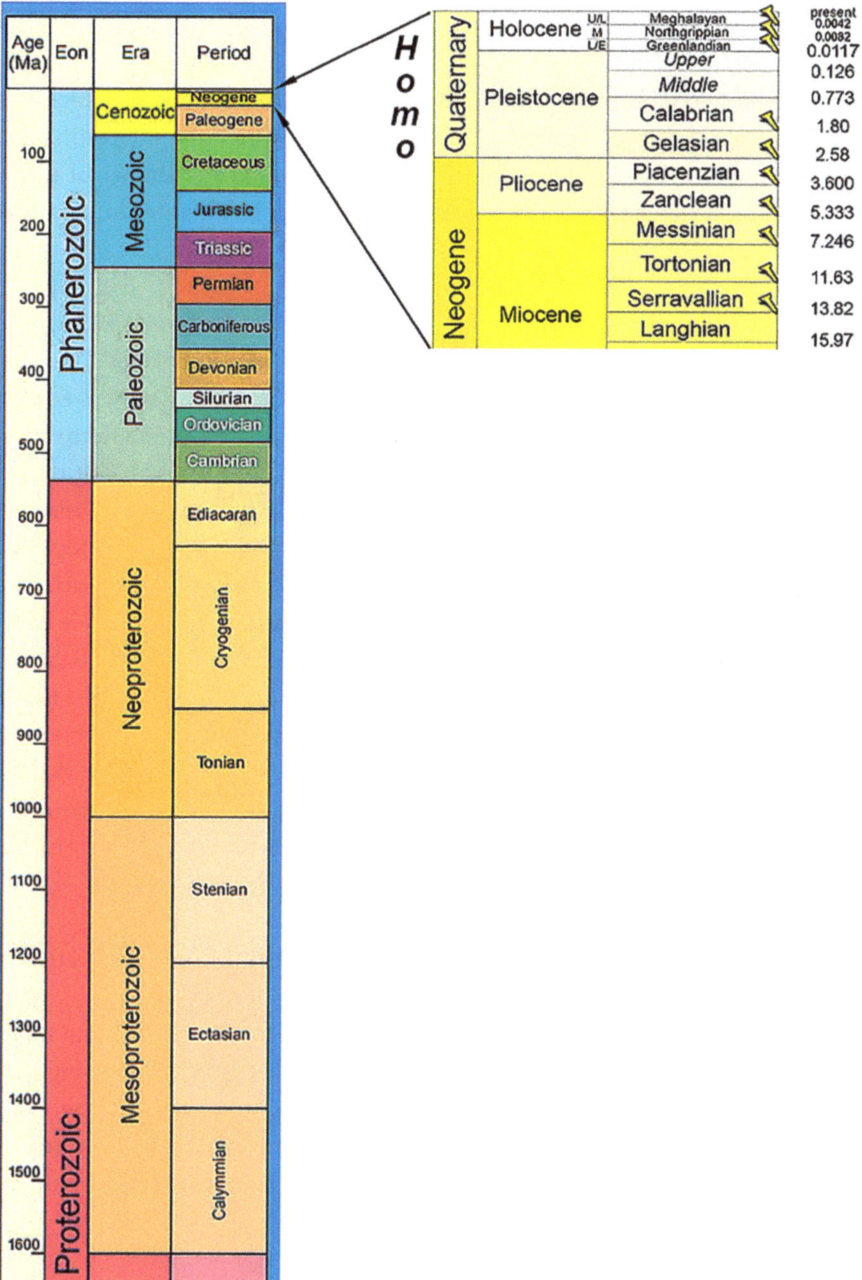

Figure 1.2. Diagram of the relationship of the Quaternary period—anthropogenic—with the Genozoic era within the Phanerozoic eon.

One thing seems foreign to this large, spherical celestial body—namely, organic life. It is limited to a certain place on the surface of the lithosphere. A plant whose roots penetrate the soil in search of food and who at the same time rises above ground to breathe is a good illustration of the position of organic life in the domain of interaction between the upper spheres and the lithosphere. One can identify another sphere—the biosphere on the continental surface, extending over both land and water . . . (Suess 1875, 158–159. Translated by I.S.)

The talent of the Austrian researcher and the significance of his work were highly esteemed in almost the entire scholarly community of earth science. The most interesting assessment of Suess is by a famous Russian geologist, geographer, and writer, member of the Academy of Sciences, Vladimir Afanasyevich Obruchev (1863–1956). In the introduction to his book entitled *Eduard Suess* (Fig. 1.3) co-authored with Zotina, he writes that Eduard Suess was

the author of a multi-volume work, *The Face of the Earth*, which is unlike anything that had appeared before or after in the scholarship about our planet. Thanks to his work, he can be considered the teacher of all geologists, not only in Europe, but also in other parts of the world, since it was translated from German into French, English, and Italian . . . [He] knew the face of the Earth not only in its present state but traced its development to its ancient history. Dissecting its wrinkles through a masterful analysis, studying both the old and young mountainous countries and plateaus, levelled down by millions of centuries and often covered by later layers, Suess identified the pattern and sequence of the formation of the existing and former mountain ranges. Through a brilliant synthesis, he reconstructed the role of each epoch in the development of the Earth's outlook by looking across all the continents. Figuratively speaking, he forced his readers to look at our planet and follow its rotation from a great height, as if taking the globe in his hands and turning it, studying its anatomy and features and removing the air and water barriers that interfere with observation. (Obruchev, Zotina 1937, 5–6; Obruchev 1963. Translated by I.S.)

This surprisingly lofty praise of the Austrian scientist suggests that one should treat his works with a special respect.

Another dimension of Suess's work bears great importance for this book's main subject. Suess had a clear opinion about the structure of Eurasia: he believed that it was a single continent, and not two, like ancient geographers and historians had believed and like the vast majority of people in the world still believe today. Even official documents of many countries state that Eurasia consists of two continents.[3] He was not the first to suggest the notion of "Eurasia" as Suess himself reported in his, perhaps, most important work, published in 1892:

---

[3] See, for example, A. Lukin (2018).

Figure 1.3. Outstanding geologists: the Russian Vladimir Afanasievich Obruchev (left) and the Austrian, Eduard Suess.

> Already the very first step in analyzing the earth's surface leads us to a conclusion, which makes it obvious how little work had been done in the right direction. Indeed, today we still talk about the Old World and the New World, as if there is no other definition for the entire vast mass of countries in Asia, Africa and Europe rather than the Old World. American naturalists felt a complete incompatibility of this term with the latest views and became the first to propose Eurasia in the place of the Old World. (Suess 1892, 766. Translated by I.S.)

Suess nobly points out to his readers that "American naturalists" were the first to put forward the notion of Eurasia, which he fully backed. He does not, however, provide the names of the naturalists, nor can I.

The concept of geospheres has become ingrained in scholarly language, and not only among geologists, but also among those involved in earth science. To illustrate this point, prominent Russian scientists—biologists, ecologists, and soil scientists—recently outlined their views on the role of geospheres in the life of our planet in the introduction to their book *Geospheres and Pedosphere*, published 135 years after the publication of Suess's seminal work:

The following geospheres can be identified: the lithosphere, pedosphere,[4] biosphere, hydrosphere, atmosphere, ionosphere, and magnetosphere. The lithosphere formed 4 billion years ago with the emergence of a basaltic sphere with islands of gneisses. Loose deposits (sedimentary rocks) appeared 3.8 billion years ago … Around the same time, 3.5 billion years ago, a magnetic field formed around Earth. The biosphere dates back to the same time … They all interact, communicate, and create the ecological conditions that determine the existence of the biosphere … (Dobrovolsky, Karpachevsky, Kriksunov 2010, 8. Translated by I.S.)

The list of geospheres, as we can see, was expanded. Below, a number of other categories often seen as having an extraordinary significance will be discussed. Among them, special importance is often given to the noosphere, or the sphere of reason, which has a tremendous effect on human life. The outstanding Russian scientist Vladimir I. Vernadsky developed the concept; we will touch on this problem in more detail in the finale of our book

1.5. The Lithosphere and the Formation of Supercontinents in the Phanerozoic

First, we must define the lithosphere and lithospheric plates. Lithospheric plates account for large areas of Earth's crust, that is of the lithosphere—the hard sphere of the earth. There are two main types of the earth's crust: continental and oceanic. The borders of each lithospheric plate are defined by zones of seismic, volcanic and tectonic activities (see, for example: Khain 2001; Bird 2003).

Narrow zones of activity and latitudinal splits subdivide the planet's lithosphere into blocks of various shapes and nature. The plates themselves are unstable in terms of their basic shapes, their spatial coverage, as well as their location on the planet. The plates are in constant motion, often colliding with each other, as they often did during the Phanerozoic. The plates served as the basis for the formation of supercontinents that later disintegrated into smaller pieces. On the modern map of the planet, they are familiar and appear as continents. More than nine tenth of the Earth's surface is covered by 13 or 14 lithospheric plates, here only the six largest are mentioned: Pacific (more than 103 million square km); North American (about 76 million square km); South American (more than 43 square km); Eurasian (almost 68 million square km); and African (more than 61 million square km) (Fig. 1.4). These six plates were the bases of the supercontinents.

Rodinia (from the Russian word *Rodina* ("Moterland")) was a hypothetical supercontinent that formed about 1.1 billion years ago and disintegrated about 750 million years ago. At that time, the planet Earth consisted of one gigantic landmass—the supercontinent Rodinia—and of the Mirovia Ocean that surrounded it.

---

[4] Pedosphere refers to deposits that were formed on top of the lithosphere—mostly, soil, even though sometimes silt layers of water basins are included there as well.

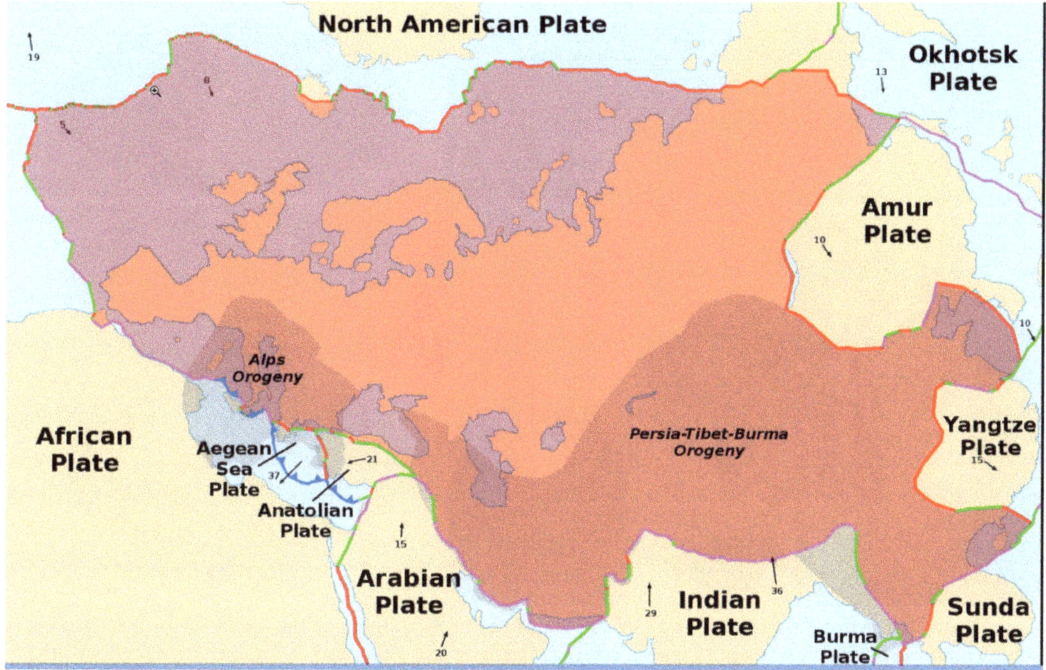

Figure 1.4. Eurasian Littoral Plate and its neighbors (USGS map).

Gondwana (or Gondwanaland) was a supercontinent in the southern hemisphere of the planet, covering the territories of modern Africa, South America, Antarctica, Australia, New Zealand, Arabia, and India. Formed as a result of the split of the supercontinent Rodinia, it existed from approximately 750 to 530 million years ago.

Laurasia was a supercontinent that existed alongside Gondwana, covering the planet's northern hemisphere. It mainly covered the areas of modern North America and Eurasia. Some time later, about 200–135 million years ago, it separated from Gondwana.

The supercontinent Pangea ("entire land") assembled from Gondwana and Laurasia 228–200 million years ago, uniting almost the entire landmass of Earth. It is believed that Pangea, the last land mass, which is fragmenting to this day, has served as the basis of the modern geomorphology of our planet, including all its continents and large islands, as well as the oceans (Wegener 1929; Khain 2010). The maps in Figs. 1.5 and 1.6 demonstrate the four stages of transformation of the supercontinent Pangea from its emergence around 228–200 million years ago to this day.

Around 1910, prominent German geographer Alfred Wegener put forth a theory of continental drift, a process associated with the disintegration of the original supercontinent Pangea. For a long time, his theory was harshly criticized by the scholarly community (Hellman 2007, Chapter 8) but is now generally accepted by geographers and geologists.

Figure 1.5. Alfred Wegener's book of 1929 and the author's idea of the splits of the Pangaea supermaterial (this map of splits is reproduced on the stone plaque of the memorial plaque on the building of the Institute in Marburg, where A. Wegener worked).

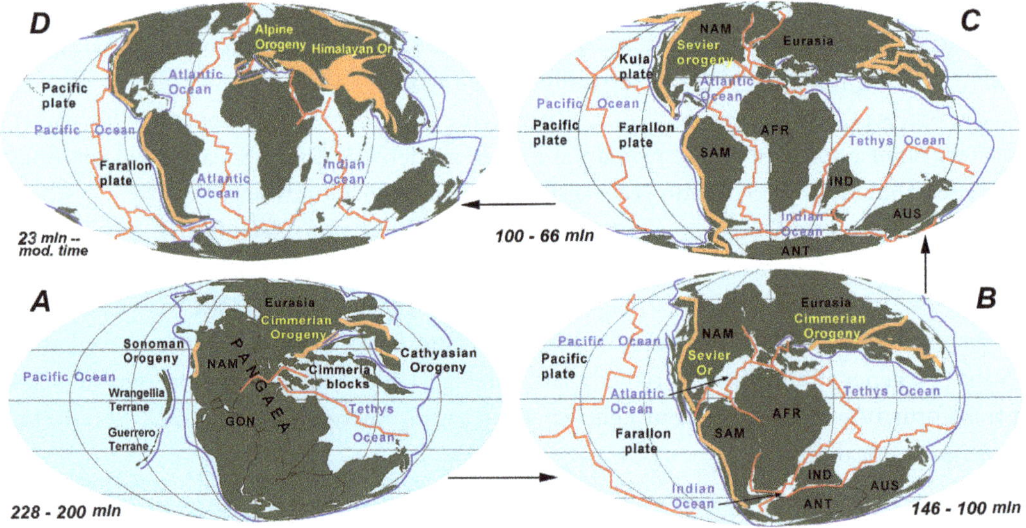

Figure 1.6. The main and currently accepted in science phases of the collapse of the supercontinent PANGAEA (https://pubs.usgs.gov/gip/dynamic/historical.html).

1.6. Metamorphoses of the Eurasian and African Littoral Plates

A quote by Russian geologist and historian of science, the author of a number of major monographs and articles with a truly global perspective, Viktor E. Khain (1914–2009) (Fig. 1.7).

The African platform is characterized by the prevalence of exposed base in the areas of development of sedimentary cover, especially in its southern part. Only in the northern part, the base appears in the form of separate massifs against the background of a wider development of the cover. The latter part of the platform can be provisionally designated as the Sahara plate, while the southern half of the platform can be called the South African shield. (Khain 2001, 309. Translated by I.S.)

The geologist's analysis about the division of the African platform into two parts—the northern Sahara plate and the South African shield—is extremely important for both archaeologists and paleoanthropologists. The South African shield, according to most experts in

Figure 1.7. Outstanding Russian geologist Victor Efimovich Hain.

these two interrelated sciences, had paramount importance since it served as the cradle of the genus *Homo*. The territory of the "shield," the birthplace of the genus *Homo*, amounted to about 20 million square km. The northern Sahara plate, with a territory of about 10 million square km, appeared to be outside the "cradle." The isolation of these two parts (even though this was not a complete isolation) continued to impact human cultures on the African continent for an amazingly long time—right up to the Modern period.

Starting from the early Paleolithic, the Mediterranean African north is appropriately treated together with the giant Eurasian mega-geoarea, but by no means with the spatially closer African mega-geoarea (Chernykh 2018a, 43–45). We have encountered an issue that is of direct relevance to archaeology and concerns the Paleolithic immigrants from South and East Africa. It would be reasonable to conclude the geological part of the book with an overview of the phenomenon of the Alpine-Himalayan orogenic belt (or geosyncline), which has had such a strong effect on both the settlement and the formation of cultures of the pioneering African human populations in Eurasia, so different from the African "cradle."

1.7. Alpine-Himalayan Orogenic Geosyncline

To begin, Khain can be quoted once again:

The Alpine-Himalayan orogenic geosynclinal area is the youngest part of the Mediterranean geosynclinal belt (Fig. 1.8), actively developing in the Mesozoic and Cenozoic, a mobile section of the earth's crust between the continental plates of Eurasia, North Africa, Arabia, and Hindustan . . . At the end of the Triassic (approximately 210 million years ago) continental plate began to split and rift formation began . . . From the end of the Jurassic (approximately 150 million year ago), Africa-Arabia and later Hindustan began to approach Eurasia; at the end of the Eocene (somewhat 35 million years ago), the continental blocks of Africa-Arabia and Hindustan—came into a direct collision with Eurasia. The process was accompanied by intense folding and thrusting with the formation of large tectonic covers, mainly to the periphery of the region . . . in the Pyrenees, Atlas, Apennines, Alps, Carpathians . . . the Balkans and the Caucasus, mountains of Turkey and Iran, southern Afghanistan, and the Himalayas . . . The most intense rise of these mountain ranges began in the Late Miocene (c. 10–5 million years ago) and continues to this day. At the same time, the formation of deep-water basins of the Mediterranean, Black and Caspian (southern part) seas, as well as of smaller seas—Adriatic, Aegean, Azov, Caspian (northern part) took place. [Along the] foothills and intermountain river valleys troughs of the Ebro, Guadalquivir, Po, Danube, Kuban, Terek, Rioni, Kura, Indus, Ganges, and Brahmaputra Rivers were formed. (Khain 1984a; 1984b. Translated by I.S.)

Apparently, by the end of the Neogene and in the Pliocene, later than five million years ago, the map of the Eastern Hemisphere gained an appearance that is more familiar to us, while the outlines of the continent of Eurasia began to appear more in line with their current state (Fig. 1.6D).

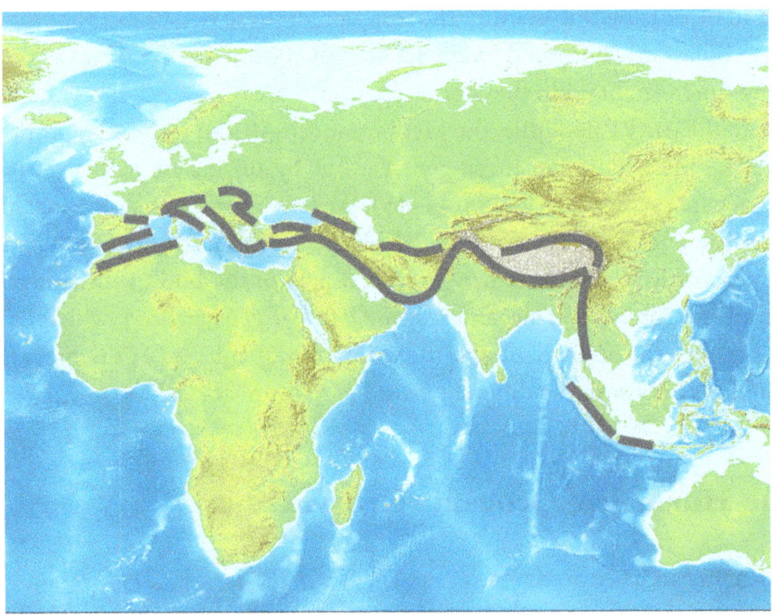

Figure 1.8. Diagram of the Alpine-Himalayan geosynclinal region within the Mediterranean geosynclinal Belt (Wikipedia).

The Alpine-Himalayan orogenic geosyncline undoubtedly caused the most significant "trauma" in Eurasia, which by that time had acquired its current outlines. The geosyncline's majestic and diverse mountain ranges stretched for at least thirteen thousand kilometers, winding from the Atlantic and North-Western Africa to the northeastern Tibet, the Quilian Mountains. After Tibet, this chain appears to descend to the south all the way to the islands of the Malay Archipelago, with its syncline reaching 17–18 thousand kilometers in length (Fig. 1.8).

Its structure reveals clear and vivid traces of unimaginable, fantastical collisions of lithospheric plates. This took place when the giant African-Arabian and Indo-Australian platforms, after moving for tens of millions of years to the north, collided with the smaller, northern Eurasian plate.

Such were the early changes along the North-South axis that affected the entire group of continents of the Eastern Hemisphere, only to become more pronounced later. No less remarkable were the changes that split the Western and Eastern Eurasia. All these processes gave rise to striking differences in human cultures so dependent on the geomorphology and geoecology of the continent.

This chapter concludes a brief overview of geological scholarship about the most significant stages in the formative history of our planet. Earlier scholarly ideas on the subject are discussed in the final chapters of this book. Next, we shall focus on an equally immense subject—both spatially and chronologically—the human cultures, to which archaeologists always pay such special attention.

Chapter 2

Humans on the Globe in the Pleistocene Era: Geology and Archaeology

2.1. What Are Our Origins?

The seminal opening sentence of the *Tale of Bygone Years*—"These are the narratives of bygone years regarding the origin of the land of Rus'"—could serve as a fitting epigraph to this chapter. In the place of "Rus'" one could insert any other country, since in the past for every ethnic group their origins always remained a mystery. There are countless theories about ethnic origins. Probably, there are as many of them as there have been ethnic groups in the history of the Earth, starting from the mysterious and bottomless depths of the preliterate history of humankind. It was during that time that various human groups began to see themselves as distinct and to ponder over their origins. Our focus shall remain on a much broader category—the first ancient humans.

> "Africa is the homeland of the genus *Homo*. This axiom is indisputable among geneticists, anthropologists, archaeologists, and representatives of other scholarly fields that study human origins," the leading Russian scholar of the Ancient Stone Age, a member of the Russian Academy of Sciences, Anatoly P. Derevyanko (2017, 435), writes, and I fully agree.

2.2. South and East Africa: The Cradle of the Genus *Homo*

Section 1.6 mentions that not all of Africa served as the "cradle" of humans. According to leading geologists, specializing in the geomorphology and geoecology of the African platform, the latter was divided into two unequal parts—the northern Sahara (near-Mediterranean) plate and the South African shield. It was the South African shield (Fig. 2.1), in the view of archaeologists and anthropologists, that was of paramount importance, serving as the cradle of the genus *Homo*. To quote Derevyanko, "researchers usually outline two global migration waves out of Africa in the Early Paleolithic but attribute them to different chronological periods" (Fig. 2.2). He believes that "the first global wave of migration from Africa to Eurasia began 2 to 1.5 million years ago. *Homo ergaster* or *Homo erectus* left the 'cradle' and moved beyond, marking the beginning of the first Great Migration and leading to an event of extreme importance—the settlement of the planet by man. The second wave, dating from approximately 1.5 to 0.9 million years ago, was marked by migrants that in all likelihood followed a route across the Isthmus of Suez and the Sinai Peninsula." (Derevyanko 2009b; 2017. Translated by I.S.)

One cannot but agree with assigning "extraordinary importance" to these waves. In the history of humankind there have been several great migrations and all of them led to dramatic and vast transfigurations of the earlier canvases of the cultures of *Homo*; this is particularly applicable to migrations within a single continent. When such migrations spread across continents, the structure of all human populations across the entire Earth underwent striking changes. The archaeological and anthropological finds from the Middle East and the Caucasus have served as the most ancient evidence of the appearance of African "guests" in Eurasia (Amirkhanov 2007).

The African migrants reached the Balkans and the Iberian Peninsula later—0.6–0.4 million years ago, as evidenced by the dates of the second Great Migration of Peoples to the north (Derevianko 2017, 467–488). Whether these waves of migrants were able to cross Gibraltar and emerge on the Iberian Peninsula, remains a controversial question. Around the same time, the so-called Peking Man (also known as the *Sinanthropus*) appeared in the Far East, although some finds might place his initial appearance in an earlier period.

The migrant pioneers encountered a very different world in northern Eurasia in terms of geoecology and climate. Across its vast expanses, they not only had to master an unusual landscape but also to fight against animals they had never seen before and which, surely, resisted the invasion of the bipeds. The area surrounding the mountain ranges and all the valleys were under massive ice sheets, making the new world of the African *Homo* strikingly different from the landscapes they had abandoned and, possibly, somehow preserved in their memory. The ability of equatorial African migrants to embrace an unfathomable and harsh new reality and adapt to it cannot but amaze us. How could they adapt to such unusual geoecology including the harsh cold climate of Northern Eurasia?

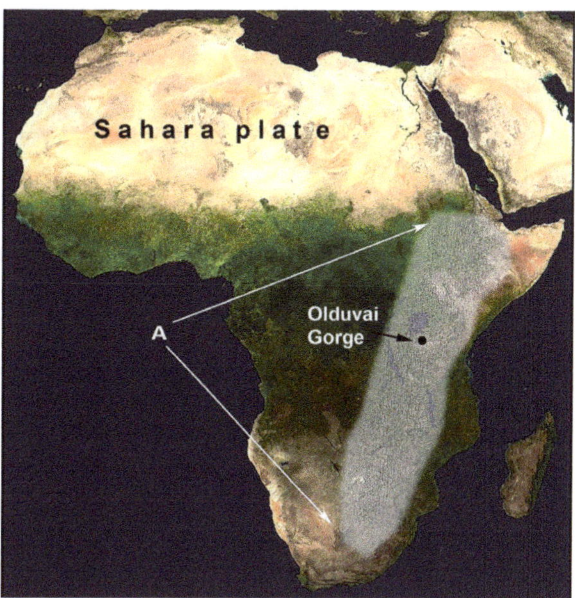

Figure 2.1. The African continent: the Sahara plate (north) and the South African shield (south). The eastern part of the continent with the Olduvai Gorge in Northern Tanzania is highlighted.

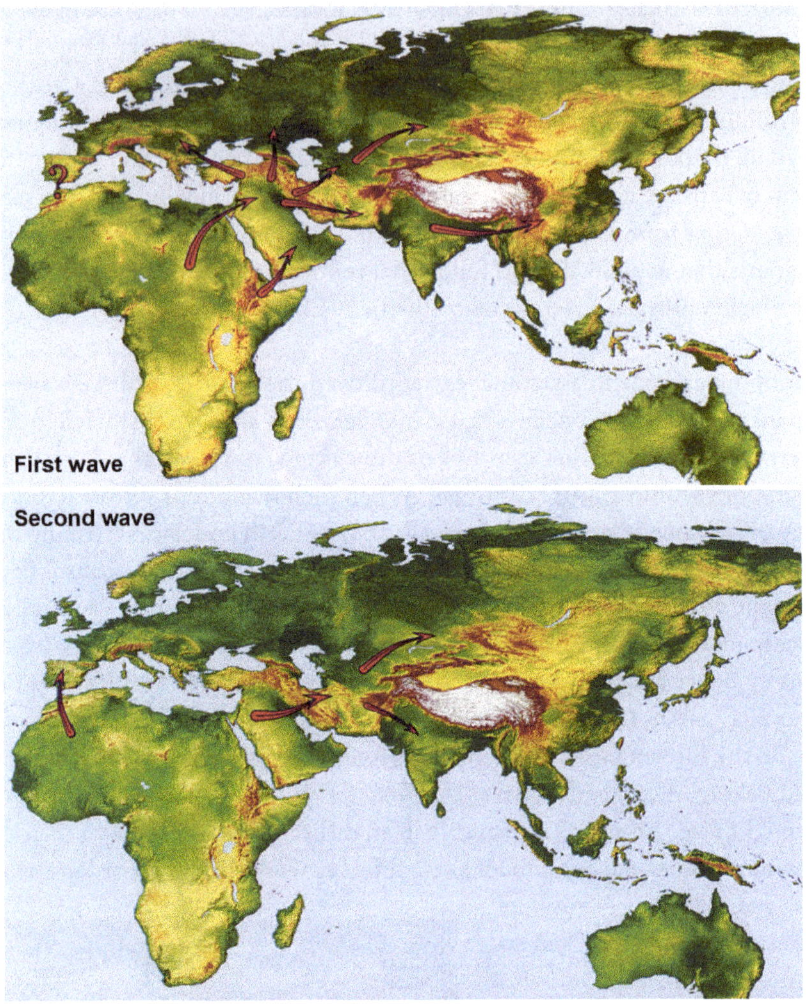

Figure 2.2. The two earliest migrations of ancient humans from Africa to Eurasia (after Derevianko 2012, Figs. 2 and 12).

The ancient humans who carried the burden of the two early great migrations were replaced by the Neanderthals or *Homo neanderthalensis*, as they are known in paleoanthropology. This species, associated with the biological evolution of the "Upright Man" or *Homo erectus*, also appeared first in eastern and southern Africa, that is, in the regions where their predecessors had lived. The first signs of the planned bioevolution date back approximately 400–350 thousand years ago. At the same time, significant changes in anthropological features were reflected only in the materials of the western half of Eurasia, mainly Europe. These materials date back to a much earlier period of time between 150 and 120 thousand years ago. Moreover, to date there have been no assured discoveries of Neanderthal remains to the east of the Altai Mountains (Fig. 2.3). Apparently, this was the third Great Migration. The African origins of European Neanderthals are undisputable, and most paleoanthropologists recognize Neanderthals as a dead-end branch in human evolution.

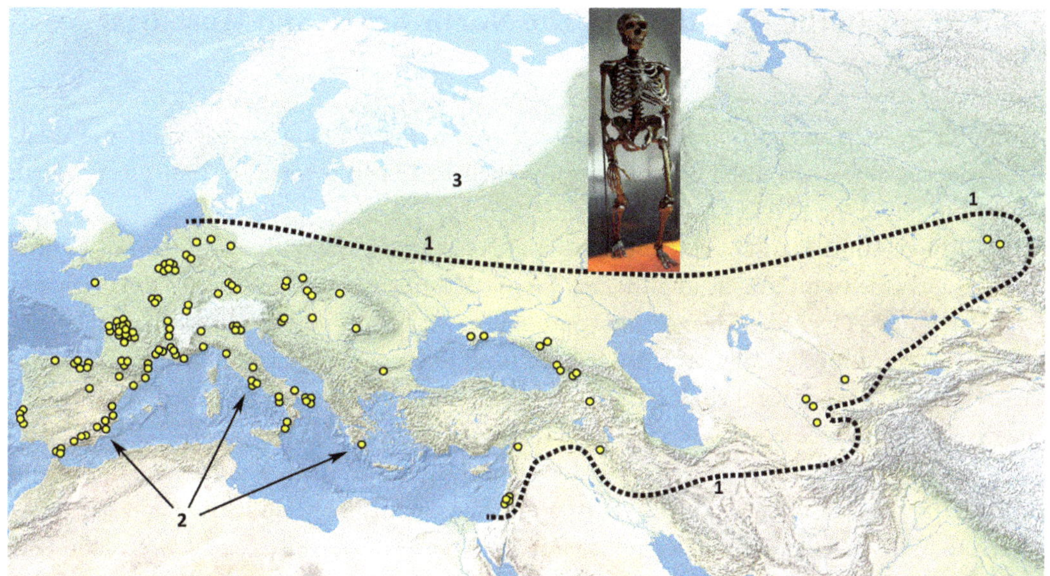

Figure 2.3. Distribution of the (1) most important Neanderthal sites (2) in Western Eurasia. Also highlighted is the glacier (3) (source: "Map of Classic Neandertal . . .").

However, there are heated debates among anthropologists about the history and relations of modern human species, often referred to as *Homo sapiens* or *Homo sapiens sapiens* ("wise men" or simply "modern men"). Where did they come from? "Of course, from Africa, and from there they came to Eurasia." This is how the majority of researchers, including both paleoanthropologists and archaeologists specializing in the subject, reply to this question. The dominant position of the "Africanists" should not be surprising, since their opinion is in full accord with the theory of Afro-centrism in the origins of *Homo*. Geneticists also seem to point to this conclusion, although data on the subject is not as clear as data about the three migration waves out of South Africa (Derevyanko 2007, 435–492; Chernykh 2019, 69). Later on, the development of anthropological families and archaeological cultures in Eurasia occurred independently of each other (Derevyanko 2012, 59, Fig. 38). The latter can be seen as the Fourth Great Intercontinental human migration.

It is quite possible that relatively new discoveries in northwest Africa will be able to confirm the leading role of the continent in the development of *Homo* into *Homo sapiens*. Indisputable remains of *Homo sapiens* were discovered in the Jebel Irhoud cave in Morocco, about 70 km from the Atlantic coast. The remains are about 100,000 years older than the previously found earliest remains of *Homo sapiens* and date back to c. 315,000 years ago (Hublin et al. 2017). This evidence supports the theory of African monocentrism and the fourth African wave of migrants to Eurasia, already associated with the "wise men." In any case, these discussions are by no means over and will carry on further.

2.3. Eurasia in the Pleistocene: The North-South and West-East Dividing Axes

The Pleistocene was part of the Paleolithic, therefore the dividing North-South and West-East axes cut across the Paleolithic Eurasia,[1] which had been formed by four great human migrations out of the African cradle.

The Alpine-Himalayan geosyncline appeared to cut across the entire continent of Eurasia from the Atlantic to the Pacific and Indian Oceans, creating two unequal parts: Northern and Southern Eurasia (Fig. 1.8). The northern part of Eurasia, larger than the southern one, covered at least two-thirds of the continent. There is no need to focus on climatic differences between these two parts here, since the contrast between the periglacial areas of the North and the rich and diverse South is obvious.

However, the differences between Western and Eastern Eurasia are more remarkable and striking. The immense ice cap of the Earth's northern hemisphere noticeably and resolutely shifted to the West (Fig. 2.4). Powerful ice layers covered North America and Greenland (the latter, by the way, is still covered in ice to the present day). Eurasia had far less ice cap with ice covering only the northwestern part of the continent, leaving almost the entire Arctic center

Figure 2.4. The last glacial period of the Pleistocene. Ice cap covering the continents and islands of the Earth's northern hemisphere.

---

[1] The division into the South and the North usually is based on geographical latitude, even though other factors (such as the Alpine-Himalayan orogenic belt) are seen as decisive in this regard. The West-East division is also usually based on geography—particularly, longitude—yet, sometimes there are other important factors, including the Ural Mountains or the valleys of great rivers such as the Yenisei River. Different approaches to the topic are discussed in other parts of the book.

and northeastern part of the continent free from ice (Walker et al. 2009; Patton et al. 2017; Chernykh 2019, 96, Fig. 4.15).

The absence of an ice cap might serve as an evidence of a significantly warmer climate in the East of Eurasia. However, this conclusion requires a further explanation. Eastern Eurasia had relatively warm summers and extremely harsh winters with either little snow or sometimes even no snow at all. The climate of the Eurasian West, on the other hand, was more balanced. Indeed, to this day, we can easily identify the frequent and obvious contrasts between winter and summer months, when comparing places in the West with places in Central and Eastern Eurasia or Asia, located on the same latitude (Chernykh 2019, 91–97, Figs. 4.13, 4.14). It might sound strange, but our generation is in a way a climatic successor of the distant Eurasian Paleolithic. Most likely, the relatively warm ocean currents of the Atlantic (a predecessor of the Gulf Stream?) made the climate in the West prone to large snowfalls. On the other hand, the cold, dominant currents of the Arctic Ocean often yielded very little precipitation.

2.4. The Two Last Waves of the Great Migration of Peoples in the Paleolithic

Now let us briefly turn to human history in the latter Paleolithic in Northeastern and Southeastern Eurasia. The drive of these people to discover the unknown prompted the fifth Great Intercontinental Migration of Paleolithic humans. Whereas the first four migrations were spearheaded by African cultures, the fifth migration was set forth by Eurasian populations.

First, let us focus on migrations in southeastern Eurasia. The fact that humans had migrated to the far southeastern borders of Eurasia was established a long time ago, upon a discovery of *Pithecantropus erectus* by the Dutch doctor Eugène Dubois. The migration of *Homo sapiens* to Australia is dated to 60,000–40,000 years ago, when they managed to cross the wide but shallow Torres Strait (Fig. 2.5).

To have a clearer idea about intercontinental routes, it is important to look at geological data.

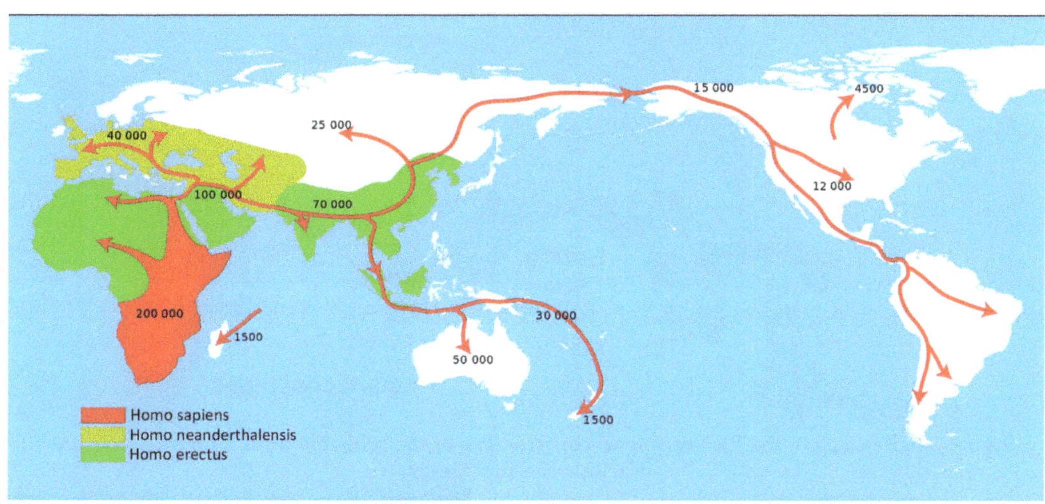

Figure 2.5. Probable routes and timelines of the distribution of different *Homo* species across the globe (source: Wikipedia).

The Pleistocene glaciations, which had such a profound effect on early human history, also impacted the appearance of the Earth's surface. The ice shells, covering many millions of square kilometers of the planet, required great amounts of water. They pumped it out from the oceans, transforming millions of cubic kilometers of water into glaciers that covered continents and islands. Geomorphologists believe that at the time the sea level dropped by many tens of meters (see, for example: Selivanov 1996; Kaplin, Selivanov 1999; Patton et al. 2017), as evidenced by abrupt changes in the coastal outlines of both continents and islands (Carter 1988; *Coastal Evolution . . . 1994*). Most probably, this also caused changes in the riverbeds, due to the connection between river mouths and seas and oceans. This process also led to a sharp decrease in the water level of the Torres Strait, saturated with small islets, which separated the mainland of Australia from New Guinea. It is believed that it was the Torres Strait that served as the main route for migrants coming from southeastern Eurasia. Reaching 150 km in width in some areas, the strait acted as a land route to the south.

However, metamorphoses in the far corner of northeastern Eurasia caused by the fifth Great Migration were of much greater importance. Two great continents—Eurasia and North America—became linked by land, in an area that was later named Beringia (Fig. 2.6). Almost all archaeologists agree that it was via Beringia that inhabitants of northeastern Eurasia crossed into unknown America. Paradoxical and truly global aspects of this migration require some explanation.

Whereas the first migration from southeastern Eurasia to Australia, who settled on surrounding islands, appears rather uncomplicated, migration from northeastern Eurasia posed some questions. On the surface, the picture seems quite clear. One can easily imagine how *Homo sapiens* crossed Beringia into the vast neighboring North American continent and became the true pioneers on a territory stretching from subarctic Alaska to subpolar Tierra del Fuego. Who if not them could have done something like that? And what is the reason behind such striking scarcity of Paleolithic monuments in the harsh Arctic areas of the Eurasian continent?

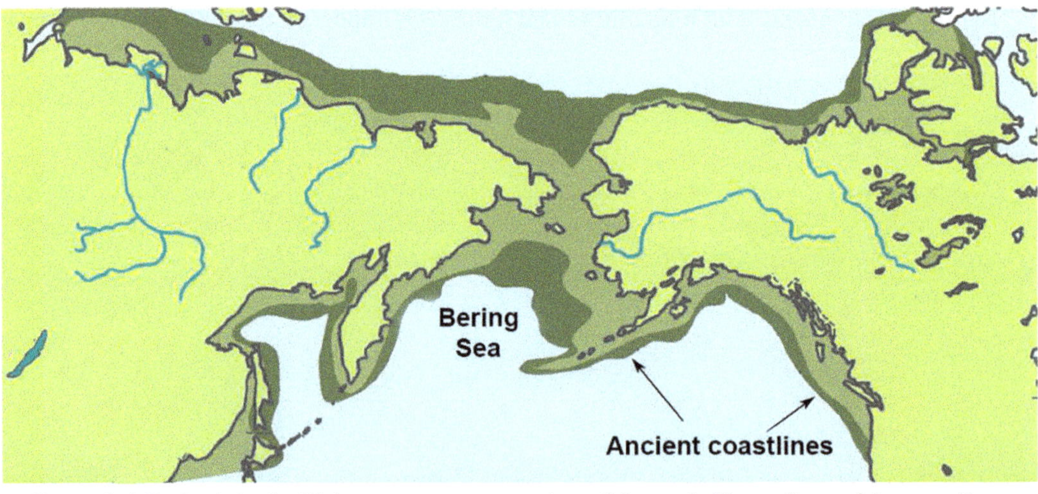

Figure 2.6. Beringia in the Pleistocene: reconstruction of the probable outlines of the coastlines of Chukotka and Alaska.

A prolific and long-time researcher of the Paleolithic in the Eurasian Arctic and Subarctic, Vladimir V. Pitulko writes, "Evidence of human presence in the Arctic prior to the last major glacier, that is dating back to 28,000 years ago, is extremely minimal." In recent years, scholars have found clear evidence of the early presence of Paleolithic migrants in these harsh conditions, dating as far back as 40,000 years ago, however, this evidence is still questioned today. To quote Pitulko once again, "Broadly, the materials discussed here might be linked to the first human settlements in the New World . . . The ancient hunters of Arctic Siberia had enough time to make use of their access route to the New World [Beringia] that laid open before them. However, at the moment it is hard to say whether they ever used the opportunity." As we can see, there are still a lot of unanswered questions. In any case, the earliest Paleolithic monuments in North America date to 20–15,000 years ago. (Pitulko 2016; Pitulko et al. 2016)

In the Holocene, abrupt climatic changes caused by another bout of cosmic shifts led to another series of cardinal turns in human history.

Chapter 3

Eurasia in the Holocene: The North-South Division

3.1. Global Metamorphoses in the Holocene

The Holocene, one of the periods in the Earth's geological history, started about 11,700 years ago and lasts to the present day (see Section 1.3 and Fig. 1.2). A period of sudden warming, called Alleröd, ended the Pleistocene and gave rise to the succeeding cold Younger Days, which were followed by the Holocene. However, the Holocene might simply be a long interglacial period of the Pleistocene, and not an independent period, as archaeologists believe.

A whole series of questions comes to mind: What will become of the Earth's climate in the future? Will the current phase of climate warming be replaced with another glaciation? Or will it cause a planetary catastrophe, for which the growing cacosphere is currently blamed (see Section 1.4)? After all, the greenhouse gas emissions associated with human activity have reached a historic maximum, as one hears in the news (Figs. 3.1 and 3.2).

Climate change, which impacts economic growth and population growth, has widespread consequences for both humans and natural systems in every country and on every continent. As air and ocean temperatures rise, the amount of snow and ice decreases, while sea levels rise. According to current forecasts, the temperature of the Earth's surface will continue to increase in the twenty-first century. This is of great concern to the world. Concerns over climate change, often seen as potentially even more detrimental than wars and global terrorism, led to the famous Paris Agreement on climate change, signed by 196 (sic!) countries in 2015. So far only 60 have ratified it, but let us hope nevertheless that the next generation will not be enmeshed in climate horrors.

3.2. Geomorphological Paradoxes and the Fifth Intercontinental Migration

One interesting geomorphological paradox is associated with the Holocene: as a result of fluctuations, global oceanic waters returned to the pre-Pleistocene levels, gradually rising by many tens of meters. The melting of glaciers transferred back to the ocean millions of cubic kilometers of water. The latter was not salt, but fresh water, which only intensified the thermal effect (Nesje et al. 2004). Climate warming soon brought the populations of both America and Australia into a virtually complete isolation, which lasted for more than ten millennia, right up to the Modern period, that is, until the fifteenth to seventeenth centuries, when they were "discovered" by European sailors.

This unexpected and surprising paradox presented researchers focusing on the history of human cultures a wonderful opportunity to effectively and efficiently study the dynamics and rates of development of cultures in each of the isolated continental enclaves. Therefore, it would be worthy to at least briefly compare the two waves of Late Paleolithic migrants of the fifth Great Intercontinental Migration—the early Australian and later American waves—which had very different fates.

The American wave of migration reached its pinnacle with the sumptuous civilizations of Central and South America, such as the Aztecs, Mayans, and Incas. The cultures in the southern periphery of the American continent greatly differed from them. The migrants from the Australian wave were very similar to the people of Tierra del Fuego, as Charles Darwin thought when he observed the latter as a young man. Native Australians froze or nearly froze in the state of development they had been in upon reaching the unknown south. There are several innovations that are attributed to indigenous Australian cultures, including the boomerang, however, both in their character and in appearance, they undoubtedly have retained their Paleolithic

Figure 3.1. Early Industrial Era: coke furnaces in England (*Coalbrookdale by Night* (1801), a painting by Philip James de Loutherbourg).

Figure 3.2. Metallurgical production in Southern Urals,
the town of Karabash and its surroundings today.
A: Wikipedia, Karabash's geoecology;
B: Wikipedia, the photo by V. Raskalov;
C: photo by the author.

features. This was how the European pioneer sailors saw them in the seventeenth century. Essentially, native Australians retained their indigenous way of life until the nineteenth century (Fig. 3.3). In any case, the "progress" of Australian aboriginal cultures appears minimal when compared with the achievements of the civilizations of Central America and South America,

Figure 3.3. Aboriginal peoples of Australia, a nineteenth-century photo (source: Wikipedia).

Figure 3.4. Gold figures; culture of El Tesoro Quimbaya, Central America
(Photo: Juan, por Revista Semana).

whose monuments are filled with gold objects—for example, Quimbaya (Fig. 3.4) in Colombia (El Tesoro Quimbaya, 2016). On the brink of the industrial era, the gap between Australian and European cultures must have seemed truly astounding.

3.3. The Dividing Alpine-Himalayan Orogenic Belt

During the Paleolithic's Pleistocene, the mountain ridges of the Alpine-Himalayan geosyncline continued to remain the main dividing line that separated Eurasia into its northern and southern parts. Stretching from the far western Atlantic borders of Eurasia and northwestern Africa up to the Tibetan or Indo-Chinese sharp "turn" to the south all the way to the junction of the Indian and Pacific oceans (Fig. 1.8), it remained unchanged. There had been no fundamental changes in such a short—geologically speaking—period of time. However, during the Antropogen, the peaks of the central ridges of the geosyncline continued to rise and still continue to rise today. Apparently, the littoral plates, especially in the area of the Tibetan-Himalayan knot, are still colliding (Trifonov et al. 2012).

In the Holocene, the most striking changes took place in the northern or northwestern part of the continent. Scandinavia and the Baltic region began to gradually (or rather rapidly in terms of geological time) free themselves from the ice cap. Thus, the Baltic region became completely free of ice about 9–8 thousand years ago, when Lake Ancylus (the future Baltic Sea), as it is known to geomorphologists, began to form. Curiously, the oceanic waters of the Atlantic did not flow into the lake; the lake's fresh water rushed into the ocean due to a rise of the continental platform underneath the lake (Monin, Shishkov 1979, 320–331, Fig. 10.4).

During that period, riverbeds and valleys also underwent changes, bringing the general outlines of the Eurasian continent closer to its modern appearance. The mighty Ob, Irtysh, Yenisei, Selenga (through Baikal and Angara), Lena, Indigirka, Khatanga, as well as Pechora and Northern Dvina Rivers in the West flowed into the Arctic Ocean. In those circumstances, the Alpine-Himalayan geosyncline served as the main dividing line, as if cutting the entire Eurasia horizontally.

Other rivers, including Dniester, Dnieper, and Don flowed into the Black and Azov seas. The Volga and the Ural Rivers flowed in the same direction (from north to south), letting into the Caspian Sea, which geomorphologists call an internal drainless basin. The Amu Darya and Syr Darya also rushed to the drainless Aral Sea, yet nowadays the rivers' weakened waters are only able to fill the deltas at the confluence of this small sea-lake.

In addition to the Amur River, the Yellow He and the Yangtze River with their sources in the Tibetan Mountains also belong to the Pacific Ocean basin. The rivers of southern Eurasia—the Tigris-Euphrates, Indus, and Ganges-Brahmaputra—are linked with the Indian Ocean.

The formation of Eurasian river valleys of Eurasia during the Holocene played a vital role in the entire subsequent history of human cultures. Indeed, it is around freshwater rivers and their valleys that the life of virtually every human society on Earth has always revolved around.

3.4. The Phenomenon of the Steppe Belt[1]

The mountain ranges of the Alpine-Himalayan orogenic belt, stretching for many thousands of kilometers, cut horizontally almost across the entire Eurasian continent. As a result, each of the regions defined by the belt had its own specific climate. Thus, clearly, northern slopes of mountain ridges and the areas adjacent to them found themselves in a colder climate. However, some of the differences in northern and warmer, southern parts of the Alpine-Himalayan geosyncline were striking. For example, there are pronounced weather contrasts in such very geographically close areas as Ciscaucasia and Transcaucasia and the Tibetan plateau and the north of Indian subcontinent. The differences in the climate of the Carpathian Mountains and other western areas seem much more minor.

Undoubtedly, the Steppe Belt, of such great significance to Eurasia, was the most striking outcome of the North-South division. Approximately 8,000 kilometers long (Fig. 3.5), it can

---

[1] The Steppe Belt has been analyzed in detail in my previous book, *Nomadic Cultures in the Mega-Structure of the Eurasian World* (Russian edition published in 2013; English in 2017), especially in Vol. 1, Chapter 2. My views on the subject have not undergone any significant changes in the short time since the publication. Therefore, this book will deal only with those visual and textual aspects not covered in my earlier work.

be divided into two main parts: the western and eastern. The famous Dzungarian Gate serves as a border between the two parts. The gate itself is somewhat of a "corridor," about 10 km wide and about 50 km long, between two mighty mountain systems—the northern or northeastern ranges of the Tien Shan (Dzungarian Alatau) and the Western Altai Mountains. Historically, the Dzungarian Gate has served as a convenient route for all sorts of exchanges—trade, military, etc.—between western nomadic cultures and the east of the Steppe Belt. Proponents of the Great Silk Road hypotheses see the gate as the clearest evidence of a reliable route between the East and the West (Figs. 3.5–3.7).

In the south, the Steppe Belt borders the massifs and ridges of the Alpine-Himalayan geosyncline, apart from two areas—the northern coast of the Caspian Sea and to the west—the northern shores of the Azov and Black Seas. The northwestern corner of the Black Sea, near the mouth of the Danube and the Eastern Carpathians (by the way, another part of the Alpine-Himalayan geosyncline) marks the western border of the Steppe Belt (Fig. 3.5). Compared to the far eastern corner of the Steppe Belt, the area close to the western edge does not measure much from north to south and looks like a peculiar dot, worthy of scrutiny. After all, the eastern edge of the Belt, located eight thousand kilometers away, is nothing like a dot, marked by a steep descent from the high ridges of the Steppe Belt to the Great Chinese (or more precisely, the North Chinese) Plain, populated by cultures very different from those of the stock-breeders of the Steppe Belt.

In this regard, one rather important circumstance needs to be pointed out: the median height of the eastern Steppe Belt is significantly higher than that of the western part (see Fig. 3.5). The differences between the ecosystems of the two parts of the Steppe Belt resulted in certain differences between the peoples of Eurasia and gave rise to a culture of mobile stockbreeders.

Figure 3.5. The Steppe Belt and its borders (1) and division of the Northern Eurasia into Western and Eastern parts (2). The arrows point to the Dzungarian Gate (3).

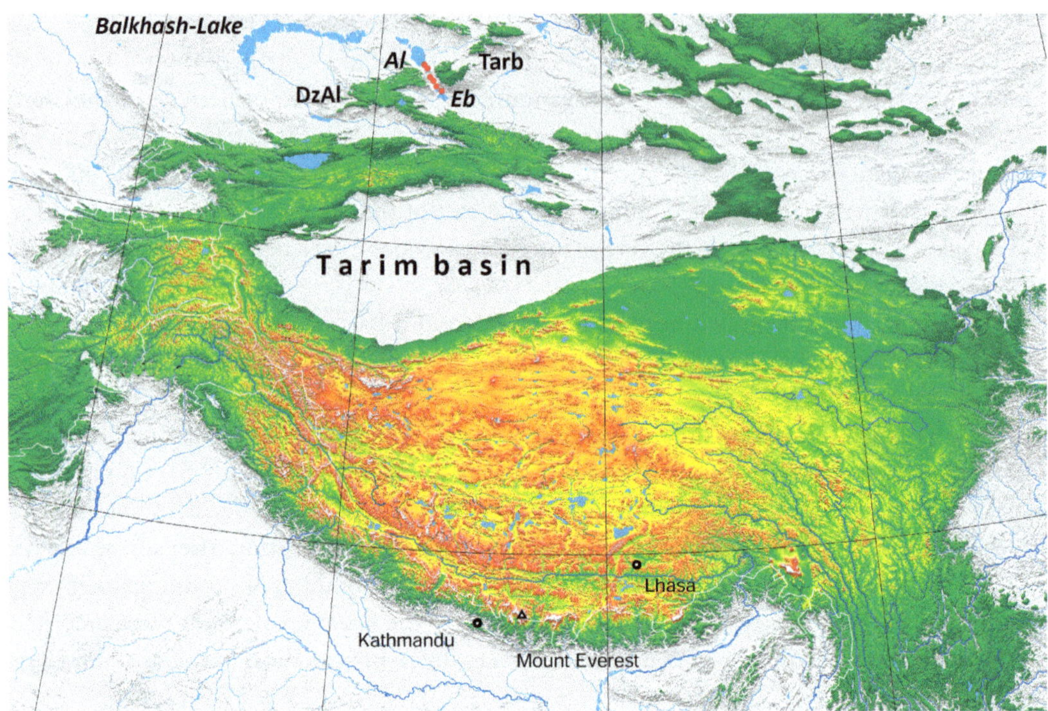

Figure 3.6. The Dzungarian Gate on the background of Tibet, Tarim basin and other different objects: DzAl—Dzungarian Alatau; Tarb—Tarbagatay mounts; Al—Alakol lake; Eb—Ebinur lake.

On the same note, what is the explanation behind the relatively short north-south extent of western Steppe Belt? The answer seems simple enough: it was caused by the glaciers of the Pleistocene. Relatively recently (in geological time), almost all of northwestern Eurasia was either covered with glaciers, or was in proximity of glaciers. The sources of the Steppe Belt—arid territories—were located at that time further to the south and to the east (*Paleogeografiya Evropy* [Europe's Paleography] 1982, maps 1–3; Chernykh 2019, 61, Fig. 3.7). At the same time, the Atlantic Ocean, with its mighty warm current, the Gulf Stream, had an undeniably powerful effect on the climate of all the western regions of the continent.

For a long time, the boundless central zones of Eurasia have served as illustrations for diverse geo climates and ecosystems on the continent (from subtropics to tundra), including in schoolbooks. This topic is dealt with in detail in a number of my earlier monographs (Chernykh 2009; 2013; 2017; 2019) and therefore will not be analyzed here. Following an analysis of Eurasia's division into the West and East, a number of paradoxes presented by archaeological finds will be discussed, which might be of special interest to some. In this chapter, several maps outlining the Steppe Belt's vast northern zones—forest-taiga, forest-tundra, and tundra—are sufficient.

Figure 3.7. Aerial shot of the Dzungarian Gate between Lakes Alakol and Ebinur. The read arrows indicate the directions of caravans' movement.

Chapter 4

The Holocene in Eurasia: Division along the West—East Axis

Let us now turn to the longitudinal division of the Eurasian continent. Whereas the latitudinal or horizontal division of Eurasia is grounded in the Alpine-Himalayan geosyncline, which passes through the entire continent, the longitudinal division of the continent is not based on such a well-pronounced, elevated borderline. Thus, the main issues surrounding the vertical division of Eurasia appear much more complex in comparison with its horizontal fragmentation.

Just as in the case of the preceding chapter, the subject of the present one has been analyzed in detail in my earlier work, *Nomadic Cultures in the Mega-Structure of the Eurasian World* (Chernykh 2017, Chapter 3). However, unlike the preceding chapter, this one includes extensive quotes from my earlier works, as well as a more detailed analysis of certain aspects, not covered in that publication. First, the almost eternal problem of establishing a border between the two essentially inseparable continents will be discussed here.

4.1. The "Border" between Europe and Asia

This issue is debated endlessly. It is discussed in detail in my earlier work, and, first, some of its excerpts will be offered here. Discussions of a border between Europe and Asia might appear something out of a school book. And yet, even if we adopt a purely geographical perspective, there is often significant disagreement about the position of this line. Following the remarkable Flemish cartographer Gerardus Mercator (Fig. 4.1), there is a consensus that in the north of the continent the division follows the line of the Ural Mountains (the Riphean Mountains[1] of earlier texts). However, from a geoecological perspective the mountains of this ancient seam between the Eastern European and Siberian cratons seem too low and diffuse to create a meaningful barrier. Further south, the precise position of the borderline becomes increasingly confused. One textbook might refer us to the Kuma-Manych depression, another to the line of the Greater Caucasus, still others to the course of the Ural River, or to Emba, flowing through

---

[1] The briefly mentioned so-called Riphean Mountains are discussed in much more detail in Section 10.5.

the Ryn Desert. This debate, like the origins of the Eurocentric Worldview, is not a product of recent political history. It has been a point of discussion for more than 2,000 years.

Herodotus, writing in the fifth century BCE, had this to say on the issue:

> For my part, I cannot but laugh when I see numbers of persons drawing maps of the world without having any reason to guide them; making, as they do, the ocean-stream to run all round the earth, and the earth itself to be an exact circle, as if described by a pair of compasses, with Europe and Asia just of the same size. The truth in this matter I will now proceed to explain in a very few words, making it clear what the real size of each region is, and what shape should be given them. . . . But the boundaries of Europe are quite unknown, and there is not a man who can say whether any sea girds it round either on the north or on the east, while in length it undoubtedly extends as far as both the other two. For my part, I cannot conceive why three names, and women's names especially, should ever have been given to a tract which is in reality one, nor why the Egyptian Nile and the Colchian Phasis (or according to others the Maeotic Tanais and Cimmerian ferry) should have been fixed upon for the boundary lines; nor can I even say who gave the three tracts their names, or whence they took the epithets. (Herodotus 4, 8–10) (Figs. 4.2 and 4.3)

A thousand years later, in the sixth century CE, the Byzantine chronicler Procopius of Caesarea (History of the Gothic War 8:2) also discussed the Cholchian Phasis (the present-day Rioni River), as the border between the two continents:

> In this land [the Transcaucasus] there are very high steep mountains, covered with forests. They stretch up to the very Caucasus Mountains. Behind them, to the east, is located Iberia extending as far as the lands of the Persarmenians. The River Phasis runs across the mountains rising high up to the sky. It derives from the Caucasus Mountains and descends in the middle of the "halfmooned" Pontus. Some people believe that in this location the River Phasis serves as a dividing line between the two continents. The lands downstream to the left are considered Asia, and lands downstream to the right, Europe. The settlements of the Lazi are located in the European part and they have no towns, fortifications or villages worth any attention on the other side. . . . According to the local legend, the Golden Fleece, which according to myths of the poets forced the Hellenes to build Argo, was located in that part of Lazica. But in my opinion it is not true at all.

Echoing Herodotus, the great Russian chemist Dmitry Mendeleev (1907, 143; Fig. 4.4) suggested that "the separation of Europe from Asia is artificial in every respect and, in the course of time, by all means will be undermined or perhaps even disappear." However, this prediction has not proved entirely correct. The desire to distinguish Europe and Asia has by no means disappeared. In fact, it remains remarkably persistent. The French journalist and philosopher

Bernard-Henri Lévy suggested that "Europe is not a place, but an idea." If this is the case, it is a very powerful idea. For modern Russians, questions such as: "Does Russia stand for the ideals of Asia or Europe?" are very familiar in the popular media. Since disputes of this kind appear to be endless, a positive contribution to the issue can only be made by approaching the problem from a different angle. In the previous chapters I have tried to relate major ecological zones with social, economic, and cultural differences in human populations. Can a similar method be applied to the borders between Europe and Asia?

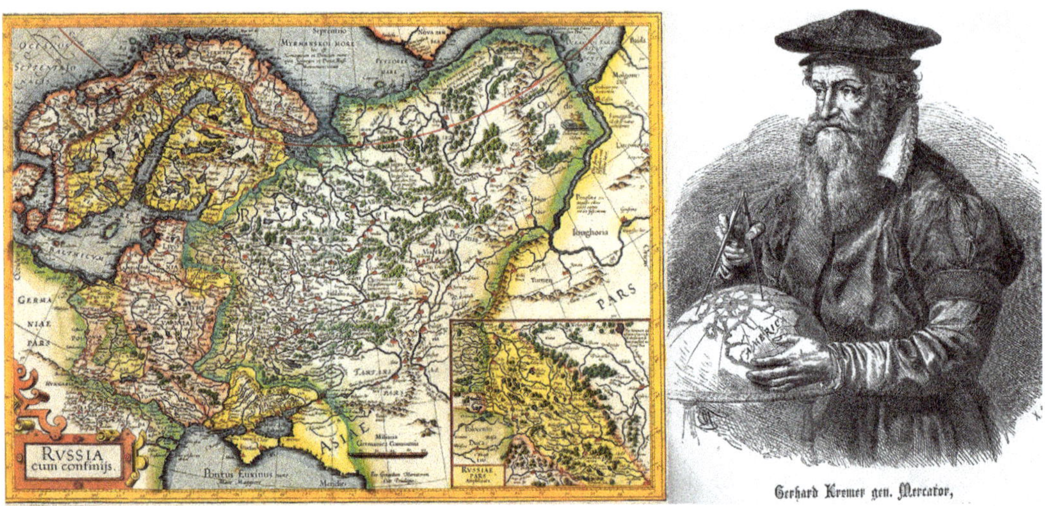

Figure 4.1. A map of European Russia by Gerard Mercator (1572),
outlining Europe's eastern border with Asia.

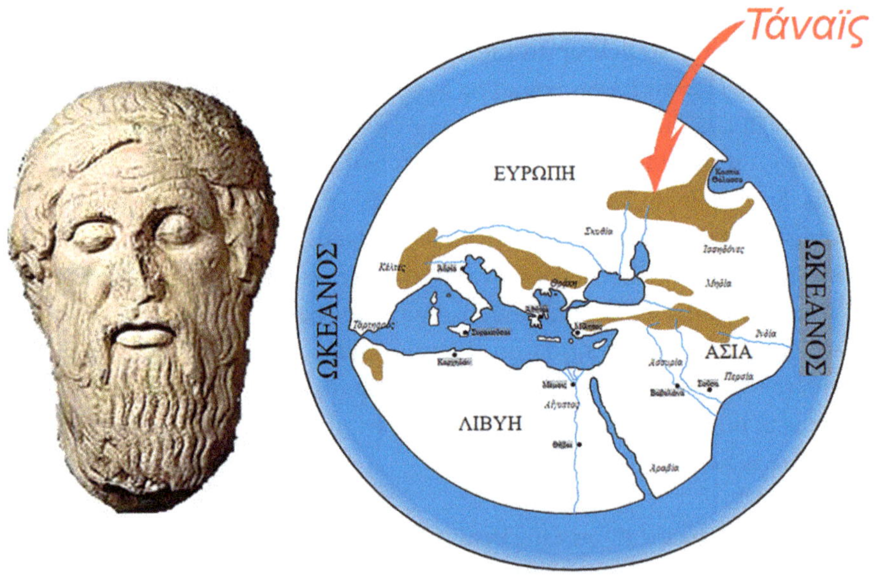

Figure 4.2. A map of Hecateus of Miletus (ca. 550–476 BCE).
Hecateus was an an early Greek historian and geographer,
even known as the "Father of Geography."

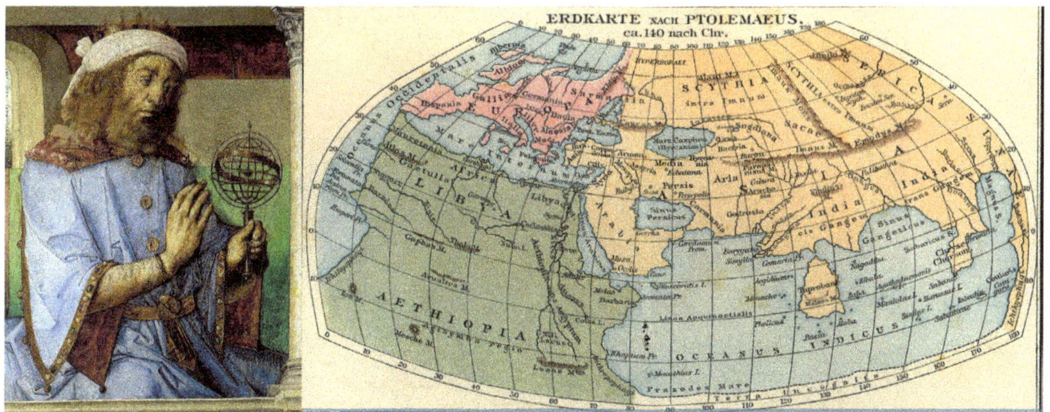

Figure 4.3. General map (~ 150–160 years) of the Land of Claudius Ptolemy with the designation of the borders of the three continents (Europe, Asia and Libya-Ethiopia/Africa) as well as the main divider of the Asian continent: the Imaus-Emodus mountain system.

We have already mentioned in passing that from our geoecological perspective, the accepted Uralian border seems unsatisfactory, its time-weathered mountains providing, at best, a permeable frontier. The same might be said of proposed divisions within the Caucasus, which offers no significant obstacle to human or animal movements. This argument can be supported by the fact that there are very few differences between communities living on either side of these questionable intra-continental frontiers. It seems reasonable, therefore, to retain the Steppe Belt as the major division in Eurasia, in spite of its latitudinal character; for example, the famous Dzungarian Gate (see Figs. 3.5, 3.6).

Figure 4.4. Dmitry I. Mendeleev, wearing the robe of a Juris Doctor of the University of Edinburgh (a watercolor painting by Ilya E. Repin (1885)).

4.2. Two Outstanding Figures: Tatishchev and Strahlenberg on the Role of the Urals

Let us now turn away from geology and focus on two figures, who influenced not only the outlook of the Ural mountains, formed in distant Antiquity, but also their character.

Figure 4.5. Portraits of Vasily Tatishchev (left) and Philip-Johann von Strahlenberg.

In the Modern period and especially under Peter the Great, the Urals (or the Stone Belt, as it was known in the past) became gradually absorbed into Russian possessions. The eternal issue of the border between Europe and Asia continued to be the subject of heated debates. When discussing the role of the Urals in this division, one often turns to the famous Russian statesman and historian Vasily Nikitich Tatishchev (1686–1750). In 1720, Peter I ordered him "to build factories and smelt silver and copper from ores in the Siberian province on Kungur and in other places, where it would be possible." Tatishchev had a complicated life, and his creative energy seemed truly boundless (Ivanov 1962).

A native of Sweden, Philip-Johann von Strahlenberg (1686–1747, born Philip Johann Tabbert; Fig. 4.5) comes to mind when discussing Tatischev and his views on the border between Europe and Asia.

He was among those few officers of the Swedish army who, after the defeat in the Battle of Poltava (1709), followed his king and fled under the auspices of Porta. But at the very border of the Ottoman Empire, Tabbert was captured by the Cossacks. The only thing that is known for certain is that he was sent to a settlement in Tobolsk as a prisoner of war. It was there, in that remote land, that the educated captive officers of the army of Charles XII were supposed to remain until the end of the war. According to Peter the Great's clever plan, they were expected to seek means of subsistence using their skills, eventually involuntarily contributing to the exploration and development of Russian Siberia. Gradually, in the course of the fairly protracted war, the city's Swedish settlement developed into the most numerous and productive Swedish diaspora in Russia.

During his Siberian exile, Tabbert, like many of his compatriots, occupied himself with research and science: he began collecting geographical, historical and ethnographic information about his country of residence, searched for

Turkic manuscripts, practiced in drawing geographical maps, copied rock petro-glyphs, compiled a Siberian dictionary, improved his knowledge of Russian and studied minerals and roads (Napolskikh, 1998).

One of Tabbert's cartographic sketches was presented to Tsar Peter the Great, earning him royal benevolence: the foreign officer was asked to stay in Russia and take an important government post. Tabbert refused this offer and chose to return to his homeland after the war. In Sweden, Tabbert was fully rewarded for the 12 years of "torment" in Russian captivity: he was elevated to the rank of nobility, awarded the rank of lieutenant colonel, and given a new surname—Strahlenberg. It was under the latter that he was destined to go down in the history of Russian and European geography.

In 1730, Strahlenberg self-published a book in Stockholm, in which he presented general facts as well as his own observations about Russia. The book contained diverse information about the historical, geographical, military, and political state of Russia. Due to the abundance of original information, it was deemed one of the most informative European sources on the subject by contemporaries. The parts describing the Northern War, biographies of Peter the Great's associates, reports about the campaigns of the Russian army, infor-mation about Siberia, Tartary and China became especially popular among his readers (Ehrensvärd Ulla, *Svenskt biografiskt lexikon*).

Tatishchev's translation of the book's title into Russian alone is interesting: "A historical and geographical description of the ancient and new state of the midnight-eastern part of Europe and Asia, or the Russian Empire, which is recognized as the midnight part among the earlier. Composed by Johann Strahlengerg and printed in Stockholm in 1730."[2]

Strahlenberg's book quickly reached Russia and right away became the sub-ject of interest among the enlightened elites. It was of particular interest to Tatishchev, who at that time was working on his *History of Russia*. As it turned out, Tabbert and Tatishchev were not only fellow scholars, but also knew each other personally, having been introduced back in Siberia in 1720. During that time, Tatishchev shared some of his historical and geographical findings with the Swedish scholar. From 1724–1726, the two men met as old friends and colleagues in Sweden. Stahlenberg noted that Tatishchev "spoke and wrote in German quite well," and did not require a translator while carrying out his

---

[2] The German title of the book is very long, however, not unusual for that time and resembles the book's thesis: *Das Nord-und Oestliche Theil von Europa und Asia, in so weit solches das gantze Russische Reich mit Sibirien und der grossen Tatarey in sich begreiffet, in einer Historisch-Geographisch Beachreibung der alten und neueren Zeiten, und vielen andern unbekannten Nachrichten vorgeatellet, nebst einer noch niemahls anc Licht gegeben Tabula Polyglotta von zway und dreyszigerley Arten Tatarischer voelcker Sprachen und einem Kalmuo-kischen Vocabulario, sonderlich aber einer grossen richtihen Land-Charten con den benannten Laendern und andern verschiedenen Kupfferstischen so die Asiatisch-Scythische Antiquitaet betreffen; Bey Gelegenheit der Schwedischen Kriegs-Gefangenschafft In Russland, aus eigener sorgfaeltigen Erkundigung, auf denen veratatteten weiten Reisen zusammen gebracht und aus-gefertiget von Philipp Johann von Strahlenberg* (Stockholm: In Verlegung des Autoris, 1730).

duties for the Russian government. Knowing that Stahlenberg was working on a book about Russia, Tatishchev wrote to him several times asking to send him the book to St. Petersburg. The author sent him a copy as soon as it was published (*Zapiski kapitana* [Captain's Notes], 1985).

4.3. The Urals as a "Divider" between the Continents

I have no desire to continue the endless discussions about the Urals as a divider between the continents. Not so long ago, I outlined my views on the subject in sufficient detail in a chapter entitled "Transitions from East to West: Across the Layers of the Eurasian Geoecology" (Chernykh 2017, 49–63). The main points from the chapter will be briefly given below.

To begin with, the Urals is not an appropriate "divider" between Europe and Asia, contrary to a long-lasting perception, which is widespread to this day. In reality, the Urals do not divide anything, except for separating the Pechora and Volga-Kama basins in the west from the Ob River in the east. The mountains extend from north to south in the tundra and the forest-tundra, mainly passing through the forest areas of Northern Eurasia. These territories, stretching from the Yenisei River to the Baltic Sea, were inhabited by Finno-Ugric peoples, who were hunters and fishermen with anthropologically and linguistically unique features. The southern edge of the Ural system (the Mugodzhar Hills), bordering the Steppe Belt, cannot serve as a divider either, since the latter is essentially a very different world (Fig. 4.6).[3]

Nevertheless, to this day the Urals (Fig. 4.7) is seen as a dividing line between the two in fact inseparable continents. This line, almost 2,500 km long, is marked in various ways—from grand (Figs. 4.8–4.9) to minor, poor and wretched gestures (Figs. 4.10–4.11)—even in tundra and on the Arctic coast (Fig. 4.9). Moreover, this imaginary border is cherished by tourists (Fig. 4.12).

4.4. A Complex Look at the Yenisei River Valley

Rather than the ridge of the Urals, the depressed valley of the mighty Yenisei River, with a basin covering 2.6 million square km, is an appropriate divider of Eurasia. The main valley of the river appears to cut it into two approximately equal parts—the east and the west—the entire bulk of northern Eurasia from the Arctic Ocean to the Sayano-Altai region. The Yenisey proper, from the confluence of its source rivers the Great Yenisey and Little Yenisey at Kyzyl

---

[3] In addition, a quote from Khain: "the Ural mountain system is only a part of the Ural-Mongolian orogenic belt, one of the largest dynamic belts of the Earth's crust, which stretches across the Asian continent and divides the East European, Siberian, Tarim and Chinese-Korean ancient platforms. It actively developed in the Late Precambrian's Paleozoic. The formation of the belt refers to the Riphean, marked by the fragmentation and faulting of the Early Precambrian continental crust (Khain 1991, 269). It should also be noted, that the geochronological Riphean period, which is as old as the Urals (the ancient Riphean Mountains), occurred between 1650–650 million years ago, long before the formation of the Alpine-Himalayan geosyncline, which played an incomparably more significant role in the history of the formation of the Eurasian continent (see, for example, Semikhatov 1984, 343–344).

Figure 4.6. Different proposed divisions into Europe and Asia: A—The borderline currently used by the U.N.; B—Along the Ural Mountains and the Ural River; C—Along the Yugorsky Strait, the Pay-Khoy ridge; the Urals and the Ural River; D—Along the border of Kazakhstan; E—Along the northern foothills of the Caucasus; F—Along the watershed of the Greater Caucasus; G—Along the southern foothills of the Caucasus; H—Along the Caucasus near the Rioni and Kura Rivers; I—Along the Lesser Caucasus and the Araks and Kura Rivers; J—Along the former border of the Soviet Union.

city (Figs. 4.13 and 4.14) to its mouth in the Kara Sea, is 3,487 km long. From the source of its tributary the Selenga, it is 5,075 km long.

The upper reaches of this gigantic waterway, about 3.5 thousand km long, are located relatively close to the iconic Dzungarian Gate, which divides the Eurasian Steppe Belt into an

Figure 4.7. The Ural Mountains. The satellite image clearly shows the contrast between the Urals and their "smoother" ridges due to old geological age and the Greater Caucasus with its ridges comprising part of the young Alpine-Himalayan Geosyncline.

Figure 4.8. A border sign between Europe and Asia near Yekaterinburg, Russia.

eastern and western part (see Figs. 3.5 and 3.6). It is also noteworthy that practically all major tributaries of the Yenisei River are located on its right bank, in the Central Siberian Upland. Yenisei's left bank—essentially the West Siberian lowland—has almost no tributaries.

For an in-depth (albeit brief) analysis of the matter at hand, one needs to examine materials from such diverse fields as archaeology, anthropology, linguistics, and ideology (religious and lay worldviews). All these features, together with specific economic models, ultimately shaped both the character and outlook of the main societies and cultures in western and eastern Eurasia.

Archaeology. The majority of Western Siberian archaeological monuments from the late Holocene epoch (that is in the Neolithic, the Bronze and Iron Ages) are settlements, testifying to the sedentary nature of local hunters and fishermen; there are virtually no cemeteries with burials. There is a completely different picture in Eastern Siberia. There are very few settlements and numerous impressive burials. East Siberian hunters and fishermen led a nomadic way of life.

Paleoanthropology. Anthropology distinguishes between two large families or races among the peoples of Eurasia: the Mongoloids and Caucasians. In the past several centuries, there have been no significant changes in the geographical distribution of these two groups, which make up the two major Eurasian races. The diverse Caucasian race occupies western Eurasia, whereas the Mongoloid race—the east of Eurasia. The latter is confirmed by archaeological and paleoanthropological finds east of the Yenisei River, in the valleys of its numerous tributaries, where representatives of the Mongoloid race are predominant in burials.

Figure 4.9. A borderline and a monument near the Promysl village on the road between Kachkanar and Chusovaya river.

Paleolinguistics. Undoubtedly, eastern Eurasia served as the domain of three major language families. Linguists recognize the Indo-European family as the largest in that part of the continent. There are more people speaking Indo-European languages than any other languages from a single language family on the planet. In the south, people speaking Indo-European languages lived together with Semites, whereas the entire northern vast forest zone of Eurasia was the domain of a relatively small population, who spoke Finno-Ugric languages.

In Eastern Eurasia, the groups who spoke the Tungus-Manchu languages settled mainly in the forest and steppe zones. The southern zone of eastern Eurasian was populated by numerous representatives of the huge Sino-Tibetan linguistic mega-family. The languages spoken by the populations of the Great Plain of China (between the Yellow He and Yangtze basins) are the most prominent in that family. Once again, the Yenisei River valley serves as a divider—this time, in terms of linguistics.

Figure 4.10. A border sign between Europe and Asia deep in the forests of the Middle Urals.

Worldviews, ideologies, beliefs. The most prominent ideological system in western Eurasia is the trio of related monotheistic Abrahamic religions: Judaism, Christianity, and Islam. Judaism was the first religion, associated with the mythical figure of Abram (or Ibrahim

Figure 4.11. The northernmost border sign between the two continents in the tundra zone near the Yugorsky Strait, between the cape and Vaygach Island (photos taken in the winter and summer).

Figure 4.12. The Middle Urals: tourists adorn the border between Europe and Asia with ribbons.

in Islam). It was him that the Almighty had chosen among all the humans to reveal himself to for the first time, according to the dogmas of these religions, and bound him with a strict Testament.

Eastern Eurasia has a different religious tradition. Its southern part, inhabited by sedentary agricultural cultures, is represented by three major distinct belief systems: Confucianism, Taoism, and Buddhism. In my opnion, only Buddhism can be regarded as a religious system. Compared to the Abrahamic religions of western Eurasia, both Confucianism and Taoism, the "competitors" of Buddhism, are hardly religions, but rather moral and philosophical teachings.

The above-mentioned ideological systems almost did not apply to the forest dwellers who believed in shamanism and animism. The cultures of nomadic stock-breeders of the Steppe Belt began to play a role in them only in or around the Modern period.

Figure 4.13. The confluence of the Great (left) and Little (right) Yenisei—
the place from which the Yenisei River properly begins in Eurasia.

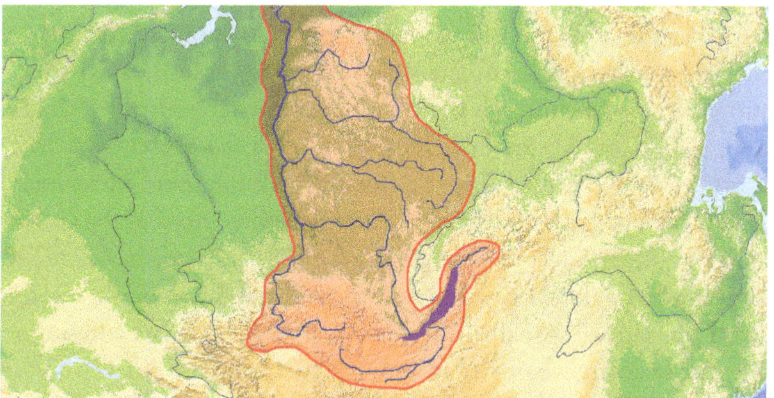

Figure 4.14. The Yenisei basin with the eastern bank's main tributaries.

4.5. Gero von Merhart and the Archaeology of the Yenisei River Basin

From the Urals, we have traveled 2,000 kilometers to the Yenisei basin, which clearly serves as a distinct dividing line between Western and Eastern Eurasia. In relation to the Urals, we have talked about Philip Johann Strahlenberg—a prisoner of war during the time of Peter I, who played a significant role in developing an understanding of the true significance of the Urals, so unique and unlike any other mountains on the continent, in the history of Eurasia.

I shall now turn to another eminent Russian prisoner of war, who was captured almost two centuries or 207 years later—Gero von Merhart (Fig. 4.15A, B, C). In 1914, almost at the very beginning of the First World War, an Austrian soldier, he was unlucky enough to find himself in remote Siberia, where his fate was very similar (down to some details) to the one of his Swedish predecessor. Just like Strahlenberg, he managed to build a name for himself as a serious archaeologist in a previously unknown and boundless eastern country. Let me quote

43

from E. V. Detlova and S. V. Kuzminykh (2019), who have examined the life of the remarkable Austrian archaeologist in a much more detailed and thorough way:

> Gero Kurt Karl Maria Merhart von Bernegg (1886–1959), in the future a famous European archaeologist and at that time only a prisoner of war of the Austro-Hungarian army, did not come to Siberia of his own free will. The scholar spent his childhood and youth in Bregenz, a town on the border between Austria, Germany, and Switzerland. . . . Merhart received a good education at a Jesuit gymnasium in the town of Feldkirch, near Bregenz. This was followed by several years at the University of Munich, which he graduated in 1913 with a degree in geology. In the same year, he began working as an assistant at the Munich Anthropological-Historical State Assembly. . . . The First World War set in motion a chain of tragic events in his life. In July—August 1914, Gero Mergart was called up for military service. . . . The first enthusiasm caused by the successes of the army gave way to disappointment, which intensified when he was captured by Russians in Galicia and was sent to a war camp for war officers in the town of Chita. However, the young scholar was much more fortunate than many of his brothers in arms. . . . he did not only survive, but also got the opportunity to return to his scientific work. In 1919, Gero Merhart became a restorer at the Museum of the Yenisei Territory in Krasnoyarsk (now the Krasnoyarsk Regional Museum), and a little later, under the Soviet regime, he was appointed head of the archaeology department. The results of his extremely fruitful work in Siberia are reflected in his numerous publications and reports in the archives of the Krasnoyarsk Regional Museum. These documents testify to the colossal amount of work that the Austrian archaeologist accomplished during less than two years that he spent in Krasnoyarsk, in extremely harsh conditions. His research of the Bronze and Early Iron Age of Southern Siberia brought him particular fame as a specialist in the field of the Early Metal Era. In addition to his most well-known works, his *Bronze Age on the Yenisei* (Merhart, 1926) remains unsurpassed and in demand, [despite its title, it does not focus on the Bronze Age alone (see Merhart 1923)]. The book comprises a huge amount of material, which includes not only the results of his own field observations, but also data obtained from various Russian museums and from a small body of scholarship that existed at the time. In the fall of 1921, Gero Merhart returned to Austria. The high point of his career came a few years later, when in recognition of his services to science he was invited in May 1928 to become the first full-time professor of ancient history at the department of the University of Marburg-an-Lann (Germany).

Merhart's role in the formation of archaeological knowledge about Southern Siberia is very significant, although his book entitled *Bronzezeit am Jenissei*, published in Vienna in 1926, is usually rightly seen as the most important publication in his career. Moreover, his account of his seven-year stay in Siberia, which fell on the particularly difficult and seminal years 1914–1921, is of great interest not only to Western but also Russian readers. A book with his perceptions of this difficult period in Russian history was published under an unusual and interesting title *Daljóko* (Latin transliteration of a Russian word which means "Far away"). The book (Fig. 4.15D) together with his other works has been rather recently discussed in a book by prominent German archaeologists (Parzinger 2009).

4.6. The Invisible "Vertical" of Division

We are finishing our story about one of the most important "verticals" of the northern half of Eurasia, or the vast Yenisei Valley (Figs. 4.13 and 4.14). The importance of this great Siberian river lies in the fact that relying on the Yenisei bed, we can divide the most massive northern part of the continent into two halves, West and East, as many researchers do.

Figure 4.15. Gero von Merhart and the cover of his book *Daljóko* (source: Merhart 2008).

Some readers may have noticed that the author has discussed the northern (Siberian) half of the continent, closely connected with the Yenisei Valley, with various data on history, archeology, paleoanthropology, and linguistics. And relative to them, geology (especially in this chapter) has faded in apparent importance. But without a doubt, it is geology that should have led this list of sciences, since geology is clearly the real foundation of existence in the megastructure of all of the continents. The author deliberately placed an emphasis on these other fields because the scientific findings of Siberian researchers is less well-known.

Returning to the topic: following the northern Yenisei "vertical," we are obliged to extend this vertical in a southern direction, to the south of the Alpine-Himalayan geosyncline heights. Admittedly, it is not easy to find among the many Eurasian rivers one that closely and obviously corresponds to the northern Eurasian vertical and continues to follow the Yenisei riverbed in the south of the continent. The author has decided to propose the very famous Indus River, whose bed also crosses huge spaces south of the Alpine-Himalayan geosyncline. Now, we shall say a few words about this river.

4.7. The Indus River and Yenisei: From Ocean to Ocean

The Indus River is a great trans-Himalayan river of South Asia. It is one of the longest rivers in the world, with a length of 3,200 km, and drains an area of about 1,165 thousand square km (Fig. 4.16). The river's annual flow is about 243 cubic km—twice that of the Nile River and three times that of the Tigris and Euphrates rivers combined. The river's conventional name derives from the Tibetan and Sanskrit name Sindhu.

The general flow of the Indus is caused by the collision of two littoral plates: the Iranian and Hindustani. Therefore, it is assumed that the central channel of the Indus coincides with the junction line of these plates (Fig. 4.17).

The Yenisei River flows into the Kara Sea of the Arctic Ocean basin, having traveled no less than 3,200–3,500 km from the confluence of its sources (see Figs. 4.13, 4.14). Similarly, but in the opposite direction, the waters of the Indus travel from their sources in the snowy ridges of western Tibet about 3,200 km into the Arabian Sea and the basin of the southernmost ocean on the planet: the Indian Ocean (Fig. 4.18).

4.8. The European Mega-Peninsula Is Europe Proper

Let us briefly talk about the so-called European mega-peninsula, which can be considered Europe proper, albeit not as the continent, but only a part of Eurasia, a special area or only as specific exclave of western part a gigantic continent.

To begin with, let us define the borders of the mega-phenomenon. The Mediterranean Sea together with its northeastern "clone"—the Black Sea or Pontus—is, in fact, a gigantic, many thousand-kilometers-long gulf of the Atlantic Ocean, which cuts deep into the body of Eurasia. Bearing many similarities to the Mediterranean Sea, the Baltic Sea in the North is not as wide and diverse. These two gulfs of the Atlantic Ocean delineate the European mega-peninsula in the south and north. The elevation of this peninsula to the rank of a mega-peninsula is justified

Figure 4.16. The Indus River. The northern section of the river bed is blocked by the snowy ridges of the Himalayas. The southern section of the river bed is located on the flat plain of Pakistan.

by the fact that this entire far western part of Eurasia has branched out into other smaller peninsulas. The latter, of course, differ greatly in terms of their territory: from the huge Pyrenees, Apennine Mountains, and the Balkans to the small Peloponnese Region and Jutland. All in all, the total area of Europe amounts to approximately 4.4–4.5 million square km (see Fig. 4.19).

This brief description alone makes it clear that the European mega-peninsula consists of two relatively equal halves: the southern and northern. This division is based not only on a proximity to the Baltic or to the Mediterranean, but also to the western ridges of the Alpine-Himalayan geosyncline.

The European mega-peninsula has experienced in full the consequences of the Atlantic greenhouse effect during the Paleolithic Pleistocene. As a result of it, the phenomenon of the Steppe Belt declined, having encountered a different geoclimate in the Eastern Carpathians and the lower reaches of the Danube.

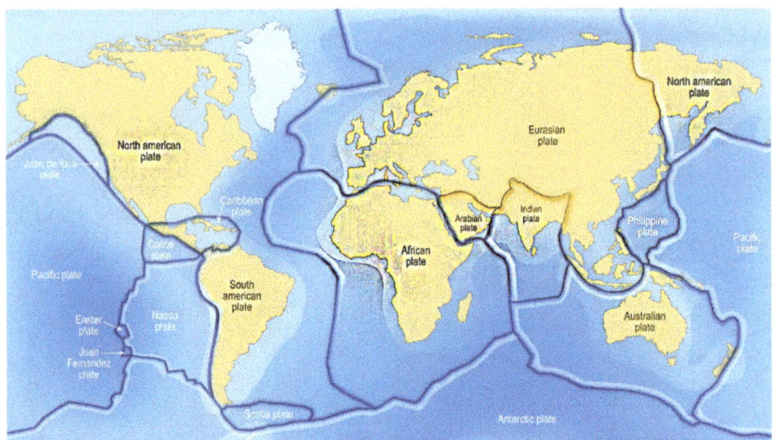

Figure 4.17. The littoral plateaus in the Eastern Hemisphere of the Earth.

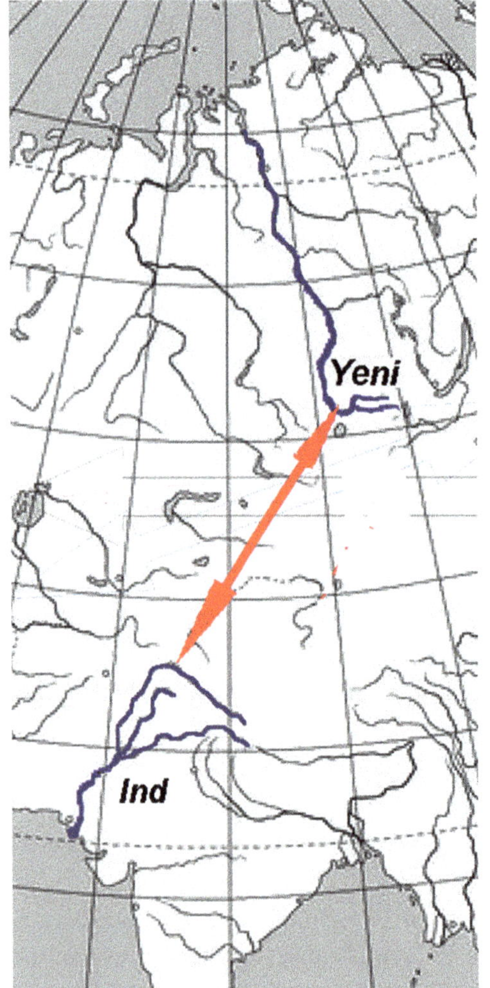

Figure 4.18. Proposed symbolic vertical division from the Arctic to the Indian Oceans. The red line indicates the symbolic connection between the sources of both great rivers.

The above-outlined approach to the continent's geography makes it possible to define the borders of Europe in a completely different way: from the edge of the Eastern Baltic up to the eastern edge of Pontus, or more precisely to the eastern edge of the Sea of Azov. The latter and not the Urals have always marked a perceived division of Eurasia into the imagined continents of Asia and Europe. However, we will emphasize once again, and very insistently, that it is not the continent of Europe, but a specific extreme western exclave of Eurasia (see Fig. 4.20).

On this geographical map, the European exclave looks quite impressive against the backdrop of the Eurasian continent, which is fundamental to it. This exclave differs from its foundation by its very small size and, perhaps, by the striking and impressively "ragged" character of its coastal outlines.

4.9. The Great Chinese Plain and the Han

Let us now make another giant leap: from the western to the far eastern regions of Eurasia. Remarkably, these two regions of Eurasia are separated by the Steppe Belt, stretching for 8,000 kilometers (Chernykh 2009; 2013a; 2017). This division is different from the one described

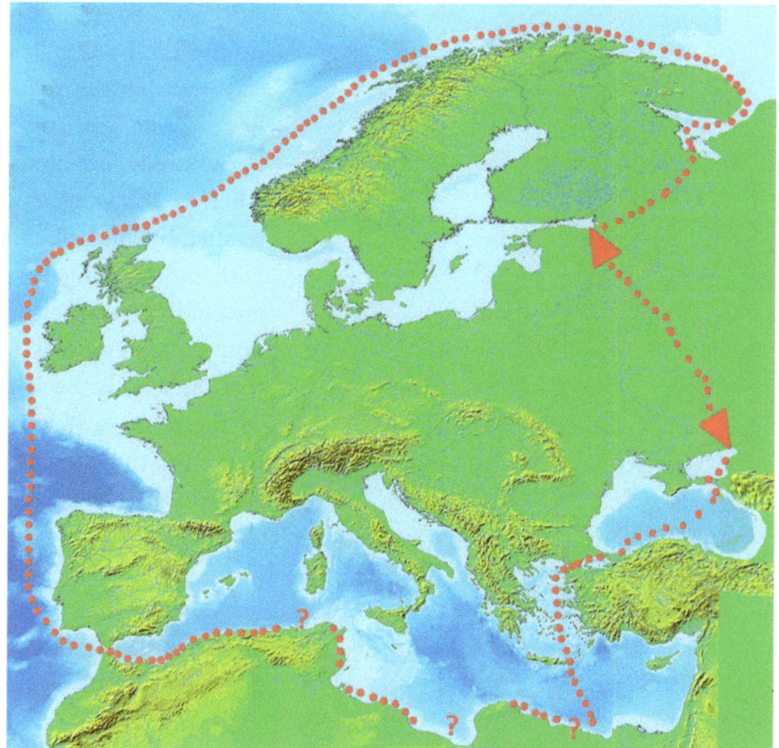

Figure 4.19. The general borders of the European exclave. The red arrows point to the extreme points of two giant gulfs of the Atlantic: the Baltic and the Mediterranean Seas.

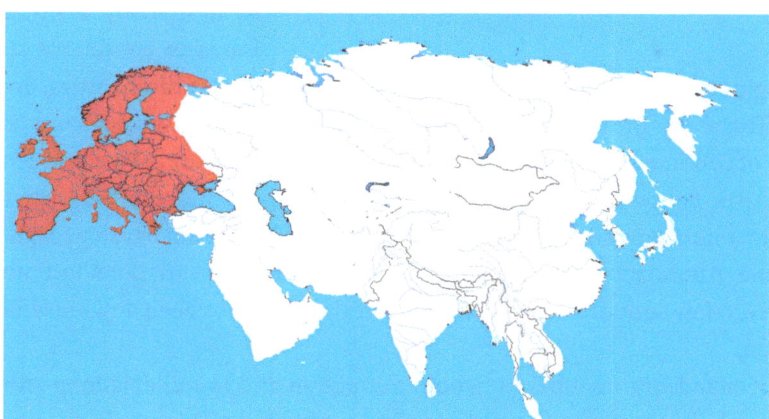

Figure 4.20. General contours of the Eurasian continent, with the European exclave in red.

in the previous chapter. While the Steppe Belt separated the North from the South of Eurasia, it also divided the continent into a western and eastern part.

The far eastern region of Eurasia is defined mainly by the valleys of the middle and lower reaches of the great Chinese rivers Yellow He and Yangtze (Fig. 4.21), even though it stretches far beyond them. The western border of the region is marked by the mountains of the Tibetan highlands (the Henduan massifs and others), through which the two famous Chinese rivers

Figure 4.21. The borders of the Great China (*Han*) Plain.

run. The eastern border is defined by the Pacific Ocean with its seas starting from the Yellow Sea to the South China Sea. The northern border of the region abuts Manchuria, whereas the southern frontier is defined by the Indochina Peninsula with the Mekong River. The total area of the Han is close to four million square km. It is worth noting that the Han is the self-designation of the Chinese people. No other regions in the vast Eurasia are mentioned under their ethnic names in this book.

4.10. Two Millennia of Discussions

I'd like to conclude this fascinating discussion with the following question: so where does the border, so desired by many, between the Eurasian continents stretch? Right away, one begins to doubt whether this subject is worth all the endless battles, even if they are verbal only. The above-proposed division based on the Yenisei River does not correspond to the principle of continental divisions, since it only serves as a dividing line between the western and eastern parts. Following that principle, it would be much more appropriate to divide the continent into a northern and southern part, with the Alpine-Himalayan geosyncline serving as a borderline. Abundant archaeological and anthropological variations in finds from these parts would certainly support such a division, but there has not been anyone willing to accept such a division.

Quite understandably, my opinion would not put an end to millenia-long efforts to define the coveted borderline between the two imaginary continents. This would not happen even if I include here one of the numerous geological reconstructions of continental lithospheric plates, elaborated by multiple world experts in the Earth's geology (Fig. 4.17). The Eurasian plate, one of the largest of all continental plates on Earth, does not reveal any rift (discontinuous) seams in the location of the supposed intercontinental borderline. The separation of Chukotka and Kamchatka in northeastern Eurasia, along the edge of the Sea of Okhotsk, and their attachment to the North American Plate would not solve the problem. After all, no one is going to regard this part of Siberia as part of the continent that is so "disliked" by many Eurasians. Therefore, the eternal Russian question of whether Russia's place is in Europe or Asia will remain unanswered for a very long time.

However, if we turn briefly to the obvious signs of mutual influence between two small plates—the Indian (Hindustan) and Arabian—then we can note the result of the "geoclash" that took place millions of years ago. The effects of this clash allow us to divide the Eurasian continent in addition to the north and south also into its two other parts: the west and the east.

It is a great pity that the idea expressed more than a hundred years ago by the great Dmitry Mendeleev about Eurasia being a single continent has remained virtually unnoticed; discussions on the subject continue to this day.

4.11. Discussions Continue

Only discussions featured in the pages of a new, peer-reviewed journal entitled *Journal of Eurasian Studies* will be examined here. The journal's editors describe it as an open access publication, issued by the Asia-Pacific Research Center of Hanyang University (South Korea). The journal is interested in studying materials not only pertaining to the whole of Eurasia, but also other regions. Alexander Lukin, a Russian researcher whose analyses were published in the magazine, has described the goals of the publications most consicely in an article entitled "Eurasia—from confrontation to partnership" (Lukin 2018). To qoute from the article:

> The idea of Eurasia per se is not nearly as old as that of its constituent parts—Europe and Asia. The latter two date back to ancient Greece, whereas, according to some accounts, not until the 1880s did the Austrian geologist Eduard Suess first coin the term "Eurasia" [See Chapter 4]. His idea was to fashion a union of the two divided parts of a single continent as a demonstration of their inherent unity—initially in the geologic and geographic sense, and later, in the social and political sense.
>
> The idea, however, that Europe and Asia are fundamentally different—rooted in European thinking since the time of Ancient Greece and later carried from there to Asia itself—is still widespread. Ever since the Enlightenment and the colonial period, Europe and the "Western civilization" founded on it have most often been viewed as "advanced," while Asia has been seen as a "backward" continent in need of being "pulled up" to the level of Europe.
>
> The concept of "Greater Eurasia" or "Greater Eurasian Partnership" differs significantly from the 19th-century geographic interpretation of Eurasia, the romantic ideas that Russian Eurasianists held in the 1920s, and the abstract constructions of classical geopolitics. It is, in fact, a very concrete and modern concept. When the collapse of the Soviet Union destroyed hopes of building a united Europe stretching from Lisbon to Vladivostok, and when the West's relations with Russia and China began deteriorating early in this century, Russian political scientists proposed the idea of "Greater Eurasia" and Russian leaders soon adopted it as official policy. The idea was that "Greater Eurasia"—based on cooperation between Russia, China, India, Central Asia, and other countries of the continent—would become one of the economic and political centers of the

emerging multipolar world, or even a conglomeration of several non-Western centers that had no proper place in the sweeping global system imposed by the West. China gave its official sanction to the concept and began concrete work with Russia towards its implementation.

This Special Issue of Journal of Eurasian Studies is the first collection of articles by contributors from Russia, China, USA, and Central Asia devoted to the concept of a Greater Eurasia. Writing from a variety of viewpoints, they examine the processes currently unfolding in the region. Many of these authors helped develop the concept of Greater Eurasia, while others discuss the approaches of major actors interested in joining or understanding the integrative processes on the continent. Despite some differences in their approaches, all look favorably on the Eurasian integration process.

The authors of the articles in this special issue examine the history of the development of this concept in detail and assess the prospects for the implementation of the "Greater Eurasia" project. It begins with an article by Sergey Karaganov, a well-known Russian international relations expert and one of the originators of the "Greater Eurasia" concept.

Since even the first part of this chapter is titled and dedicated to the "almost eternal issue of a border between Europe and Asia," a view that there is no end in sight to this ever-lasting issue would hardly surprise anybody.

4.12. Four Main Parts of the Continent and Their Spatial Shares

We will conclude this chapter with a brief description of the spatial characteristics of the four main parts of the Eurasian continent, divided North-South and West-East (Fig. 4.22). The total territory of Eurasia, according to the Great Russian Encyclopedia, is 53.6 million square kilometers, including all the islands connected to the continental landmass, which is about 3.45 million square kilometers. That is, the continental part is close in area to about 51 million square

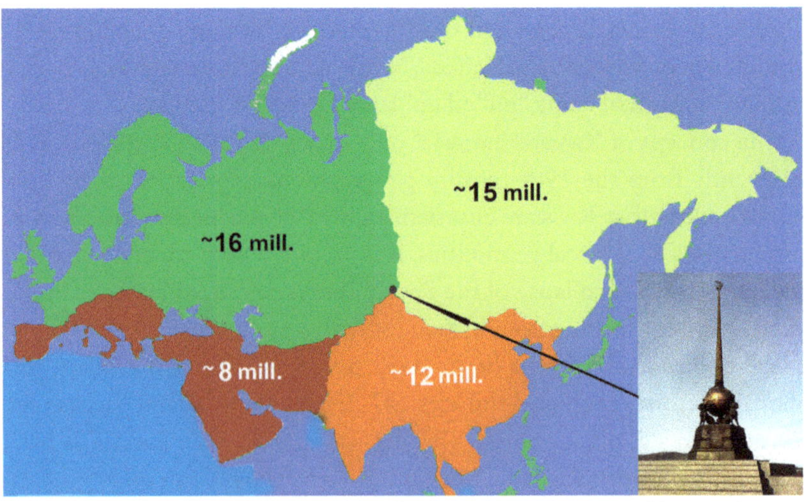

Figure 4.22. The general "quarters" of the Eurasian continent (the numbers indicate the area of the territory of each of the "quarters" in square kilometers).

kilometers. In studying the Eurasian georegions, which will be discussed below, the author has mainly relied on the characteristics of the continental landmass, although sometimes it was also necessary to use materials from a number of islands close to the mainland; for example, the British Isles. During the division of the Eurasian continent from north to south, when the Alpine-Himalayan geosyncline served as a basic support, it was possible to record a picture of extremely pronounced spatial differences in the data of both halves of the continent.

Geographically, the northern half of the Eurasian landmass is sharply predominant, occupying about 31 million square kilometers, or about 61% of the total land. The south of Eurasia is represented by no more than 20 million square kilometers of land, that is, about 39%. The picture looks completely different when studying the territorial proportions of West-East: here the shares of each side are, in fact, quite close to each other: about 24 million square kilometers (West) and 27 million (East); that is, about 47 and 53% on each side (Fig. 4.22). The spatial differences recorded are certainly important in constructing a general picture of Eurasian geographic zones.

Chapter 5

Important Geoareas and Enclaves of the Eurasian World

The structure of Eurasia, the largest continent on the globe, is complex. Most likely, its complexity is the main reason behind the unceasing debates about the border between Eurasia's Europe and Asia, which have been ongoing for more than 2,500 years. The giant Eurasian lithospheric plate served as the foundation of the Eurasian continent, as was discussed earlier in Chapter 4 (Fig. 4.17). Rejecting hypotheses about the plate's bicontinental nature, it has been stressed earlier in this book that Eurasia is a single continent. This monograph's main goal is to develop concepts about the megastructure of the Eurasian world by identifying areas that are most demonstrative of most important global composite parts.

5.1. Sixteen Important Geologo-Geographic Areas in Eurasia

Our research on the division of Eurasia along the lines of North—South and West—East (Chapters 3 and 4) allows us to outline sixteen geoareas for the Holocene epoch on our continent (Fig. 5.1). We will list them very briefly below, together with data on the approximate land area covered by them (in parentheses, in millions of square kilometers).

A. Northwestern Eurasia, five geoareas:
1—Northern Europe and the British Isles (~2.8),
2—Fennoscandia (~1.7),
3—Western Urals (~3.6),
4—Eastern Urals or Western Siberia (~3.4),
8—Western part of the Steppe Belt (~4.5).
Total area — approx. 16 million km².

B. Northeastern Eurasia, four geoareas:
5—Eastern Siberia or Yakutia (~7.0),
6—Chukotka (~3.3),
7—Kamchatka peninsula and the Kuril Islands (~1.2),

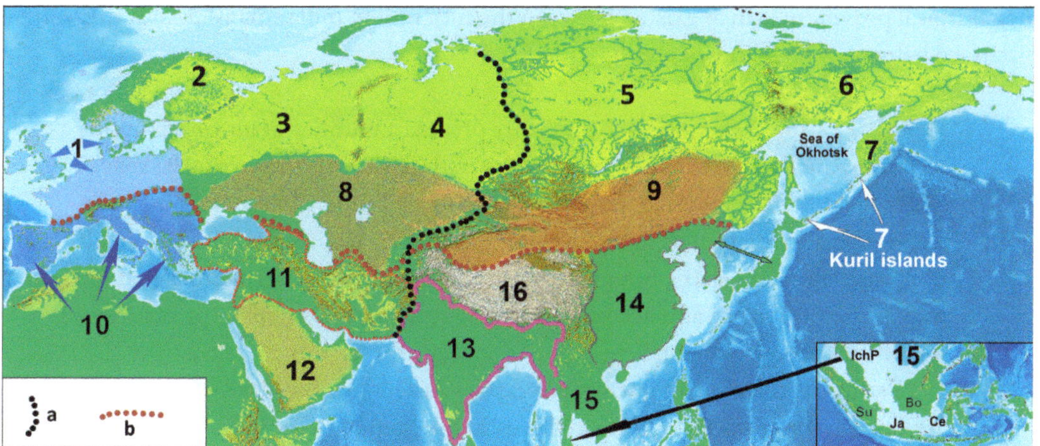

Figure 5.1. Borders of the 16 most important geoareas of Eurasia. Symbols of the sections of Eurasia along the lines West—East and North—South.

9—Eastern part of the Steppe Belt (~3.5).
Total area — approx. 15 million km².
Northern Eurasia, total area — approx. 31 million km².

C. Southwestern Eurasia, three geoareas:
10—Southern Europe (~2.0),
11—Anatolia and Iran (~3),
12—Arabian Peninsula (~ 3.0).
Total area — approx. 8 million km².

D. Southeastern Eurasia, four geoareas:
13—Hindustan (~3.0),
14—North China Plain (~4.0),
15—Indochinese Peninsula (~2.8),
16—Tibet (~2.2).
Total area — approx. 12 million km².
Southern Eurasia, total area — approx. 20 million km².

Their total area is close to about 51 million square kilometers. At the same time, the area of geoareas of the northern half of Eurasia reaches about 31 million square kilometers, while the southern geoareas are sharply inferior to the northern ones, with their coverage close to 20 million square kilometers (Fig. 4.22).

Each of these areas had distinct geomorphological, geoecological and also geoclimatic features. The latter were products not of men, but nature (as the first part of the preceding chapter states). Perhaps, one of the most interesting research areas is the intersection between the enclaves' nature-made and man-made components or the effect of human cultures and their different economies (e.g. the appropriating and production economies) on nature. Another interesting problem is the interrelation between the enclaves and each of the ten Eurasian metallurgical provinces, which existed between the fifth to first millennia BCE. But this will be discussed below—mainly in Chapter 6.

Here, the author is inclined to necessarily present a number of his observations regarding the limit or the limit of accuracy in determining the territorial spaces of the geoarea. The figures given above in millions of square kilometers are given in some approximation or non-variable rounding. In reality, it is hardly difficult in all cases to establish the boundaries between neighboring formations that are true in their certainty and accuracy, and it is clear that this applies primarily to formations on the mainland.

Perhaps this feature of uncertainty is manifested in the analysis of the northern border, for example, of both enclaves of the Steppe Belt: there, a wide strip of forest-steppe does not allow us to determine any accurately and definitely a "hard" border in the zones of the "junction" of the Steppe Belt and forest-tundra geoareas, which is why the territorial estimates we have proposed are largely conditional. But also in other cases, such dividing lines could relatively rarely be determined with proper accuracy and certainty, unless they were connected, for example, with maritime borders. It was very often possible to observe fluctuating/undulating border shifts, especially characteristic of, say, the Han enclave. For this reason, the author was forced, in fact, to round the values, and in both directions—both plus and minus. The corresponding figures of spatial coverage are given here, but in such an approximate form. Moreover, we note, however, that it is unlikely that such approximations will have any significant effect on the general picture of our study.

5.2. Significant Differences between Eurasian Geoareas

Each of the sixteen Eurasian geoareas outlined here differed from the others—neighboring or relatively distant—fifteen areas by quite definitely expressed specifics of geomorphological, geoecological and, accordingly, geoclimatic features. However, it hardly makes sense to discuss this topic in any detail in this chapter of our book. A detailed (or even a cursory) description of these issues in the most diverse literature is likely.

However, it is quite well known to readers, and therefore the author does not see any sense in attracting additional attention to these issues. Nevertheless, in this series it is necessary to outline one of the problems to which different researchers paid very little attention, as if in passing. This problem is connected with the position and real dimensions of the geoareas outlined in our book, and this has been extremely noticeable, and probably affects the processes to this day the history of the historical development of various Eurasian peoples.

The territorial dimensions of each of the 16 geoareas are not always similar to each other. Sometimes the differences between them appear as extremely significant. Here, for example, are two geoareas located relatively close to each other: the first is Yakutia/Sakha, covering about 7 million square kilometers, and the second is Kamchatka with the Kuril Islands, the length of the entire ridge of which is close to 1,200 km, but with a total approximate coverage of only 1.2 million square kilometers. However, for us, perhaps, the characteristics of the average geoarea values in the northern and southern parts of the Eurasian continent are of much greater interest.

The northern half of Eurasia is more than one and a half times larger than the southern half (see the previous chapter, Section 4.11 and Fig. 4.22). In terms of the number of geoareas, the northern half of the continent also seems to surpass the southern one, but not at all as expressively as in the general area: here there are ten geoareas against seven in the south. However, when comparing the data of the territorial coverage of the compared areas, the differences between them look significantly more impressive. In the north, on average, each geoarea occupies about 3.75 million square kilometers, and in the south—almost a million less—2.86.

The question is obvious: what are the reasons for such discrepancies? And how can this question be answered? Here again we must remind the reader about the incessant movement and collision of giant continental plates, about the pressure on the Northern Eurasian plate, two southern plates — African and Australo-Indian. These constant collisions still lead to surface distortions of the junctions of all three littoral plates. Compared to the relatively flat—even in gigantic spaces—"landscape" of Northern Eurasia, the southern half looks as if it is broken, even warped. It is for this reason that in the south of the continent it is much easier for us to both mark and delineate the boundaries between individual geoareas.

Now we will move on to another aspect of these complex issues: on any map of Eurasia, even an extremely small-scale one, it is not difficult to notice that on the southern edge of the continent, there are three strange, massive, and seemingly hanging "pockets" to the Indian Ocean (for example, see Fig. 2.2 or 5.1). In cartography, such hanging "pockets" are usually referred to to the category of peninsulas, and such in this case are three: Indochina, Hindustan and Arabia. The history and roots of this kind of pockets are of interest to us, of course, extraordinary. However, we note at the same time that to Indochina. This does not quite apply to China. The situation there seems quite clear: this extreme southeastern peninsula in Eurasia is formed by its formation. About the earthly origins of the Eurasian world is due, as it were, to the expressive breaks of the Alpine-Himalayan geosyncline hanging down to the south of Tibet and facing the Indo-Eurasian Islands.

The sources of the other two peninsulas located west of Indochina look much more indicative for us. Here, the distinct signs of endless collisions of intermaterial littoral plates can serve those geoareas that we can consider alien to Eurasia and which, as it were, "stuck" to it from the south. And such strange Eurasian "geoadopters" can be considered two geoareas, Arabia and Hindustan. Note at the same time that both formations are very different from each other.

5.3. Arabia as Part of the Sub-Saharan Littoral Plate

The alienness of the Arabian Peninsula and its corresponding enclave is clearly visible already against the background of other formations located to the south Alpine-Himalayan geosyncline and closely pressed either to the ridges of this geosyncline, or to the shores of the depths of the Black and Caspian Seas (see, for example, Fig. 5.1:12). The peninsula itself is adjacent to the

Figure 5.2. The Arabian Peninsula, view from the south (satellite imagery, based on materials from Wikipedia).

Iranian-Anatolian geoarea, or rather to Palestine and Mesopotamia. The desert geoecology of Arabia is very contrasting with the northern and Irano-Anatolian Highlands. Equally expressively different from them are the most important features of its geomorphology: border frames of the peninsula are the narrow slit-shaped Red Sea and the Persian Gulf (Figs. 5.2 and 5.3). All these important details of Arabia say that this semi-island is, of course, the northeastern, sub-Saharan part of Africa, rather than the southern "hanging pocket" of Eurasia.

But only during the breaking of the supercontinent Pangea, the system of rift (slit-like) articulations went "not as it should have." And if the space forces had broken the platform of that supermaterial through the gap of the Persian Gulf, then Arabia would have retreated to Africa.

However, all our observations and conclusions turned out to be very consonant with the views of geologists on the relationship of Arabia with both continents: this peninsula undoubtedly has obvious African roots. Here are only brief excerpts from V. E. Khain's book "Tectonics of Continents and Oceans," where the author considers these issues against the background of general planetary phenomena and in connection with them:

Figure 5.3. The Red Sea and its most important bays, the Nile River, the Dead Sea (satellite imagery, based on materials from Wikipedia).

In the Oligocene, Africa and Arabia entered the neotectonic stage of theirdevelopment. Soon the giant East African rift system began to form, and in the Late Miocene [23–5.3 million years ago] the Gulf of Aden and the Red Sea opened up with simultaneous spreading in the first and Pliocene [5.3–2.6 million years ago]. . . . The late Miocene and the beginning of the Pliocene were the time of a new activation of endogenous processes. Its most remarkable expression was the spread of the spreading axis The Arabian-Indian ridge . . . with the formation of the first eponymous cauldron-the fault, and then the Gulf of Aden and, finally, the rift Suez and Aqaba grabens with the emergence of the East African Rift system. These events were associated with the separation of the Arabian Plate from Africa and the beginning of its movement to the north, along the transform fault passing through the Dead Sea (Khain 2001, 336, 584).[1]

---

[1] The professional language of geologists is often incomprehensible to an untrained reader, so it may be useful to supply here some explanations to the text of the above book by V. E. Khain: A *rift* is a linear depression in the earth's crust formed at the site of a crust rupture as a result of its stretching; a *rift system* is a complex of such depressions and ruptures; *spreading* is stretching or expansion; a *graben* (here) is a bay connected to the rift gap.

It is also important to add that the entire so-called Mediterranean or Saharan shield of North Africa (see Section 2.1 and Fig. 2.1) can be considered a special region that played a very significant role in the formation of the megastructure of Eurasia, although its base rested on a different (African) littoral plate—though it also rests on the Arabian one.

5.4. Hindustan: The Northern Part of the Australo-Hindustan Plate

Hindustan is a part of a large, so-called Australian-Indian tectonic plate, which for many millions of years has been steadily advancing in a northerly direction. It is and on the present time is in incessant collision with an even larger massif of the Eurasian plate (Figs. 5.1:13, 5.4).

The Hindustan subcontinent is the only fragment of Gondwana [also a super-continent, but earlier—E. Ch.], which turned out to be entirely in the Northern Hemisphere during its collapse and disappeared about 40 million years ago into the new supercontinent of Eurasia. . . . After Collisions with Eurasia, India's movement to the north did not stop, but it slowed down. The territory of the subcontinent began to experience an increasing rise (Khain 2001, 338, 348).

In essence, the pressure of the tectonic plates—the Australo-Hindustani plate in conjunction with the African plate—on the Eurasian platform led to that huge uplift of the Alpine-Himalayan geosyncline, for many thousands of kilometers. After all, it was she who literally cut, albeit unevenly, the whole of Eurasia into northern and southern halves. The appearance of giant plates

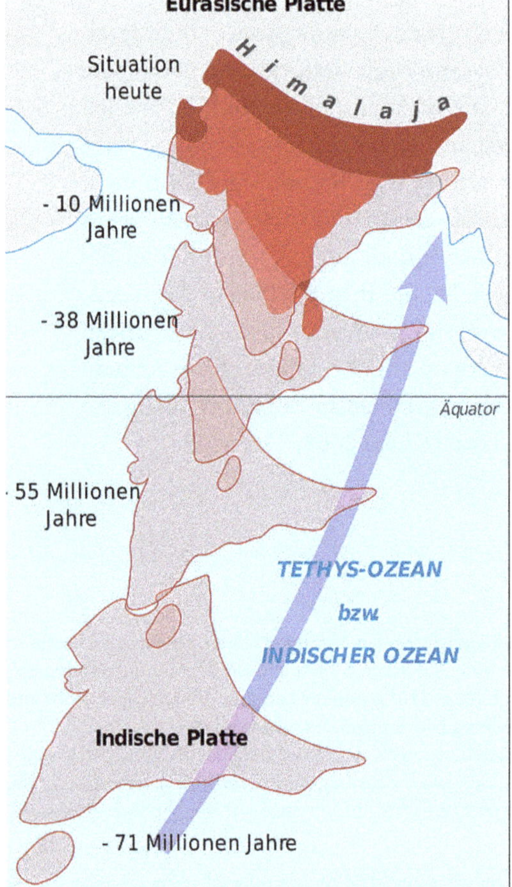

Figure 5.4. Diagram of the movement of the Hindustan littoral plate with the formation of the Himalayas in the zone of junction with the Eurasian continental plate.

Figure 5.5. The Hindustan plate, the Hindustan Peninsula and Tibet.

Figure 5.6. The Tibetan Plateau and the Himalayan ridges in the Far Pan (based on Wikipedia).

on the ocean surface of Hindustan proper (Figs. 5.4 and 5.5), and, according to cartographic terminology, we will call this geoisland a peninsula. In the same series of geological events that transformed the appearance of the Earth, as one of the most important consequences, we will certainly also mention the addition Tibet is the Tibetan plateau, so different both in its gigantic scale and amazing heights from other high-altitude objects of our planet. It is believed that the Hindustan plate is "crawling" or "pressing" under the Eurasian plate, which is why the plateau of Tibet has risen to a five-kilometer height, fenced off from the low Hindustan by the ridges of the Himalayas (Figs. 5.1 and 5.6). In the end, archaeologists and historians do not care enough whether it is native to the Eurasian or alien to this continent Hindustan and the Hindustan geoarea combined with it. However, even in the case of alleged indifference, we are unlikely to be able to ignore that massive of difficult-to-explain mythological paintings and narratives that turned out to be associated with the subcontinent of India (but more on this below and in other chapters).

5.5. Chukotka and Kamchatka with the Kuriles: Are Geoareas Alien to Eurasia?

It is also impossible to ignore the rather widespread opinion that the entire vast north-eastern edge of the Eurasian continent—that is, Chukotka and Kamchatka with the extended ridge of the Kuril Islands are, in fact, only part of the littoral North American plate (Fig. 5.7).

But from our position it is not so! It is quite likely that much earlier—according to the geological account, say, in the so-called Carboniferous period of the Paleozoic, and this is about three hundred million years—the supposed unity of the littoral plates was observed. However, later there was quite an obvious problem between them—a crack or a gap, as reported, by the way, in his famous book Alfred Wegener, the founder of the theory of the mobility of tectonic plates or the so-called "mobilization" (Wegener 1929, 18, 19; Figs. 4 and 5). For us historical archaeologists, of course, the current localization of both continents—Eurasia and North America—is much more important, when a clear territorial "gap" began to grow between their littoral plates. It was then, quite interesting that in the Quaternary geological period both truths were formed—but the Eurasian geoarea is Chukotka and Kamchatka (Figs. 5.1 and 5.8). However, we—archaeologists, historians—cannot, and hardly have the right to actively interfere in highly professional disputes of geologists regarding the connection of the geoareas highlighted in our works with the tectonic "monolism" of littoral plates. In this kind of search, our primary task is, of course, the correct interpretation of geomorphology data for a desirable and equally correct interpretation of the history of various bio-social formations of *Homo* against the background of various geo-changes on our planet.

We are nearing the end of our geomorphological discussions. All these various shifts and changes on Earth we attribute, of course, to the category of natural and therefore not man-made phenomena. There is no doubt that people—especially in recent centuries—are trying to participate in and influence these natural, and therefore non-man-made additions in some way to the best of their abilities. However, if *Homo* is reflected in the structure of our planet, then it is unlikely that these changes can be too noticeable.

Nature has prepared for the genus *Homo* very diverse and somewhat non-repeatable geoareas, which have served as an absolutely necessary foundation for humanity for its million-year existence. And without such a base, the existence of humanity, apparently, would simply be impossible.The history of the initial settlement of all the continents of the planet Earth was completed during the so-called fifth Intermaterial Migration of Peoples (see Chapter 2, Section 2.4). It was preceded by the fourth migration, when obvious traces of *Homo sapiens* of the late Paleolithic were discovered and from scientists are already in the subarctic zone of the extreme northeast of Eurasia. That is at least 20 thousand years ago, it could be assumed that all the spaces of Eurasia, and therefore all the geoareas we planned, were inhabited by various *Homo* groups. At the same time, the Paleolithic cultures of the Eurasian continent had the honor to become pioneers in the development of the gigantic spaces of North and South America—from Alaska to Tierra del Fuego.

And all these vast expanses of the earth were inhabited by very diverse groups of *Homo*. In victorious battles with the most diverse representatives of the terrestrial biosphere, the entire land, and later even the water element of the planet was under the rule of the human race. But this applied, first of all, to those representatives of the biosphere who were easily visible people and distinguishable by them. But that's when—and even to this day—humanity does not really manage to master and cope with the microscopic world of the biosphere that remains invisible to the human eye. And this world remains in many ways very harmful and insidious for *Homo*. We have allocated only sixteen geoareas for Eurasia. Are there many or few of them? The answer to this question may lie in the general task that a particular researcher sets for himself. Yes, and by the way, in many ways, all decisions of this kind are distinguished by an approach that is not objective at all, but subjective. Let's say you set yourself the task of determining the megastructure of the continent when dividing it only into two or, say, into four parts. And then what we have done and completed in chapters three and four will be quite sufficient for you (see Fig. 4.22). In the first edition of this book, nine geoareas were presented, but now such a picture seems insufficient, which is why we expanded the range of our divisions to 16 geoareas. . . . Probably if the task of the divisions was limited, say, only to the European mega-peninsula, where two geoareas were planned (Northern Europe

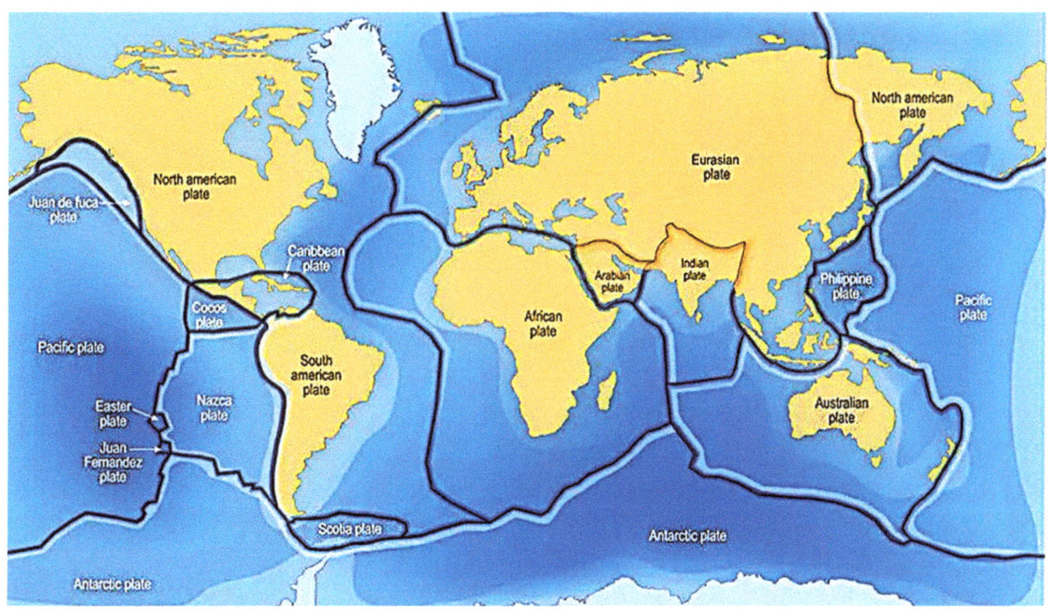

Figure 5.7. The circular arrangement of various continental and oceanic plates on the globe; the authors of such constructions are confident in the relation of the northeastern part of Eurasia (geoareas Chukotka and Kamchatka with the Kuril Islands) to the North American littoral plate (based on Wikipedia).

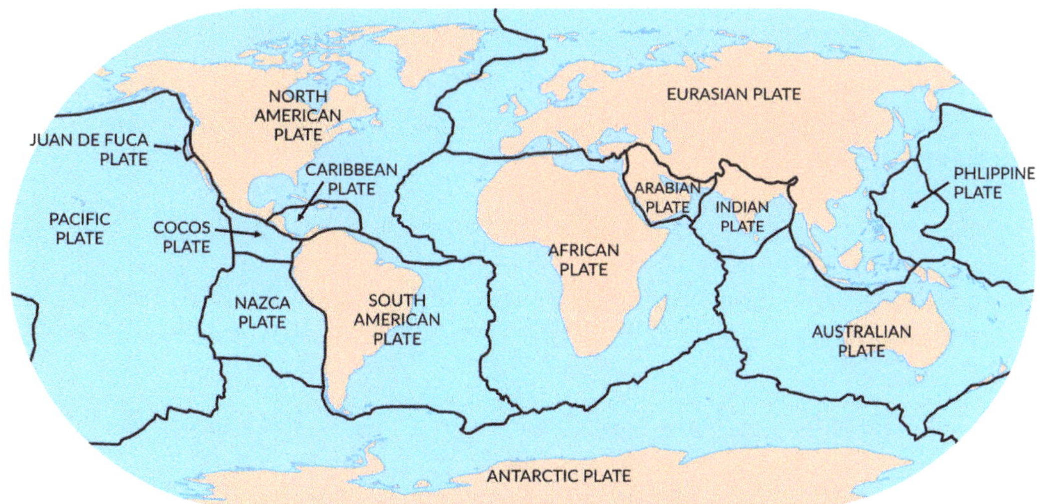

Figure 5.8. The circular arrangement of various continental plates on the globe, which the author of the book considers more realistic.

and Southern Europe), then in such a case, and the separation would require greater fragmentation—and this seems quite obvious even with a visual analysis of geoareas ... (see, for example, Figs. 4.19, 5.1:1 or 5.1:10). Beginning in the next chapter, creations will come to the fore which are man-made; it is with them that such key forms of *Homo* as enclave, ethnos, and culture are closely intertwined.

Chapter 6

Geoareas and Enclaves: Difficult Problems of Definitions

Suppose that a certain person has to work in parallel in the fields of sciences of the natural science cycle on the one hand, and on the other, the humanities—and now this is not such a rare variant of such conjugacy. It is likely that such an imaginary person, most likely, could have caught the eye of those significant differences in the crown definitions of both scientific fields, which largely determined the basic foundations of those sciences. After all, the formulations of such definitions could sometimes sound almost polar in such comparisons, if you focus on their well-established clarity, universality of sound and consistency with the definitions of "neighboring." A special contrast was perhaps the sound of definitions in such sciences as mathematics, physics, on the one hand, and on the other—disciplines aimed at understanding the foundations of the social and multifaceted history of mankind and, perhaps, we emphasize specifically—philosophy. Against the background of natural science definitions, the humanitarian formulations of such definitions were very often distinguished by their apparent vagueness and uncertainty, as well as by their frequent saturation with noticeable contradictions.

However, the author of the book does not have the slightest desire and intention to invite the reader to analyze all kinds of interdisciplinary dissonances in definitions—this cannot be the task of our book in any way. Unfortunately, we will not be able to completely avoid evaluating such comparisons either. Therefore, we will try initially and consistently to clarify and clarify the interpretation of the concepts of "enclave," "ethnos," "culture," as we discussed in the final paragraph of the previous chapter. Let's start with the concept of enclave.

6.1. Defining an Enclave

Against the background of the other two definitions just mentioned, the enclave is far from being one of the most common concepts in the field of humanities. Here, for example, is how its definition sounds in the Great Russian Encyclopedia:

> ENCLAVE (a French *enclave*, from the Late Latin *enclave*—to lock with a key)
> is the territory of one state (part of it), surrounded on all sides by the territory

of another state (other states). Enclaves that have a seashore are often called semi-enclaves. There are, for example, the Republic of San Marino, surrounded by the territory of Italy, and semi-enclaves—the Kaliningrad region; also the Russian Federation, surrounded by the territories of Lithuania and Poland and washed by the waters of the Baltic Sea, the Republic of the Gambia, surrounded by the territory of Senegal and washed by the waters of the Atlantic Ocean.

However, the author was not able to find in the countless archaeological materials of Eurasia anything directly corresponding to this definition from the Great Russian Encyclopedia. I thus had to turn to the works of Eugene Yu. Vinokurov (born 1975; Fig. 6.1), who has relatively recently published in Russian and translated English a capacious book called "Theory of Enclaves," in which he sets out his understanding of the enclave. At the same time, the author touches on a wide variety of subjects, which seem to be closely related to this topic:

> The theory of enclaves should be comprehensive and comprehensive. Its task is to study at least three facets of the enclave's existence—their political, economic and social life. It should cover a wide range of issues. The first level—issues related to the phenomenon of enclaves and exclaves. Are they really specific? Do they have any common characteristics that allow us to consider them as a single class of spatio-political objects? A number of issues concern the emergence, development and disappearance of enclaves. Why and how do they arise? How do they develop and build their internal politics and economy, as well as relations with the outside world, especially with surrounding and mother states? . . . The term "enclave" is used quite widely. It is usually used to determine the existence of a foreign fragment inside a certain environment. It is very typical to use the term in geology to denote the existence of a rock fragment. In fact, looking through databases of scientific publications, you can find much more work on enclaves in the field of geology than in the field of politics. In church law, the term traditionally defines the territories of one diocese surrounded by another, which is far from a rare case. . . . It is also widely used in sociology and other social sciences, meaning a compact settlement that differs significantly from the surrounding territory in terms of national, political, socio-cultural or other characteristics. . . . The term is also used in military affairs, in agriculture and in the process of land surveying, in industry. Let's end with the fact that the word "enclave" is widely used in fiction and in everyday speech to

Figure 6.1. Evgeny Yuryevich Vinokurov (born 1941), author of the book "Theory of Ethnos."

characterize the state of isolation of an object, group or any phenomenon from the outside world. (Vinokurov 2007, 13–19)

But that's not all. The structure of the phenomenon itself is complicated by E. Vinokurov due to the "addition" of so-called exclaves to the enclaves:

> There are three types of exclaves. Firstly, a large number of exclaves are simultaneously enclaves in relation to the state surrounding them. . . . Secondly, there are exclaves that are simultaneously semi-enclaves . . . And, thirdly, these are pure exclaves, that is, territories surrounded by more than one foreign state and, therefore, are not enclaves in relation to them. . . . Pure exclaves may or may not have access to the sea. The decisive factor in determining their status is the separation from the main territory of the mother state on land. (Vinokurov 2007, 19)

According to E. Vinokurov, an exclave is an area characterized by the specifics of its culture, localized within the enclave, and at the same time relatively isolated and sufficiently isolated from the enclave surrounding it. True, the exclave for our purposes is a relatively unacceptable category: it corresponds incomparably closer to political systems, when some politically foreign formations are considered as if included in the body of another state (examples: the Republic of San Marino or the Vatican on the Apennine Peninsula, politically belonging to Italy). Examples closer to, say, Russians are associated with the small republic of Adygea in the North Caucasus, completely surrounded in the Kuban basin by other regions of the Russian Federation, or with a much more extensive steppe Kalmykia.

Perhaps significantly more expressive examples relate to high-altitude regions, such as, say, Tibet. There, the specifics of—albeit few—cultures clearly differ from neighboring ones by signs of material existence and production. A different language dominates there, and all these cultures are associated with certain areas reflected on maps (Fig. 6.2). It is likely that similar patterns could have been observed and, of course, were also observed in deep—even Paleolithic—antiquity, when similar anomalies of this kind were manifested within a certain mega-culture or a large cultural community.

However, I must admit that the above excerpts from the methodological guidelines of E. Vinokurov, the author of the book "Theory of Enclaves," did not particularly captivate me. The reason for this is quite simple: the definitions proposed by him were very awkwardly combined with those objects that archaeologists and ethnologists usually take as the most important objects of research. Therefore, the author—in this case, also the reviewer of the book by E. Vinokurov—prefers to use his own definitions.

The geoareas we have identified themselves, by their fairly clear isolation from each other, could certainly meet the requirements, if not E. Vinokurov, then at least an Encyclopedia. But geoareas are, in the original version, primarily geological and geographical objects, but not related to *Homo* cultures. Therefore, we will call each of the sixteen geological and geographical areas of Eurasia outlined here an enclave, only in case of its indispensable coverage and settlement by any *Homo* formations.

Figure 6.2. Ethnocultural areas on the plateau of Tibet.

So, geoarea is inhabited, and we move it to the category of enclaves. But then the following question will also arise: what is the formation of *Homo* that occupied this geoarea? How to determine the ethnicity of these "uninvited invaders"?

6.2. The Phenomenon of Ethnicity

Ethnos against the background of the enclave, the phenomenon on which we touched in the previous section, is incomparably more widespread in the ethnological and historical literature of various countries. Therefore, the sections of this chapter devoted to the ethnic problems of *Homo* cultures will occupy much more space in our presentation. As well as in the chapters of the book devoted to geology, the author will prefer to initially give the floor to our leading (mainly Russian) specialists in the field of ethnology and social anthropology, and not only the history of Eurasia.

Let's start with an attempt to decipher the original concept of ethnos, that is, going back to venerable ancient Greece. In other words, this concept has been known in the world for at least two and a half thousand years. This is how the Russian academician Julian V. Bromley (Fig. 6.3) defined the meaning and interpretation of this word:

Figure 6.3. Academician Julian Vladimirovich Bromley (1921–1990).

Dictionaries of the ancient Greek language give about ten meanings under this word, including: people, tribe, crowd, group of people, class of people, foreign tribe, pagans, herd, genus. Etymological analysis of this word makes it possible to see in it an indication of any set of identical living beings having certain common properties (customs, habits, appearance, etc.). A sequential examination of the use of the word ethnos in ancient Greek literature from Homer to Aristotle shows that at the archaic stage of the existence of this word, the meaning of "flock, swarm, group" prevailed. However, the meaning of "tribe, people" already exists. In the historical literature of the V–IV centuries BCE, this meaning becomes dominant. (Bromley, 7–8)

But here is the opinion of another also very significant Russian historian, ethnologist, social anthropologist, and academician Valery A. Tishkov (born 1941; Fig. 6.4).

There is no generally accepted definition of ethnicity. In modern science, the concept of ethnicity is most often used as a category denoting the existence of distinctive identities and ethnic groups formed on their basis. In Russian ethnology, the term ethnos is used in almost all cases when it comes to the people and even the nation. This concept presupposes the existence of homogeneous, functional and generally separable characteristics that distinguish a group from others with a different set of similar characteristics. (Tishkov 2003, 59–60)

One of the most significant scientific works of V. Tishkov was a book with a very intriguing-sounding title: "Requiem for Ethnos" (Fig. 6.4). And in this book, the author reports that the main task of the book with its almost mysterious name:

It is formulated simply: if I, along with many other ethnologists and historians, share the opinion that the basis of the modern world process is not just the evolution of world systems and organisms, but the dense and constant interaction of factors of global integration and local autonomy . . . then why not consider the main thing in the ethnographic method of included observation the installation to see "everything and in everything." . . . Thus, it is possible to radically expand the horizons of the discipline, overcoming the niche of ethnicity and the group categorization of the subject of scientific study ("only ethnos and nothing but ethnos").

However, the wording proposed in this book by the author as simple—as he evaluates it—does not look simple at all. For example, those studies remain completely unknown to me, in which their creators explain all the processes of world development by a simple "evolution of global systems and organisms." And V. Tishkov himself thinks that "having overcome the niche of ethnicity," and in fact, having rejected the concept of ethnos entirely—hence, by the way, the title of the book "Requiem for Ethnos":

We will embark on a much more promising path of cognition **of cultural diversity and cultural meanings of human activity and the various social coalitions created by it** [text highlighted by V. Tishkov] compared to the study of discrete units of anthropological (ethnological) analysis and their relationships with each other.

The continuation of this text by the author of the Requiem is also very remarkable:

However, this attitude of modern interpretive anthropology is difficult to perceive in the Russian scientific community and therefore forces me to enter into polemics with my critics, although I have tried in every way to avoid this all these last years, counting on gradually coming understanding. (Tishkov, 7)

Understanding, however, does not seem to have increased, and "Requiem for Ethnos" author V. Tishkov remained in a clearly lonely position against the background of a huge mass of ethnological and historical not only Russian, but also in other countries. I will also note that the author of the present book—he is also its unwitting reviewer—is also among those who still "hardly understand" the basic foundations of Requiem.

6.3. More on Ethnicity: Paradoxes in the Fate and Work of Sergei Shirokogorov

The fate of Sergei Mikhailovich Shirokogorov is really filled with a fair number of diverse and rather contradictory details. His name was hardly mentioned for a long time in the Russian (Soviet) scientific and related to ethnography/ethnology. However, in recent years, in some Russian publications of an encyclopedic and reference nature, he has finally been presented

Figure 6.4. Academician Valery Alexandrovich Tishkov (born 1941), author of the book "Requiem for Ethnos" and others.

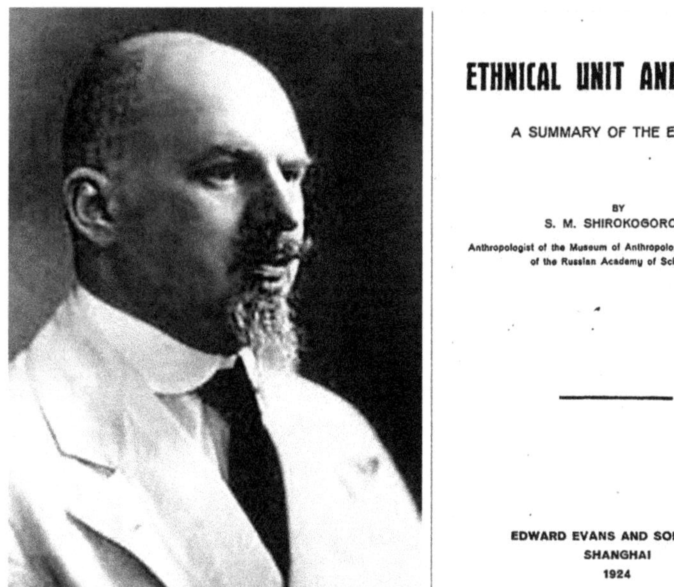

Figure 6.5. Ethnologist Sergey Mikhailovich Shirokogorov (1887–1939).

as a Russian orientalist, anthropologist, ethnographer, ethnologist. His fame was due to the authorship of the term "ethnos" in its modern meaning, as well as the ancestor of the theory of ethnos itself. S. M. Shirokogorov (Fig. 6.5) in the view of his colleagues in the scientific profile of recent years was

> a scientist of a huge scale, who was ahead of his time in many of his discoveries, lived in isolation from the main centers of European science and, nevertheless, had a considerable influence on European ethnology, being the creator of entire schools and directions in it, S. M. Shirokogorov only in Russia so far occupies a place somewhere on the periphery of the history of science. His creative legacy needs to be mastered and made available to the Russian reader. Then our outstanding scientist-compatriot will receive recognition worthy of his name not only in the world, but also in Russian science. (Revunenkova, Reshetov 2003, 119)

That is why in our book we will tell in more detail than about other scientists about the life path of S. M. Shirokogorov, which in all probability, also affected his scientific views. He was born in 1887 in the small Russian city of Suzdal in the family of a pharmacist. His parents' family, most likely, according to local estimates, belonged to the number of intellectuals, although the category there is extremely small. After graduating from a private German gymnasium in Tartu in 1904, he moved to Paris in 1905. In France, he immediately immersed himself in an extremely wide and very diverse learning process: the Faculty of Philology of the University of Paris, classes at the Higher School of Political Economy, the Anthropological School. After returning to Russia in 1910, he continued his studies in Russia, but already at the Physics and

Mathematics Faculty of St. Petersburg University, although five years later he moved away from this direction. At the same time, he was also interested in courses at the Archaeological Institute.

In 1910, Shirokogorov was enrolled as an employee in the Museum of Anthropology and Ethnography at the Academy of Sciences of St. Petersburg (Kunstkamera), and in 1916 was designated a junior anthropologist and even a curator of the Museum in the Anthropological Department. From the same year in 1910 he became an active participant in ethnographic expeditions in Eastern Siberia and the Far East.

The outbreak of the civil war caught Shirokogorov in the Far East, and he stayed to work in Vladivostok, where he took part in the organization of a private historical and philological faculty, and then the State Far Eastern University. The years 1918–1922 were marked by the victory of Soviet power in the Far East, and these events found him already in China. Then Shirokogorov found himself, as it were, in forced emigration to this country, where he began his work in educational institutions in Shanghai, Aomin, and then Beijing. In Beijing, he taught at Catholic University and Tsinghua University. In exile, S. M. Shirokogorov wrote many of his works, published mainly in English, and partly in German and French. Among them, two works should be mentioned:

- *The place of ethnography among the sciences and the classification of ethnic groups: Introduction to the course of ethnography of the Far East, delivered in 1921–1922 at the Far Eastern State University* (Vladivostok, 1922).
- *Ethnos. The study of the basic principles of changing ethnic and ethnographic phenomena: A print from the LXVII Proceedings of the Oriental Faculty of the State Far Eastern University* (Shanghai, 1923).

Both of these works were reprinted much later in the series *"Selected Works of Shirokogorov."* We will also point out a small but very important article "Ethnic Unit and Milieu," translated by Shirokogorov himself into English and published in Shanghai in 1924 (Fig. 6.5) and we will mainly refer to its text below. China, in fact, became a second homeland for S. M. Shirokogorov, where he remained until his death in 1939. In China, he also acquired, as it were, his new name 史祿國, which in English sounds like Shi Luguo.

6.4. The Work of S. Shirokogorov on Ethnic Unit and Milieu

This major article S. Shirokogorov himself translated into English from his Russian-language work of 1922 and promptly published it in Shanghai in 1924 (Fig. 6.5). However, at the very beginning of its text, the author formulated his definitions of terms and concepts related to the ethnos.

The term "ethnos" and its meanings

The author introduces a new term "ethnos" (from ἔθνος), which means a group of people who speak the same language, recognize their common origin, have a set of customs and a social system that is consciously maintained and explained as a tradition and differs from other groups. It is an ethnic unit in which the processes of variation (growth and decline) of all ethnographic, linguistic and anthropological phenomena take place.

The term "ethnic" is used as adjectives to "ethnos," and "ethnography" is used as the name of a science that studies ethnoses from the point of view of knowledge previously collected by ethnoses and led to some technical, social, mental and psychological systems, so that this science includes sociology, psychology, technology, history and science. Again, "anthropology" serves to denote a science dealing with the physical, anatomical characteristics of a person as an animal species (or genus). Finally, "ethnology" means a science that studies the process of changing ethnic phenomena, which is a science that combines the conclusions of anthropology, ethnography, linguistics, and so on.

Ethnos and its interaction with the bio-geographical environment

The process of changing these phenomena is a function of human biology, which adapts to the geographical environment (climate, soil, other animal species, etc.) by changing its mental, mental and physical abilities that ensure its existence in these physiographic conditions. From this point of view, culture, or, in other words, human knowledge and experience acquired by previous generations and inherited, is the product of a purely biological function, and all phenomena of the physical, social, mental and technical order are concrete manifestations of this process.

Ethnos and its interaction with other ethnic formations

The life, origin, decline and growth of an ethnic group occur in certain given geographical conditions, which may be favorable or unfavorable for it. There are three types of phenomena: phenomena given by nature, and phenomena created by man (ethnos), and, finally, phenomena arising as a result of communication or, rather, interethnic relations. No attempt is made here to analyze the phenomena of the first two groups, the relations of the third group, or the ethnic environment considered only. . . .

The degree of pressure from neighbors and the degree of mutual influence in all these cases are different. The possibility of merging and assimilation of these ethnic groups by their neighbors is no less different.

Thus, for an ethnic group, other ethnic groups form an environment that is also subject to the mutual influence of this ethnic group. Therefore, the analysis of these relations is necessary to understand the origin, culture and position of ethnic groups. . . .

Social institutions, philosophy, even technical culture, for example, the Buryats (Mongols of Transbaikalia), are incomprehensible without a thorough study of the Mongols and their neighbors, Russians, Chinese and Tungus, just as the history of Russia is not at all clear if you omit the history of the Mongols, Turks and Chinese.

Thus, if we recognize that the factor of the ethnic environment is of fundamental importance in shaping the position of an ethnic group, then: (1) the degree of influence of an ethnic group on others leads to various combinations of interethnic relations and conditions characteristic of this ethnic group, and (2) the strength and power of ethnic groups (which surround this ethnic group) determine mutual the relations of neighboring ethnic groups and the intensity of the ethnic environment. . . .

An ethnic group is always fighting for its existence, and if it is able to resist other ethnic groups and win victories, which often leads to the expansion of its territorial coverage, and this is one of the external manifestations of its growth. Thus, the ethnos faces adaptation to the primary environment as the first problem; the creation of a secondary environment and self-adaptation to it as the second problem. This leads to some internal organization of the ethnos (social culture), if such is necessary for adaptation to the primary and secondary environment. . . .

Finally, the ethnic environment gives the ethnos another task to solve—to create interethnic relations that can take various forms. These forms range from friendly—"cooperation" up to "parasitism." If, as a result of these relations, an ethnic group is completely destroyed, it can be assimilated, forcibly merged with other formations or replaced by other ethnic groups. But it depends solely on the biological strength of the ethnic group.

Ethnos and the number of its population

What kind of power is this? The ethnos, like all biological species, pursues the main goal—to preserve its position among other animal species and ethnic groups, to preserve its right to exist. . . . The forms of adaptation to the primary and secondary environment are diverse, the forms of adaptation to the ethnic environment are no less complex and diverse. The external manifestation of the ability to self-adapt to these conditions is, first of all, the numerical value of an ethnic group relative to other ethnic groups. . . . If there is any difference in the quantitative or qualitative value of culture, numerical superiority loses its significance. For example, a numerically large ethnic group that does not have a developed technical culture cannot resist a numerically small ethnic group with a highly developed culture; a highly developed culture allows the latter to oppress the former, even destroy it, as evidenced by the history of numerous colonizations. . . .

If ethnic groups have almost the same number, cultural superiority can also give superiority in the struggle for existence. If we take, for example, an agricultural ethnic group and an industrial one, then the latter will have superiority, because it will put the former in a state of dependence by its production activities, means of communication, and so on. At the same time, the superiority of agricultural machinery gives the ethnic group that uses it an advantage over the nomadic ethnic group.

Thus, if there are two ethnic groups that have similar and equal cultures, then the one with a higher population is the best adapted of them. You can take, for example, Buryats and nomadic Tunguses in Transbaikalia, whose numerical values are about 300,000 and 3,000, respectively. The numerical superiority of the former undoubtedly gives them all the advantages, and they assimilate the nomadic Tungus.

The second condition—the preservation of the population—should also be explained. There is a very widespread opinion, accepted by sociologists, economists, politicians, historians and even ethnographers, who claim that an ethnos (or nation, or people) should grow numerically, and those that do not grow are in a state of degeneration and decline.

Taking into account the interethnic relations that can be observed in Europe, this opinion from a psychological point of view can be understood as quite natural, although it is absolutely wrong. In fact, if in a given ethnic group its culture is stable, then the limit of numerical growth is determined thereby. If the territory covered by an ethnic group and cultural conditions provide a means of existence for a certain ethnic group, then it cannot increase its number without improving its culture. Consequently, with an increase in the population exceeding the possibilities of nutrition, the excess population should disappear or the birth rate should be regulated by some means—medical, artificial, social, etc. We know several customs, such as, for example, legal, i.e. recognized by the state or society: the murder of children or the elderly, abortions, birth control and the like. . . .

If we take as an example the murder of girls—parents who, thereby limiting the growth of the population, seem to follow the interests of society and therefore regard it as an act of real virtue. . . . Among Russian peasants, for example, who do not have preventive methods of abortion, and being Orthodox Christians, therefore infanticide is not practiced, the role of a birth rate regulator is played in their environment by a huge infant mortality. The custom of late marriages has the same biological significance, which leads in some European countries to an increase in the average age of brides to thirty years.

Thus, all ethnic groups, applying some method of preventing population growth or allowing only its slow progress, practice it in one form or another, acceptable and understandable in their cultural state, but they do it completely unconsciously.

For example, infanticide can take the form of religious sacrifice; the regulation of conception and the practice of abortion are explained by civilized Europeans as having their own motives in caring for the best conditions for other children born and born; late marriages are explained by the need to create some financial basis for a new family; rejection of marriage takes various forms—moral, religious, and so on finally, indifference to child mortality among Russian peasants finds its soothing formula in the supreme will of God. . . .

Ethnos and its spatial coverage

The territory necessary for the existence of a population whose only occupation is hunting must, of course, be larger than that occupied by the same number of people living by cattle breeding. In addition, agricultural people need less territory than pastoralists, and industrial people need less than farmers. In other words, the more intensive the exploitation of the territory, the more the population can feed this territory.

Population density maps provide many illustrations of this principle. However, one should not lose sight of the fact that the territory suitable for habitation is unevenly distributed, and population density maps should indicate the size of the territory that can be inhabited and which is not allowed due to its topographical features, deserts, rocky mountains, and so on. Thus, the population density depends on the degree of culture of a given ethnic group, so if the territory and culture are not subject to change, the excess of fertility over mortality is constant and equal to zero. It can also be expressed as follows: population density ratio, which is the ratio of population to area.

It turns out that all these dependencies of the territory of settlement of an ethnic group, population density and the nature (degree) of culture of this ethnic group can be quite clearly and definitely established and expressed using various mathematical formulas. However, such an extraordinary approach to ethnic themes is very difficult for untrained readers. In addition, it is very difficult to comment on it since it is so unusual in a humanitarian context. Therefore, the reviewer—and this is the author of this book in this case—wished to limit himself only to photocopies of sections of the text with formulas from the specified English-language publication (Fig. 6.6). At the same time, perhaps, it is already possible to sum up some of the main provisions of Shirokogorov's works. It is also interesting that at the very end of his article, he seems to attempt a prophecy regarding the existence of the entire genus *Homo*, which is also quite remarkable:

If we try to determine what period of time can be expected for the existence of mankind, we must draw our conclusion from the past. A person in conditions of very intensive growth inhabits the whole earth, but in the current state he has been living for about 6,000–8,000 generations. The period of the most intense activity (since the beginning of the Metal era) lasted only about 300 generations. Based on this, we can assume that most of the way to absolute population density (if we take into account the intensity of growth) has already been passed. Based on the above, it can be assumed that humanity is currently very close to its climax. If it cannot produce a new species (or genus) that will be able to adapt to the new conditions, it will be doomed to perish without leaving offspring, as, for example, the ammonites died.

We have already given the formula of ethnical equilibrium and the constant ω. Now let us consider this question more in detail and analyze the consequences resulting from it.

If the quantities of q, S and T are in a state of stability, the ethnos is always in a *stationary state*. What must happen, if there is variation of q, increase or decrease of it? In order to maintain the constant ω the quantities S and T or one of them must also be varied, so that *with the variation of q there appears an impulse of variation of S and T or one of them*. Then the quantity $\frac{q}{ST}$ will change proportionally to the relation of original q to the difference of original and present q_1 $(q_1 - q = \Delta q)$ so that *the impulse of variation of S and T* (it may be positive or negative) will be $\pm i_{st} = \frac{q}{ST} \cdot \frac{q}{\Delta q} = \omega \frac{q}{\Delta q}$.

Figure 6.6. This is how the formulas by which S. Shirokogorov tried to calculate data on the number of members of certain ethnic groups against the background of their spatial coverage look like (from the publication in the article "Ethnic Unit and Milieu").

In some ways, this text by S. Shirokogorov may even remind us of some points of the famous theory of Thomas Robert Malthus (1766–1834; Fig. 6.7) about the so-called excessive "over-population" of the earth (Fig. 6.8); however, Karl Marx and, following him, Soviet Marxist scientists unconditionally attributed this theory to the category of misanthropic.

6.5. The Creativity of S. Shirokogorov in the Ratings of the "Reviewer"

This author ("reviewer," in this case) has no doubt that S. Shirokogorov can be classified as a very prominent ethnologists of the twentieth century. His main and unconditional achievement was a strict and clear formulation of the theory of "ethnos" (see Section 6.4), when the author determines what can and should be included in the framework of this concept, so important for the science of our time. Quite often, various authors in different countries have positively assessed Shirokogorov's position, and the "reviewer" joins their opinion.

However, the "reviewer" is much less enthusiastic about those areas of Shirokogorov's research when he rushes to the problems of "Ethnos and the number of its population" or "Ethnos and its spatial coverage." In these cases, speculative features begin to dominate, which becomes especially noticeable against the background of the lack of specific data for certain ethnic groups. Or when the author seems to completely avoid referring to geographical maps, which always dramatically reduces the significance of his conclusions.

At the same time, the speculative nature of the broad conclusions proposed by S. Shirokogorov clearly increases when he, trying to strengthen his argument, turns to methods of mathematical statistics. And again, even in these cases, the materials of real ethnic groups remain somewhere

Figure 6.7. Thomas Malthus and his book on the principle of the [existence] of the *Homo* population on Earth.

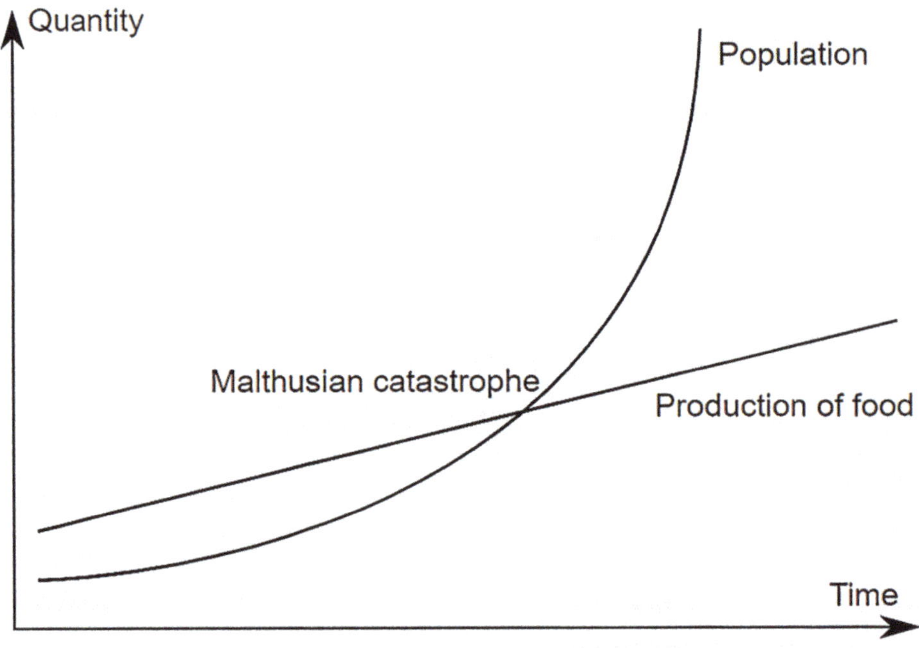

Figure 6.8. So according to Malthus, the proportional relationship between food production and the size of the *Homo* population leading to a catastrophe looks like.

aside without any references to their specifics. Unfortunately, it turns out that at first the author tries to formulate definitions of the most general nature, and already tries to "fit" the specifics of certain ethnic formations to these general constructions.

I had an interest: how often and how positively did other specialists in ethnology treat these sections of S. Shirokogorov's theory, already outside of his general views on "What is an Ethnos?" Alas, there are very few of them. Here, for example, is the work of R. I. Yakupov

"Introduction to General Ethnology" (2016: 89), that in some cases, with the help of certain "... letter values, you can try to build some mathematical formulas for the stability of the culture of an ethnos. Such attempts were made at the time by S. M. Shirokogorov." And that's it!

However, the author of this book, who has identified himself as a "reviewer" of the works of S. Shirokogorov, wishes to end the discussion with these words—after all, such a topic should not be considered important and significant for our book. And it is hardly possible to fully share the almost sublime assessments of his other reviewers, the staff of the "Kunstkamera" of St. Petersburg—E. V. Revunenkova and A. M. Reshetov (see Section 6.3)—that "his creative heritage needs to be mastered and made available to the Russian reader. Then our outstanding scientist-compatriot will receive recognition worthy of his name not only in the world, but also in Russian science."

Chapter 7

Culture: A Mysterious Phenomenon

7.1. Culture and Its Three Basic Facets

Now we turn now to the phenomenon of "culture"—the most complex of the triad we have outlined, which includes, in addition to culture, the "enclave" and "ethnos" already analyzed in the previous chapter. Culture in its original sound corresponded to the Latin word *cultura*. It is believed that its author is Marcus Porcius Cato Censorius, or the Elder (234–149 BCE). The term itself at that very remote period from us meant only agriculture and its agricultural products. However, with an extraordinary and difficult to explain speed, the concept of culture acquired a truly all-encompassing global significance, reflecting various characteristics of the results of human activity in any sphere—from material to spiritual. However, along with this, it will be extremely difficult for us to find an acceptable consonance in the formulations defining the phenomenon of "culture" in the world scientific practice.

Let's start with the definition of "culture" in the Great Soviet Encyclopedia, where this concept is associated with the Latin word:

> **Cultura**—cultivation, upbringing, education, development, reverence, a historically defined level of development of society and man, expressed in the types and forms of organization of life and activity of people, as well as in the material and spiritual values created by them.
>
> **The concept of culture.** It is used to characterize the material and spiritual level of development of certain historical epochs, socio-economic formations, specific societies, nationalities and nations (for example, ancient culture, socialist culture, Mayan culture), as well as specific spheres of activity or life (labor culture, artistic culture, everyday life culture). In a narrower sense, the term "culture" refers only to the sphere of people's spiritual life.
>
> **Pre-Marxist and non-Marxist theories of culture.** Initially, the concept of culture implied a purposeful human impact on nature (cultivation of the land,

etc.), as well as the upbringing and training of the person himself. Education included not only the development of the ability to follow existing norms and customs, but also the encouragement of the desire to follow them, formed confidence in the ability to meet all the needs and requests of a person. Such two-dimensionality is characteristic of the understanding of culture in any society. Although the word "culture" itself came into use in European social thought only from the 2nd half of the 18th century, more or less similar ideas can be found at the early stages of European history and beyond.

From this very verbose and extremely vague definition, it only becomes clear that there is no any clearly expressed concept regarding the phenomenon we are interested in. we won't get it. However, a similar conclusion will also follow from a much later publication in Wikipedia:

Culture—from Latin *cultura*—cultivation, later—upbringing, education, development, reverence—a concept that has a huge number of meanings in various areas of human life. Culture is the subject of study philosophy, cultural studies, history, art studies, linguistics (ethnolinguistics), political science, ethnology, psychology, economics, pedagogy, etc.

Basically, culture is understood as human activity in its most diverse manifestations, including all forms and methods of human self-expression and self-knowledge, the accumulation of skills and abilities by a person and society as a whole. Culture also appears as a manifestation of human subjectivity and objectivity (character, competencies, skills, abilities and knowledge).

Culture is a set of stable forms of human activity, without which it cannot reproduce, and therefore exist.

Culture is a set of rules that prescribe a certain behavior to a person with his own experiences and thoughts, thereby exerting a managerial influence on him.

The source of the origin of culture is thought to be human activity, cognition and creativity.

It seems that Wikipedia even surpasses the Great Soviet Encyclopedia in verbosity and uncertainty, and this encourages us to search for other ways to develop an acceptable definition. For this reason, the author of this book prefers to formulate his own understanding of the phenomenon of "culture." At the same time, however, the proposed formulations do not pretend to be widely used, but reflect, first of all, what the author used in the course of his research. So:

Culture is a model or way of existence of a certain *Homo* group—quantitatively very diverse, and from small to great in size. Each group of people of this kind necessarily correlates with a certain ethnic group or nation or with some kind of voluntarily or involuntarily connected set of ethnic groups. The model of existence of each group is based on three supporting, closely interrelated and mutually dependent basic blocks—on their triad or triangle.

a) **The technology of production,** where the basis is the most important of all the signs characterizing the model or method of physiological and technological life support of groups of *Homo*-carriers of a particular culture or society. This block is significantly more closely connected with the surrounding ethnos nature; it is mainly due to the laws of natural development, and its structural features depend on it (Fig. 7.1A).

b) **Social structures or structure of society or socio-structural block.** This covers all models of social formations—from a herd very close to the animal world up to the constructive peaks of *Homo* formations such as empires, monarchical formations, parliamentary republics, etc. (Fig. 7.1B).

c) **The ideological and ideological block.** This block covers the worldview or understanding of the world in which, firstly, a particular society or ethnos exists; secondly, it necessarily includes the representation of each society or ethnos about the place that it occupies in the surrounding and conscious world, as well as about the role that he himself believes he plays in this world around him (Fig. 7.1C). Against the background of the other two blocks, its sometimes complete isolation and even isolation from the natural environment is very noticeable. Here, the speculative methods of reconstructions and other often global constructions almost completely dominate; the crown changes in the ideological and ideological sphere are already associated with the Era of Modern Times.

Perhaps it is possible to discuss briefly this idea. According to a number of experts, this basic trihedron of any culture may also appear as a kind of square. Some believe that these three facets should certainly be "enriched" with a fourth—a linguistic facet or a linguistic block. For example, such prominent researchers as David Christian or the Nobel laureate and American neuroscientist and psychologist Eric Kandel (Kandel 2006) are inclined to this. They consider the linguistic block to be an extremely significant factor in the development of *Homo* cultures, primarily due to the exchange of information between the most diverse ethnic groups. At the same time, they also refer to the opinion of "the great historian William McNeill, who built his classic work on world history 'The Rise of the West' around this idea—after all, historically significant social changes occur due to interaction with foreigners who have new, hitherto unknown skills" (Christians 2019, 226–227; McNeill 1990, 2).

The linguistic factor, indeed, can be significant if we are sufficiently well aware of all the most important linguistic characteristics of interacting cultures.

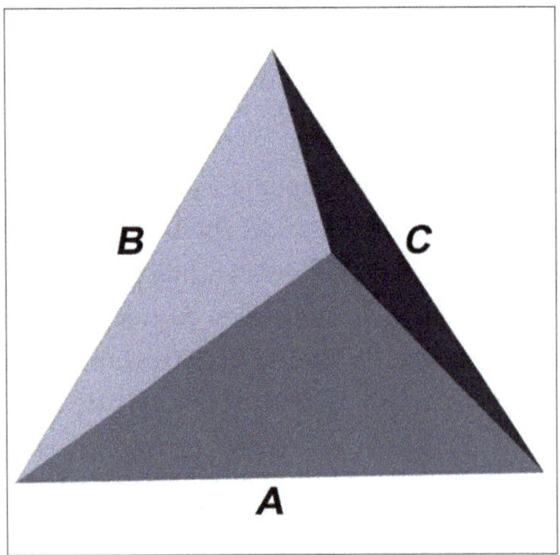

Figure 7.1. The triangle of closely interrelated and interdependent basic blocks or facets of *Homo* cultures: A—physiological and technological, B—socio-structural, C—ideological.

However, for archaeological antiquity of this kind, correct linguistic definitions can be considered for the most part only a rarity, moreover, most often exceptional. Maybe that is why all the author's attempts to "enrich" the three basic structures with a language block did not lead to anything definitely positive, and in this book the author preferred to focus on the basic trihedron of faces/blocks presented by him earlier. (Chernykh 2019, 109)

We will certainly also add that culture, in fact, is the face of a particular ethnic group, which we can judge. And that it is precisely the most important features of a person identified by various methods that representatives of various sciences, and not only the humanities, try to thoroughly study in order to develop correct judgments about a particular ethnic group.

7.2. The Interrelation and Interdependence of Facets/Blocks of Cultures

The assertion of the interrelationship and interdependence of facets of cultures, according to the author, hardly requires lengthy discussions: they are always and certainly closely interrelated, interdependent, mutually dependent (Fig. 7.1). However, approaches to this problem are completely heterogeneous, even sometimes polar opposites. The deepest differences in the assessments of this problem are rooted in the definition of the initial crown block, and the differences here relate mainly to two facets/blocks—physiological-technological and ideological-ideological, that is, in the clear majority with the religious block. Marxist materialists recognize the technological block as the initial and defining one—after all, it is directly connected with our breadwinner Earth, and everything literally depends on it—that is, on nature. Idealists of various religious interpretations are sure of the opposite: absolutely all prescriptions are conditioned by the will of the Almighty, including the connection with the "Nurse Earth," which is also completely subordinate to the Almighty.

However, we will touch on this problem in more detail already in the chapters of the second part of the book, after a closer acquaintance with the main *Homo* religions; indeed, this problem itself certainly deserves special attention.

7.3. In Trying to Understand the Essence and Forms of Belief

Many chapters of the second part of the book will be closely linked to the sphere of ideological and ideological problems, that is, the one that is essentially the most difficult to decipher in the entire basic trihedron of any culture and any ethnic group. In fact, in all cultures—up to modern ones—this block was personified mainly by religions. And if the latter are viewed through the prism of millennia, then their diversity on such a giant canvas can hardly seem extraordinary. Therefore, we must first turn to them in order to understand the essence of this basic block. And if we want to understand, we must ask for a clear definition of religion. In asking this we immediately encounter a strange accumulation of more or less "blurred" definitions, and their mass will look clearly excessive. Therefore, it was not possible to choose the most appropriate definition for the purposes of our book, which is why the author himself had to try to work out a fairly broad and acceptable approach to this, since it became a very difficult problem.

The author was extremely curious to compile a specific catalog of definitions of religion, for which he undertook a kind of run through the relevant works mentioned in the multilingual sections of Wikipedia—from Russian and English up to Chinese. Perhaps the clearest answer to these searches for definitions was in the English version; however, there is no scholarly consensus over what exactly constitutes a religion. In a fairly similar way, although most often more verbose, explanations were presented in all those languages which the author addressed.

But it is in this connection that the iconic words with which Mircea Eliade (Fig. 7.2), who was, without a doubt, a classic and the world's largest connoisseur of ancient and, of course, began his three-volume epic about religions and myths, could perhaps manifest themselves modern beliefs:

> For many years I have cherished the idea of a small, capacious book that can be read in a few days. Because reading "in one breath" gives, first of all, an idea of the fundamental unity of religious phenomena and, at the same time, of the inexhaustible novelty of the ways of their expression. The reader of such a book would have the opportunity to approach the Vedic hymns, the Brahmanas and the Upanishads just a few hours after the ideas and beliefs of the Paleolithic era, Mesopotamia, Egypt appeared before him; he would have discovered Shankara, Tantrism and Milarepa, Islam, Joachim of Florence or Paracelsus the morning after thinking about Zarathustra, Gautam Buddha and Taoism, the Hellenistic mysteries, the rise of Christianity, gnosticism, alchemy or the mythology of the Grail; he would have met German enlighteners and romantics, Hegel, Max Muller, Freud, Jung and Bonhoeffer soon after Quetzalcoatl and Viracocha, the twelve Alves and Gregory Palamas, the early Kabbalists, Avicenna or Eisaya.
>
> Alas, this compendium has not yet been written. So for now I will have to be content with a three-volume work in the hope that someday I will still be able to put it into one volume of 400 pages. (Eliade 1976; Eliade 2002: Introduction to Volume 1)

I was ready to follow Mircea Eliade's definition, but alas! What is curious is that I could not find even in this classic (!) a brief, global formulation: what are the signs that we—and in particular, the author of this book—will consider weighty in order to attribute certain doctrines to religions? Then we should set ourselves the goal of identifying at least those metas that, in the author's opinion, can serve as an indispensable and obligatory feature of any religious doctrine, ranging from extremely simple, up to very complex—for example, monotheistic? What conditions and requirements for this seem necessary, at least for the subject of this book? Let's try to formulate them.

a) Mandatory recognition of the existence of some powerful and inaccessible (to *Homo*) Higher Power (HP), initially directing and shaping all our earthly existence, elevating it into reality not only for *Homo*, but also for the whole world around us.

b) It is hardly possible to identify a specific image of the HP and one can only guess about this, but the HP does not always welcome even such attempts.

c) It is necessary to firmly understand and assimilate the requirements of the HP regarding the daily behavior of *Homo*.

d) It is necessary to firmly and clearly define and identify deviations in the daily line of behavior, and they should concern not only a specific person, but also groups of people; such deviations are especially unacceptable after warnings received from the HP or even some hints about the undesirability of such acts.

Figure 7.2. Mircea Eliade (1907–1986), an outstanding historian of world religious movements.

e) One should have a clear understanding of the nature of those spiritual and material sacrifices that are required to gain favor from the HP and atone for one's own sins and miscalculations.

f) It seems indispensable to take care of the future; namely, extraterrestrial existence, which should be prepared in advance, during earthly life, and for each individual personally.

g) In each group of *Homo*s, there are necessarily persons close enough to the HP; it is they who carry out the absolutely necessary connection between the HP and ordinary *Homo*s; it is they who are able to understand the special language of the Sun, convey the meaning of its commands to ordinary mortals and guide their daily lives; such roles fall to the lot of ordinary shamans to the highest hierarchs of the church.

However, one should not take the definitions proposed here as strict and indispensable requirements. These, of course, are not requirements at all, but only a systematic presentation of the author's own approach to this problem. Here are only recommendations of an approach to a very extensive list of various doctrines professed by a wide variety of societies. In the following chapters, we will shift the focus to the comparison of two essentially polar types of beliefs: the ancient ones, inherent, in fact, exclusively in the Greco-Roman world, and the truly world doctrines of the Abrahamic sense. The differences between them, without any doubt, look extremely impressive, although both doctrines certainly belong to the category of religions.

It should also be noted that in the collective, in the *Homo* group, where this kind of doctrine is born, there must certainly be a key person in whose head the initial questions are awakened: What is the world around us? And what are we in this world? Let's call this set of questions, starting from the concepts of modern mental language, a hypothesis. Usually, such prophetic ideas arise in the minds of only an individual and, in extreme cases, a maximum of two or three people, but not a large collective. This is how Benedict (Baruch) Spinoza defined almost four hundred years ago (Fig. 7.3) the role of such a person, calling her a prophet:

> Prophecy, or revelation, is a known knowledge about some thing revealed to people by God. The prophet is the one who interprets the revelation of God to those who cannot have a true knowledge of the objects of divine revelation and who therefore can accept the objects of revelation only on pure faith. (Spinoza, 6)

The group can accept or reject the emerging hypothesis. In any case, it is possible to assume that the founder of the idea itself lives by it and is truly immersed in it. He painfully nurtures it until he finally considers this hypothesis a reality. And then the creator himself must catch some long-awaited signals coming from the HP, from the Sun or Moon, or from forces unknown to him before: *The eternal truth has been revealed to you, you have comprehended it and now you must become its Prophet—go and create the truth.*

After the completion of the development of a sufficiently definite and systematic outline—dogmas and postulates—in the understanding of these views, and of course, if the canons of this system seem unconditional, then, as a rule, a group of like-minded people arises, associated with a clan, tribe, ethnos, or even—in the most difficult to comprehend variant—with a group of ethnic groups. A prerequisite for the existence of such a group is the recognition of such views as an absolute.

It is impossible to question and challenge this absolute in any way. But if suddenly such a drama is brewing and bursts to the surface, then all adherents of these canons will certainly find themselves in captivity of grave misfortunes: after all, that Higher Power, which people recognized as the Absolute, will never forgive them. Therefore, any doubters of the high and immutable truth of the Absolute revealed to people from the collective of like-minded people must be harshly expelled, or better yet, simply destroyed—they are traitors! Otherwise, something much worse will come: after all, the holiest thing in the existence of *Homo* is true and boundless faith! And such an approach to heretics—fighters with the Absolute—has been repeated in the history of *Homo* and is being repeated, in fact, endlessly.

Figure 7.3. Baruch (Benedict) Spinoza (1632–1677), author of the famous books "Ethics," "Geological and Political Treatise" and a number of other works.

Chapter 8

Subsistence Systems of Cultures: Geology and Archaeology

8.1. Two Basic Subsistence Models

In Chapters 3 and 4 we focused on the factors and situations that remained largely beyond the control of the cultures of *Homo*. It was all about the division of Eurasia into specific—horizontal and vertical—zones. Each of these zones and their intertwining was different to some degree or another. It is Earth, every part of it, that nourishes the members of a human group that chooses to live in a certain place. However, in the end everything depends on the ability of a culture to receive from its domain all the necessities for a normal existence.

Now we will focus on *man-made* subsistence technology—that is, the one that serves as one of the three main supporting blocks that are essential for any culture (See Section 3.3). Indeed, it is not only the availability of food for physical and physiological existence that depends on the nature and structure of this supporting block, but the integral structure of any *Homo* culture.

We can identify two basic, fundamental models of life subsistence or economies in human cultures: appropriating and production economies. Let us start with the first one.

8.2. Appropriating Economy

The appropriation model is a purely biological phenomenon inherent to all mammals. The model functions without significant obstacles only if there is a good balance between:

a) the quantity of edible plants,

b) the number of herbivorous creatures that feed on these plants,

c) the number of carnivorous creatures that feed on the flesh of the herbivores.

If the soil is low in fertility—as is the case, for instance, in tundra or forest-tundra, or in the desert or semi-desert—then the range of edible plants growing on it is very limited. It means that there will be a natural cap on the number of herbivorous animals on a given plot of land at a given time. This limit, in turn, will adjust the quantity of carnivorous animals—there will be strictly as many carnivores as there is food for them, such as the herbivores.

Human beings, bipeds, belong to omnivores, that is, to both herbivores and carnivores. The appropriative economy model was the one dominant among Paleolithic cultures of the Pleistocene epoch. In the Holocene, not all societies renounced it. It completely dominated the vast northern zones (taiga and tundra), covering almost 40% of Eurasia. Still, if one counted the number of inhabitants adhering to this model, the figure would be incomparably lower. A similar model survived right up to the Modern period as a strange rarity in several peripheral regions of technologically developed Eurasia.

8.3. Production Economy

Now let us move on to discussion of the production economy. First of all, it is conventional to apply the notion of the production economy to early *Homo* cultures. Within the framework of this model, human beings appeared to make an "agreement" with nature to "tame" it at least partially. However, humans were not in the position to rebel against nature since they were completely dependent on it. Human beings could understand its rhythmic changes and adapt to them. As a result, they could obtain from nature at least the minimum required necessities for their biological existence. Looking many million years back, we can see whether a community used the production or the appropriating model with the help of a few uncomplicated research methods.

The experience of many generations helped to develop these skills. It took several thousand years to transition from the appropriating economy to the production economy model. Moreover, these processes vary in nature and length in different regions of Eurasia.

Based on an examination of numerous Eurasian cultures of the Holocene, two seemingly polar types of production economy can be highlighted.

The first (A) is characterized by a combination of agriculture and cattle-breeding; the second (B) is characterized by a mobile, nomadic, or semi-nomadic lifestyle and cattle-breeding.

The agricultural and pastoral cultures of the A-type model were located mainly to the south of the Alpine-Himalayan geosyncline, and their ridges served, as a barely surmountable wall dividing them from the northern pastoral communities. In this category there were also exceptions or anomalies that appear unusual due to their localization to the north and east of the Alpine orogeny, on the polar flanks of Eurasia. The cultures of the Han block were in the east, concentrating mainly in the Yellow and Yangtze basins, bordering on the steep southward bend of the Eurasian orogeny on the outside. The Han cultures had a peculiar polar western counterpart—a block of cultures that populated the European mega-peninsula north of the Eurasian orogenic ridges and adjoined the Baltic Sea basin (see Figs. 4.19, 4.21).

The pastoral cultures of the B-model were mainly associated with the Steppe Belt of the continent. The huge and deserted Arabian Peninsula that shifts to the south represents an apparent anomaly in some ways like the Han cultures. It was dominated by cultures of camel-riding nomads to be discussed later.

One should not assume that the prevalence of any of the production economy models implied that in proximity there could be no other model. For example, it can be argued quite definitely that the A-type model was predominant in the regions of the so-called Middle East. Still, at the same time, in those very same regions, the cultures of mobile herders played a significant role. The Arabian pastoral cultures of Bedouin nomads (see, for example: Losleben 2003) seem to be the most explicit example of coexistence of Semitic-speaking peoples. Also, it is almost certain that the Semitic-speaking Jews, whose origins are reflected in the book of Genesis, in the third—early second millennium BCE originally were stock-breeders. The routes of their wave-like and incessant early migrations covered areas from Palestine to Egypt and back.

Moving to the north from the Eurasian orogeny we will find ourselves on the lands of the nomadic/semi-nomadic herders of the Steppe Belt who were undoubtedly predominant in that region. We will also find traces of agricultural activities, albeit in the form of a few grains of cultivated cereals found in relatively undifferentiated cultural layers of settlements (see, for example: Lebedeva 2005; Spengler 2015). However, we still cannot entirely understand how these cereals got to the steppe cattle breeders. Did people living there practice farming somewhat similar to oasis farming in the desert areas of the south? Or was there—as R. Spengler liked to put it—any specific form of agropastoralism? Or are these grains a product of exchange with other groups distant from the steppe nomads?

Structures of cattle-breeding societies of the B-model appear less complex when compared to the ones described in the previous paragraph. For example, they did not have to maintain a balance (albeit naturally regulated) between the needs of farming and tending to animals. The focus of many cattle-breeding cultures was exclusively on animals. In essence, the mobile and semi-mobile herders had to meet natural demands like the ones discussed above. In their case, the mobility of domesticated animals required certain conditions for the herds and their grazing territories. The latter were certainly much larger and diverse in terms of plant life than the grazings of sedentary cultures. Moreover, the families of mobile cattle breeders representing this model rarely constructed permanent dwellings, making do with light tents similar to yurts. Such dwellings are very difficult to find by the archaeologists.

8.4. Metallurgical Cultures

At last, it is necessary to address, albeit quite briefly, a third model for a production economy—the M-type model—characterized by the development of metallurgical production to produce enough food to survive. Indeed, the final history of mankind turned out to be conjugated with metal. It was in those millennia that the ascent towards the industrial era began, the origins of which are usually associated with the so-called Early Metal Age (EMA), dating back to the fifth millennium BCE and associated at first with the mining and metallurgical production of copper, and with various kinds of bronze, and much later with iron. Such production was most often associated with the production economy of A- or B-type models (the latter was much less common). Very rarely, and in insignificant ways, it was associated with appropriating economies.

The first variant—sedentary cultures associated with mining and metallurgical production or only with metal processing—can be designated as A-M. For the second variant—nomadic or semi-nomadic cultures—the abbreviation B-M will be used.

Metallurgical provinces (MPs) played a special role in the EMA. MPs are defined as extensive complexes of metallurgical and metal-working centers, closely interconnected throughout multiple centuries and occupying territories up to several million square kilometers. These territories were marked by networks of metallurgical and metal-working centers. Products of each metallurgical province were unique both in territorial and chronological terms. The differences were most clearly reflected in the typological and morphological characteristics as well as in the type of technology that was used (from forging to casting, using metal alloys or "pure" copper or bronze, etc.).

Each metallurgical province was usually represented by two main types of production and by closely interconnected centers—metallurgical and metalworking ones. The metallurgical centers focused on extracting ores, smelting metal from them, and forging/casting products from the smelted metal. The function of metal-working centers was much simpler: they forged or cast metal from a particular metallurgical center into products of certain characteristic shapes. The metal was obtained through various means—exchange, appropriation, etc. Usually, the products of metal-working centers replicated, or were quite similar to the types of products associated with particular metallurgical centers. Even so, in most cases it is easy to see differences in metal production in metallurgical and metal-working centers. Finally, let us note the third way of metal distribution: often metal products ended up outside the boundaries of metallurgical and metal-working centers through exchange, appropriation, or penetration of certain cultural groups into foreign, remote regions.

> Perhaps the most striking example of the latter is the now widely known Seima-Turbino phenomenon that manifested itself in Eurasia at the turn of the third to second millennium BCE. The phenomenon itself, with its impressive monuments and bronze objects, reflected an extremely remarkable and mysterious route, used by groups of metallurgists and herder-horsemen that stretched for many thousands of kilometers—from the mountain steppes of Mongolia and the Dzungarian Gate up to the eastern Baltic region (see, for example: Chernykh, Kuzminykh 1989; Chernykh 2009, 264–286; Chernykh 2017, 234–249; and other works). This kind of phenomenon was not at all common in Eurasia.

8.5. From Proto-Metal Age to Metallurgical Provinces

The key subject of our research is the megastructure of the Eurasian world of which the metallurgical provinces are an important part. In all of Eurasia, so far it has been possible to identify only ten metallurgical provinces of the EMA falling within a wide period between the fifth millennium BCE and the late second to early first millennium BCE.

The Early Metal Age was preceded by a Proto-Metal Age, associated with the most ancient sites where metal has been found. The earliest known man-made metal objects were found in Anatolia, Levant, Northern Mesopotamia, and Western Iran (Fig. 8.1). These rather simple copper and lead objects were found in layers dating back to the ninth or maybe even tenth millennium BCE. Nevertheless, archaeologists have no definitive proof that these metals were smelted from mineral ores. Therefore it is usually believed that all those objects were made out of pieces of native metals processed by blacksmith pioneers. The sites where those metal objects were discovered are traditionally dated to the early New Stone Age or Pre-Pottery Neolithic. The most striking characteristic of these sites is, of course, stone and clay architecture (Çatalhöyük, Çayönü Tepesi and others) and, to a degree, unique stone sculptures found in temples (Göbekli Tepe).

The intricacy of objects and the artistic culture of archaeological sites dating back to the Proto-Metal Age makes one wonder whether this early phenomenon should be seen as the true *cradle* of post-paleolithic cultures of Eurasia as most specialists agree. It would make sense to connect these achievements in domestic and religious stone architecture as well as mining and metallurgical technology to this cradle. In this regard, we encounter an initial paradox.

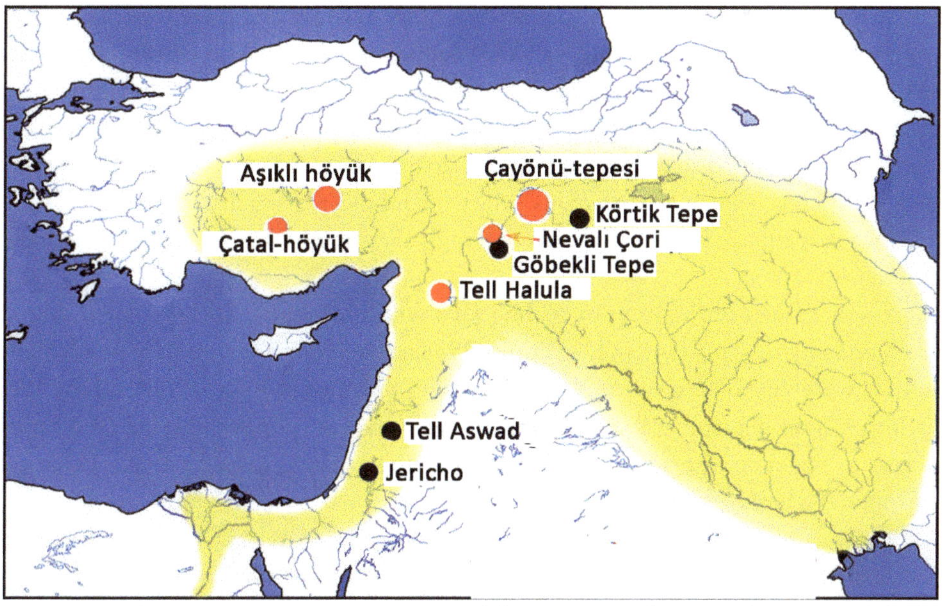

Figure 8.1. The most important monuments of the Proto-Metal Era. Settlements with finds in their layers of metal (copper and lead) are indicated in red.

8.6. The Carpatho-Balkan Metallurgical Province

The fifth millennium BCE was the time of a rapid rise of the most ancient metal production complex with a cohesive cluster of mining and metallurgical centers. Its emergence did not take place in the "Middle Eastern cradle" of universally significant technological revolutions where one would expect to find the origins of Eurasian metallurgy. In fact, metallurgical centers sprung up at a sufficient distance from it in the north of the Balkan Peninsula and in the Carpathian mountains. It was an actual metallurgical "outburst" that took place in the middle of nowhere at a time when there was no notion of metallurgical production in the "sacred cradle." The outburst gave rise to the phenomenon of the Carpatho-Balkan Metallurgical Province, the oldest MP in Eurasia.

This Copper Age cluster (there were no copper-based alloys there at the time) dates to the fifth millennium BCE and spans 1,500,000 square kilometers, stretching from the Adriatic Sea to the Middle and Lower Volga River regions (Fig. 8.2). The leading metallurgical centers of the Carpatho-Balkan Metallurgical Province (CBMP) were in the mountainous ore-bearing regions of the Balkans and the Carpathians. One of the most striking mining sites was the copper mine Ai Bunar. Large, heavy chopping copper tools and weapons were its main products. In addition to copper weapons, one finds gold jewelry; the number of these finds is close to 4,000. Likely gold was mined by washing river deposits in the mountainous regions of the Balkans such as the Rhodope Mountains (Tsintsov 2018).

Metal-processing centers were located to the east (Fig. 8.2B, C), in the steppe and forest-steppe in oreless territories of the northern Black Sea and the Volga River basin mainly in the domain of mobile and semi-mobile tribes of the western Eurasian Steppe Belt. Copper from Balkan ore sources discovered in the burials of the Volga stock-breeders was transported for at least 1,500–1,700 kilometers and probably 2,000 or more along the road. There are very few copper tools among steppe finds. Smaller copper jewelry such as beads, tubular strings of beads, and pendants are predominant, and their total number cannot but impress: *there are more than 5,500 of them*. However, in terms of total weight and significance, these trifles were certainly inferior to the "heavy" Western artifacts. Trading Carpatho-Balkan gold was probably forbidden. Hence, the precious yellow metal did not reach the steppes. Given that the province that replaced the CBMP was strikingly different in this respect, this fact needs to be stressed.

Still, this presumed prohibition of trading gold with the steppe stock-breeders should not outmatch one rather significant and, again, paradoxical fact: large parties of copper smelted in the Balkans were not to be given to the neighboring societies in Asia Minor and the Western Alps, which followed a similar life sustenance model. Instead, copper traveled far away to the north-east, to the alien nomadic stock-breeding peoples, reaching as far as the Volga region. What formed the basis of this paradox?

Most likely, it was in the Carpatho-Balkan Metallurgical Province that the practice of the international division of labor developed for the first time in history, although it still remains unclear what kind of labor the steppe-dwelling stock-breeders took upon themselves. Today, this practice is still of great significance and determines relations between very diverse societies across the globe. It is also then that the professional separation between mine-workers and metal-makers on the one hand and "ordinary" members of societies took place.

It is possible that even in this early, initial period when ore mining and smelting were just emerging, mine-workers and metal-makers either instilled in others respect or caused repulsion, as was the case, for instance, in Africa at the beginning of the Modern period. Of course, this was observed much later, but such a pattern could have occurred earlier and often in human cultures (Chernykh 1972, 192–194).

The revolutionary circumstances of the origins of the Carpatho-Balkan Metallurgical Province were without doubt a paradoxical surprise. Its demise, marked by a complete and sudden disintegration of its powerful and striking complex as well as of its numerous achievements, was also somewhat contradictory (Figs. 8.2, 8.3). In the Carpatho-Balkan region it was replaced by new production centers of the Circumpontic Metallurgical Province. This did not happen right away, but only 1,000 (sic!) years later, and the change itself was rather unconventional.

8.7. The Proto-Circumpontic Metallurgical Province

The origins of the Circumpontic Metallurgical Province (CMP) could be seen as a long-awaited debt paid by the "sacred Middle Eastern cradle." The earliest metallurgical centers of the province were associated with the mining regions of Eastern Anatolia, with the Tigris-Euphrates river system. It was there and in the nearby mountain ranges of the South Caucasus that copper and polymetallic mining, as well as copper smelting and tool forging from arsenical bronze began. Several scholars have gone as far as to call this period the Arsenical Bronze Age.

These facts made the Circumpontic Province quite different from the disintegrated Carpatho-Balkan Province in the West. The Circumpontic Province was named after a designation that the Greeks used for the Black Sea, *Euxinus Pontus*, since at one point the sea was surrounded by centers of the CMP.

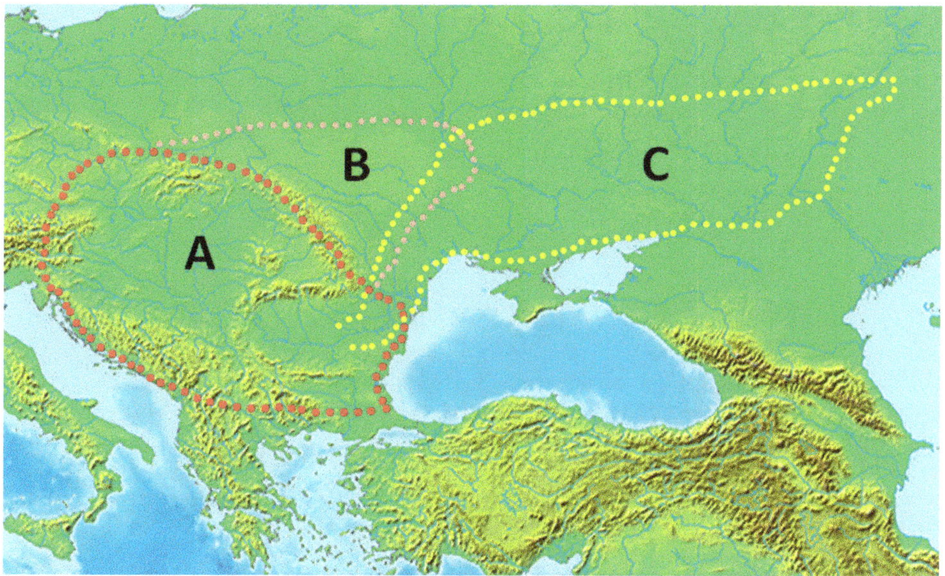

Figure 8.2. The area of the Balkan-Carpathian metallurgical province: A—zone of leading metallurgical foci; B—zone of metalworking foci in agricultural and pastoral cultures; C—zone of metalworking foci in nomadic and semi-nomadic pastoral cultures.

The history of the province is divided into two chronological stages. The early stage, known as the Proto-CMP, is associated with the Early Bronze Age, and spans the whole 4th millennium BCE (Fig. 8.4A). At the time, both metallurgical and metal processing centers were located close to the sea, but only in the east and southeast. The province's territory amounted to approximately 1.7–1.9 million square kilometers. The metal-working centers of the Proto-CMP were in Mesopotamia, which was deprived of ore deposits, as well as in the North Caucasus, where the cattle-breeding peoples of the Maykop culture, famous for its kurgan burials of nobility, played the most important role. The size of the kurgans of the Maykop culture and their unique funerary artefacts have caught the attention of archaeologists. It was there that they found many thousands of gold and silver items in burial complexes together with arsenical bronzes.

The Maykop kurgans contain yet another perplexing mystery. Metals, such as copper, gold, and silver were mined in southern regions where (unlike in the Balkans) rivers were mostly used to extract precious metals. One such mine was the ancient Sakdrisi gold mine (Stöllner, Gambashidze 2011) discovered in Transcaucasia (Georgia). It was also in the south that many thousands of different kinds of gold objects were forged and cast.

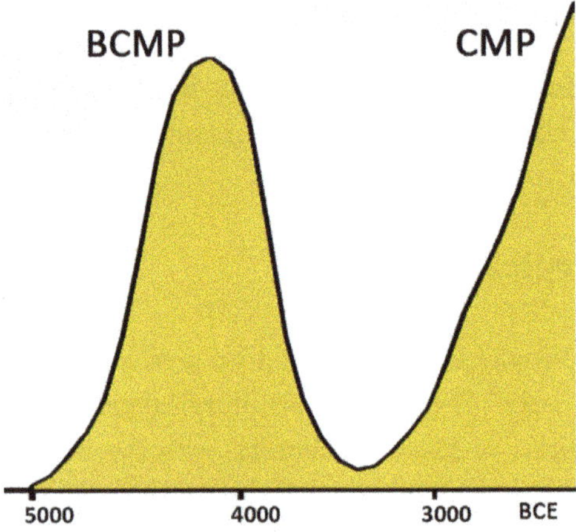

Figure 8.3. Balkan-Carpathian region: a diagram of the pulsation of production in the metallurgical centers of the Balkan-Carpathian and Circumpont provinces; a sharp failure in the production activity of the provinces between 4 and 3 thousand BCE is obvious.

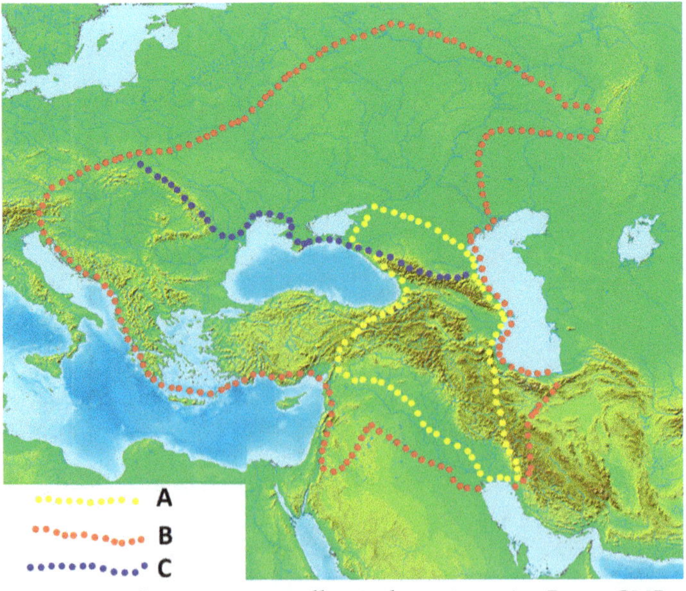

Figure 8.4. Circumpont metallurgical province: A—Proto-CMP area; B—CMP area; C—separation the boundary between the zones of metallurgical and metalworking centers (South and North of the province).

Archaeologists discovered all those objects mostly in burial mounds located in the north, a place inhabited by stock-breeding cultures where there are no known centers of metal production. Moreover, southern sites of the Proto-CMP account *only for 5–6 % of the striking amount of all the metal objects* produced by that system—nearly 10,000 objects (my database includes 9,500; see Fig. 8.5, left).

What is the reason for this phenomenon, as no examples of barter trade are known there? There is clear evidence of the destruction of Arslantepe, one of the largest east Anatolian hill dwellings composed of numerous layers. Right after its destruction, the hill became the site of a grave of the chieftain and his retinue (Chernykh 2013a, 172–177; Chernykh 2017, 152–157). Such an unusual way of constructing a burial complex might well be seen as a peculiar imitation of North Caucasian (Maykop) noble kurgan burials. It is likely that the first indications of military superiority of the fast-paced steppe riders over the settled village inhabitants date back to that period, especially since the steppe nomads had likely mastered horse riding by that time. It is entirely possible that this was the earliest episode in the pre-history of the victorious cavalries of the Huns, Turks, and Mongols, who would tread on cultures unknown to them on the vast Eurasian territories 3,000–4,000 years later.

Also, there is yet another puzzling pattern that has manifested itself in an extremely significant number of human cultures, distinguishing them among others, and illuminating a special type of religious practice. I shall call this pattern the "irrational syndrome," which encouraged human cultures to secure the afterlife of its representatives with material goods.

8.8. The Circumpontic Metallurgical Province

The second period in the history of this province spans the entire third millennium BCE and is characterized by an evident development of metallurgy and metal-working techniques as well as the types of metal products. An undoubted characteristic was the rootedness of this province's metal production in the previous era. The main achievement of this later stage was a drastic expansion of the province's territory, making it fully consistent with its name, Circumpontic. As a result, the province engulfed the Pontus of Euxinus, making it its center (Fig. 8.4B).

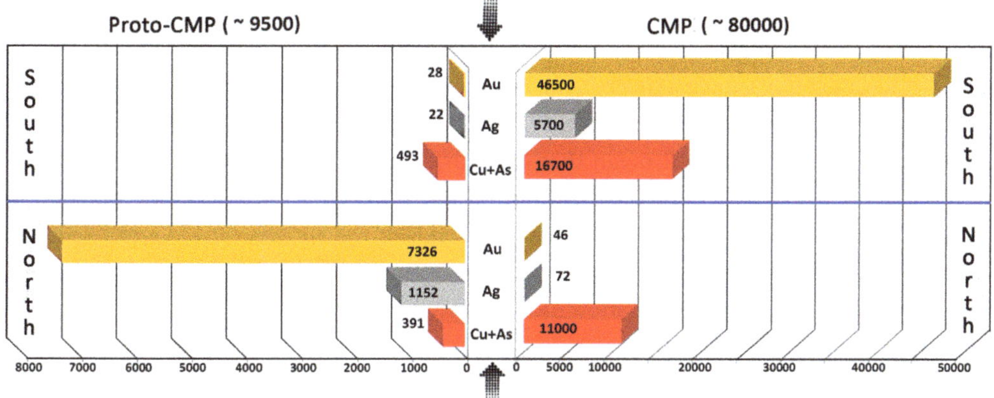

Figure 8.5. Proportional ratio between the number of gold, silver and copper/bronze (Cu+As) products in the monuments of the South and North for the periods of Proto-CMP and CMP.

At the time, the enormous territory of the CMP amounted to 4.5–4.7 million square kilometers, at least 2.5 times larger than before.

By that point, the centers of the CMP had completely absorbed the former, already forgotten CBMP. This "absorption" was accompanied by a simultaneous expansion of the CMP into the Southern Balkans and the Aegean Sea region (cf. Figs. 8.4 and 8.2). The CMP's western expansion was followed by the advancement of Circumpontic borrowings in the steppe and forest-steppe regions to the Urals. It was there that the main metal-processing centers working with imported Transcaucasian and Carpatho-Balkan metals were located. The bewildering intrusion of "entrepreneurial" migrants into the South Urals led to a rapid discovery and exploitation of copper ores with the Kargaly deposits as the most significant mining and metallurgical center.

The "watershed" between the areas of distribution of the two major types of CMP's centers—metallurgical and metal-processing—is depicted on the map (Fig. 8.4C). In this regard, the proportional ratio of the shares of three metals—gold, silver, and copper or bronze—in the collections of artifacts from the metallurgical south and from the metal-processing north and south is noteworthy. During the later period, the number of metal products increased to 80,000-almost ninefold in comparison with the earlier stage—yet another paradox (Fig. 8.5). The above-mentioned proportional ratio in the province's early history is completely different from the proportional ratio of these three metals in the Proto-CMP. The northern stock-breeding cultures account for less than 14% of objects in the database. Moreover, this percentage could have been much smaller if the chemically pure copper from the Ural's Kargaly were excluded, which cannot be considered southern. However, if in the previous case one wondered why the metal (especially the precious ones) produced in the south concentrated in the north, the CMP seems to present a more "normalized" situation, since all the major metals, produced in the south were discovered in southern sites. How can one explain the speed of these abrupt, drastically different transformations?

One thing is clear: the steppe dwellers did not renounce precious metals on their own; they were denied it, evoking the practices that had been common in the Carpatho-Balkan Metallurgical Province. Southern societies became more consolidated and resistant to attacks by northerners and they were likely to have their own cavalry. The most important reason for the consolidation of southerners and the growth of their military power was probably an ideological revolution which occurred in the third millennium BCE. All objects of significance, particularly those made from gold and silver, dating back to that period and onwards, have been discovered by archaeologists in temple and funerary complexes of the South. In the previous epoch, there was a difference: all the gold and silver from the south accumulated in the kurgans of noble stock-breeders in the north.

What does it tell us? Does it mean that the worldview of the steppe stock-breeders, who believed that life began with one's death, prevailed across the large Circumpontic Metallurgical Province? After all, it was this mindset that dictated that the most valuable and long-lasting goods had to accompany deceased noblemen. In any case, these complex questions are particularly interesting and require special attention.

The Circumpontic Province, undoubtedly, was the most striking flagship for the formation among other metallurgical provinces. That is why most traces of its presence—both direct as well as hidden and hardly detectable ones—were preserved in mining and metallurgical production of the chronologically successive metallurgical provinces up to the time of the first uses of iron.[1] This unique phenomena fell into decline during the last third of the third millennium BCE.

8.9. The End of the Early Metal Age: Eight Metallurgical Provinces

The collapse of the Circumpontic Province marked a transition of the Eurasian cultures into the Late Bronze Age and coincided with the final centuries of the EMA. From the end of the third to the beginning of the second millennium BCE, for three or four centuries Eurasia underwent a striking, all-encompassing "explosion" leading to the formation of eight metallurgical provinces. The metal-producing centers of these dissimilar, newly formed structures spread across an enormous territory between the Atlantic and the Pacific Oceans (Fig. 8.6). The explosive effect becomes apparent when we compare this geographical scope with the territory of the earlier metallurgical provinces. They were more than four times smaller (Fig. 8.7). The last provinces of the EMA were active all throughout the second millennium BCE up to the turn of the second millennium and early first millennium BCE, marking another turning point in the history of Eurasian societies—the Iron Age.

This chapter allowed us to touch briefly only upon a few of the most important aspects of the gigantic "eight-headed" phenomenon. First, the development of these new metallurgical provinces followed major, common patterns. As a rule, each province was characterized by a significant development of metallurgical and metalworking technologies, by an expansion of

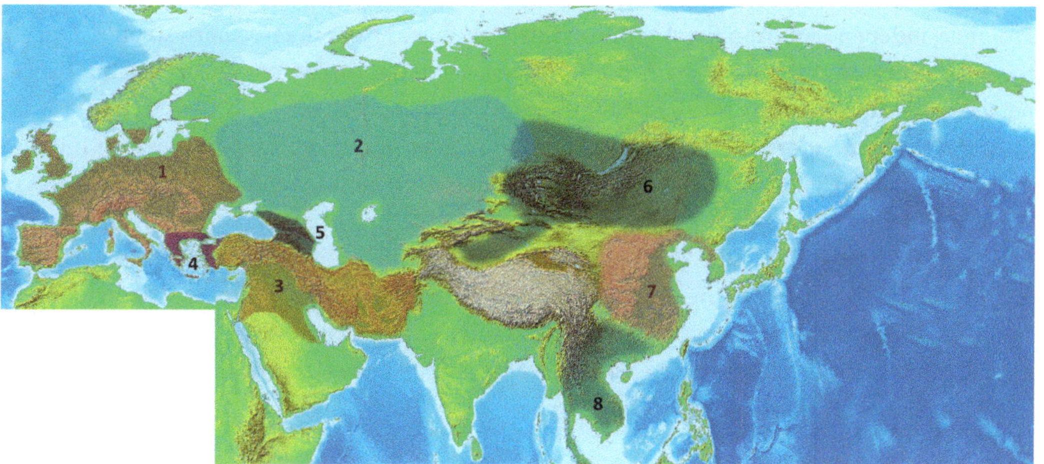

Figure 8.6. Distribution of metallurgical provinces in Eurasia: 1—European, 2—West Asian, 3—Iranian-Anatolian, 4—Aegean, 5—Caucasian, 6—East Asian, 7—Ancient Chinese, 8—Indochinese.

---

[1] Publications from my 1977 article up to the latest book published in 2019 (Chernykh 1977–2019) include references to various characteristics and details of production in numerous metallurgical and metal-working centers of this huge phenomenon.

major types of metal products (not only tools and weapons, but also jewelry), as well as by the development of new mining and metallurgical centers and their early and active exploitation.

> The most impressive among these centers was the gigantic Kargaly copper ore complex in the South Urals. Copper mining and smelting began there already in the fourth millennium BCE, during the era of the Circumpontic Province. Kargaly reached its pinnacle in the second millennium BCE, when it became one of the key centers of the West-Asian Metallurgical Province. Up to 35,000 (sic!) remains of different shafts were discovered there on the territory of 500 square kilometers, with hundred-kilometer deep workings. Excavations of one of the multiple settlements of the Kargaly miners and metallurgists revealed a huge mountain composed of 2,500,000 bones of domestic animals that had been brought to Kargaly in exchange for ore and metal. All of these remains were unearthed at one site, and excavations are still ongoing there. Kargaly's ore and metal spread all the way up to the Sea of Azov and the Lower Dnieper basin. In exchange, Kargaly received cattle from these regions.

The history of the Carpatho-Balkan, Circumpontic, and West-Asian Metallurgical Provinces presents striking cases of the actual international division of labor. During their 3,000 year-long history, there were continuous, active exchanges between the metallurgical centers of the south and the northern steppe cultures, although these relations varied greatly in terms of causes and incentives at times. The Caucasus played the role of a permanent, long-lasting *bridge* between the south and the north and had a particular importance for the Circumpontic Metallurgical Province. With the general situation changing, the Caucasus witnessed the emergence of impressive new centers of metallurgical production, a *sui generis* Caucasian Metallurgical Province, independent from the Anatolo-Iranian south and from the northern steppes. During the entire second millennium BCE the Caucasus served as an almost unsurpassable barrier between the cultures of the north and the south.

And finally, one more curious trait of these provinces of the late Bronze Age is of interest. These provinces exhibited a high degree of interpenetration between the contrasting cultures of western and eastern Eurasia. These interpenetrations were governed by a *pendulum-like rhythm*: first from the west to the east, and then back. However, these oscillations were much more manifest in the Iron Age, as the next chapter demonstrates. One of the most striking demonstrations of this penetration across many thousands of kilometers from the east to the west are the monuments of the so-called *Seima-Turbino phenomenon* (Chernykh, Kuzminykh 1989; 2010 (in Chinese); Korochkova et al. 2020).

8.10. The Iron Age: Apparent Changes in the Eurasian Jigsaw Puzzle

The first millennium BCE witnessed striking shifts and changes in the general Eurasian pattern. Step by step, but quite quickly, the systems of all the eight aforementioned metallurgical provinces of the late Bronze disappeared. It was they that, in fact, in the second millennium BCE represented the rather rigid foundation for the general structure of the Eurasian world. These eight systems covered a territory of about 23,000,000 square kilometers (Fig. 8.7). It seems that no less than nine-tenths of the continent's population lived within those boundaries, because outside of them there were only sparsely populated forest-tundra and tundra. It means that one need not doubt that the shifts and changes of the early centuries of the first millennium BCE influenced virtually all the main Eurasian communities.

The Bronze Age ceased to be with the demise of the metallurgical provinces, never to return, and was replaced by the Iron Age. The first, even though quite clear signs of metallurgy and iron-working—iron was new to the cultures of *Homo*—became apparent already in the final centuries of the second millennium BCE. These discoveries and the appearance of a new technology were a shock to those who witnessed the change, but they never expected these innovations. Probably that is why the assessments of these processes varied significantly. Here, for instance, are the maledictions of the Greek poet Hesiod who lived in the seventh century BCE:

> Thereafter, would that I were not among the men of the fifth generation, but either had died before or been born afterwards. For now truly is a race of iron, and men never rest from labor and sorrow by day, and from perishing by night; and the gods shall lay sore trouble upon them . . . [yet] even these shall have some good mingled with their evils.

Already six hundred years after the desperate lines of the Greek poet, the Roman philosopher and poet Lucretius spoke about those shifts in quite neutral tones:

> *Then by slow degrees*
> *The sword of iron succeeded, and the shape*
> *Of brazen sickle into scorn was turned:*
> *With iron to cleave the soil of earth they 'gan,*
> *And the contentions of uncertain war*
> *Were rendered equal.*

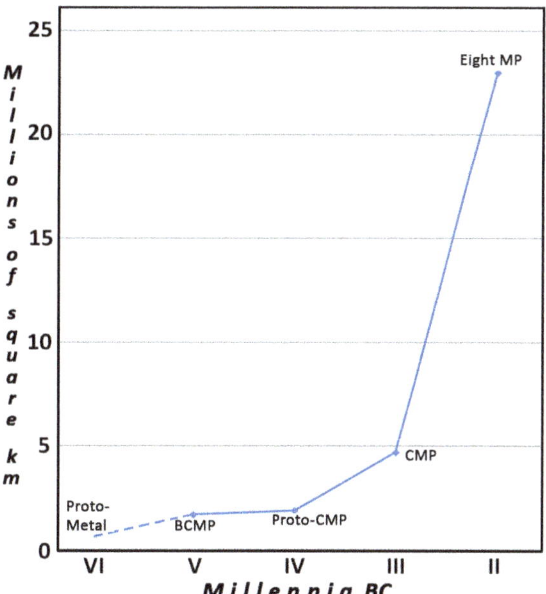

Figure 8.7. Dynamics of spatial coverage of metallurgical provinces of Eurasia depending on chronological positions.

One can assume that the most striking changes that characterize the structure of the world, including, of course, the Eurasian world, are a consequence of the great intercontinental migrations (see Chapters 2 and 3). It is true, though, that the mass intracontinental movements of human groups caused the changes. The movements that took place do not appear to be large-scal, but the changes that occurred in the main look obvious. The metallurgical provinces of the Bronze Age that represented a rather rigid foundation of the general Eurasian structure disappeared. It is quite difficult to say precisely what *structural* changes emerged at the beginning of the Iron Age as the metallurgical provinces of the Iron Age have not been yet studied in detail.

What is important here is something else: the structural shifts did not apply to all areas of the metallurgical provinces such as in the Far East of Eurasia or the Aegean Sea region in the Eastern Mediterranean or the cultures of the sub-enclave of Palestine that had been integrated into the structure of the Anatolo-Iranian province of the Bronze Age. They will be discussed in the second part of this book, when focus turns to a comparison of these three systems.

Chapter 9

Radiocarbon Chronology of "Pre-Written" History in Western Eurasia

The problems of the absolute chronology of Eurasia actually constantly worried archaeologists, although the original paradigm of this science for the largest continent of the Earth seemed to be well-established, as if even unshakable in principle approach. The dogma declared many decades ago Ex Oriente Lux or "Light from the East," when all, as well as all the crown discoveries were connected with the so-called Middle East, that is, first of all, with Mesopotamia and, perhaps, with the countries adjacent to it. It followed from this that all phenomena outside of this rather artificially defined region should be dated to a later time.

This approach was quite clearly reflected in the specific research of many archaeologists and historians, although a number of researchers simultaneously had a feeling of something artificial in this approach and therefore far from incomprehensible. In addition, many Paleolithic problems remained outside the rigid framework of this seemingly immutable axiom, which practically turned out to be unsolvable on the basis of the canons of this "irresistible truth." Yes, however, also the antiquities of the Copper-Bronze, as well as the Iron Ages, often presented archaeologists with many surprises and difficult to solve issues. However, much more vivid insights awaited the science of the age of fossil antiquities already in the decades after the end of World War II.

Initially, it was dendrochronology, and after it appeared age definitions in fossil objects based on the analysis of the isotope ^{14}C. Then we tried to combine the results of radiocarbon age determinations of archaeological antiquities with dendrochronological ones. And this, perhaps, was the beginning of the crystallization of a fundamentally new direction in science, which allowed solving very difficult and extremely important problems of the absolute chronology of the ancient stages of human development in a fundamentally different way (Chernykh 2019, 33–37; Pearson Ch. L. and others 2021).

However, it hardly makes sense to repeat here—even in the briefest retelling—the story about the main stages of the development of this direction and its successes. It is better to turn directly to the results of processing those materials that turned out to be directly related to the main tasks of our research.

9.1. Radiocarbon Dating Database

The "age frame" of our chronological research is based on the general database of dates for ^{14}C, which has been formed over the past six decades in the Laboratory of Natural Science Methods of the Institute of Archaeology of the Russian Academy of Sciences (Chernykh, Orlovskaya 2021). The cumulative and not yet fully systematically processed array of this database now exceeds eight thousand definitions of the age of archaeological objects. However, so far, a total of almost 6.3 thousand absolute age definitions associated with more than 1400 archaeological sites have been subjected to systematic analysis (Fig. 9.1, Table 1). The spatial coverage of the definitions that make up this database is close to 20 million square kilometers (Fig. 9.1).

The main focus of the database turned out to be aimed at archaeological communities, cultures or individual significant complexes of the Early Metal Era of Western Eurasia (ERM). However, in addition to this, attention is also paid to the cultures and complexes of the Proto-Metal epoch associated with this epoch in time and space, Neolithic communities of northern, forest, areas, as well as steppe communities of the early Iron Age. The results of the research are the basis for deciphering and recreating a wide—chronologically and spatially—canvas of the formation and development of ERM cultures from the earliest signs of human use of copper and its alloys, up to the apogee of the mining and metallurgical production of copper and bronze and, finally, to the Early Iron Age, that is, within nine to ten thousand years—from ten to one thousand BCE. The spatial coverage of the definitions that make up this database turned out to be close to 20 million square meters. extending from the eastern border between the eastern and western parts of the continent—the basins of the Yenisei and Indus—up to the Balkan-Carpathian region and the lower reaches of the Danube (Fig. 9.1).

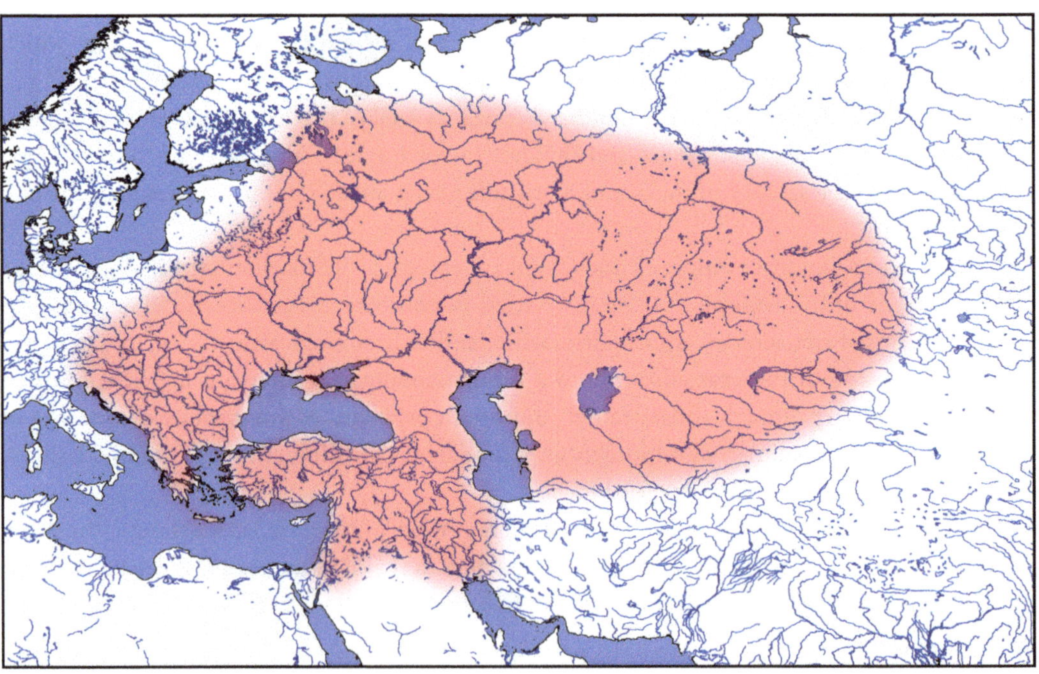

Figure 9.1. General map of the distribution of systematized radiocarbon dating.

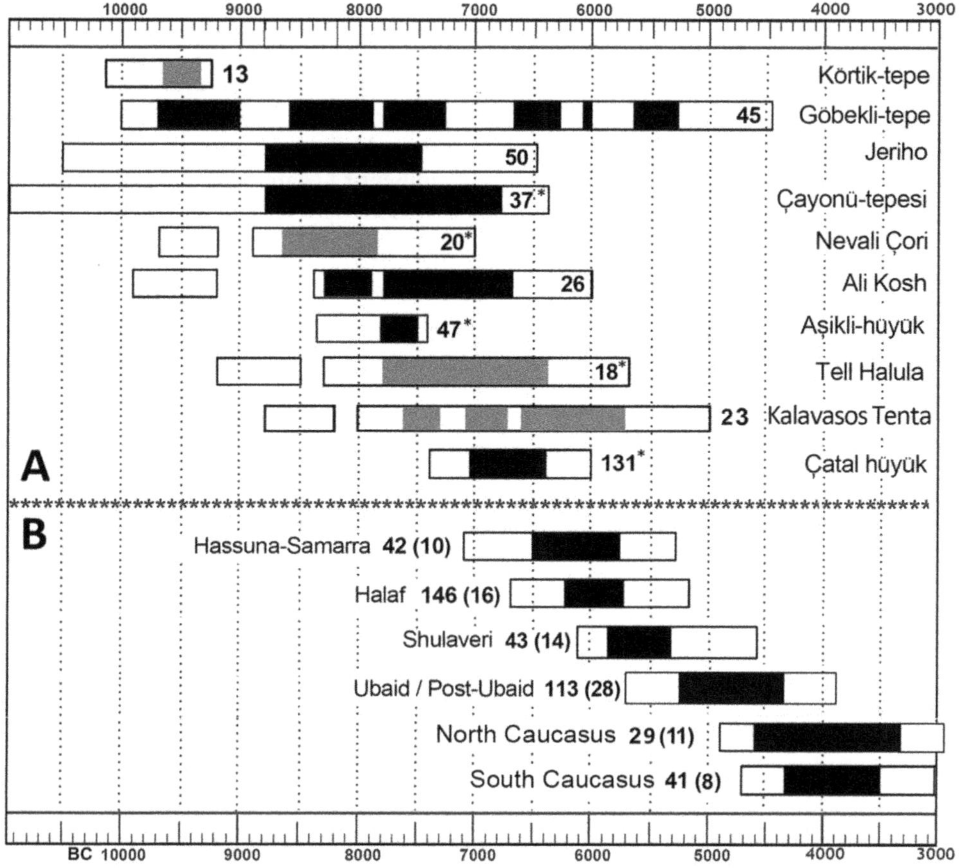

Figure 9.2. Chronological ranges of pre-Ceramic Neolithic (A)
and Ceramic Neolithic (B) monuments.

The purpose of this article is to give an idea of the most important, nodal series and aggregates of radiocarbon dating, which were the true foundation in the process of distributing most communities and cultures on an absolute calendar scale. At the same time, the main attention was paid to the dating of antiquities that were part of the systems of large metallurgical provinces of Eurasia: the Balkan-Carpathian, Circumpont and West Asian. In addition, data on the Proto-Metal epoch preceding these systems, as well as subsequent steppe cultures of the Early Iron Age were involved (Table 1).

9.2. Systematic Analysis of Radiocarbon Dating

System analysis radiocarbon dating assumes an indispensable set of the following actions:

1) Calibration of the age definitions of each of the analyses taken into account; calibration is carried out in the laboratory according to the program of the Oxford Laboratory (Research Laboratory for Archaeology and the History of Art, University of Oxford (Version 3.10 and later);

Table 1. General database: information on the sums of systematized ¹⁴C datings, the number of sites to which the recorded datings are related, as well as the area covered by the datings for each of the eras, metallurgical provinces (MP), archaeological cultures or communities (AC) and individual areas

Epochs of the provinces/regions	Phases/areas/AC and their blocks	Number of ¹⁴C dates (number of sites)	Total number of dates and sites	Coverage area of dates in million square kilometers	
				Local	General
Proto-Metal	Pre-Pottery Neolithic	422 (10)	843 (98)	0.5–0.6	1.2–1.3
	Pottery Neolithic	421 (88)		1.2–1.3	
Carpatho-Balkan MP	Central	873 (159)	1240 (279)	0.7–0.8	1.6–1.7
	Tripolye-Cucuteni	121 (52)		0.16–0.2	
	Steppe AC	246 (68)		0.6–0.7	
Post Carpatho-Balkan MP	Baden and others	114 (39)	114 (39)	0.35–0.37	
Proto-Circumpontic MP	South (Anatolia, Mesopotamia Levant, Western Iran, Transcaucasia)	443 (84)	554 (121)	1.4–1.6	1.7–1.9
	Ciscaucasia, Maykop AC	111 (37)		0.3–0.35	
Circumpontic MP	South (Anatolia, Mesopotamia Levant, Western Iran)	700 (104)	1789 (424)	1.4–1.6	4.0–4.5
	Balkan-Carpathian, Aegean	254 (59)		0.9–1.0	
	Steppe AC: pits, catacombs, etc.	835 (261)		1.7–1.9	
Altai Enclave of the Circumpontic	Afanasievo	84 (31)	138 (48)	0.15–0.25	0.25–0.4
	Chemurchek	54 (17)		0.1–0.15	
West Asian MP	Phase I: Abashevo-Syntashta-Petrovka AC	112 (27)	730 (197)	1.0–1.2	5.5–6.5
	Phase II: Construction Andronovo AC	406 (105)		3.0–3.5	
	Phase II: Forest Variants and n/a	68 (25)		0.8–1.0	
	Phase III: AC of Volchikova Pottery	144 (40)		4–5	
Forest AO of the West Asian MP	From Karelia to the Trans-Urals	244 (105)	244 (105)	1.3–1.5	
"Scythian World"	From the Lower Danube to the Sayan-Altai	624 (109)	624 (109)	5.0–6.0	
Total Number of ¹⁴C dates (sites)			**6276 (1420)**		

2) The most definite connection of a specific analysis to determine the age with the complexes of a specific Archaeological Culture (AK), or an Archaeological Community (AO), or with a specific archaeological monument (AP);

3) Determination of the geographical coordinates of each of the definitions for accurate indication on the map.

The results of the system analysis are presented on the general database; all information is reflected in the diagrams (Figs. 9.2, 9.7–9.9, 9.11, 9.14, 9.18, 9.20).

Each of the diagrams corresponds to a specific and, as a rule, a large section—the seven most important chronological and territorial systems of Eurasia: 1—The Protometallic Epoch; 2—the Balkan-Carpathian Metallurgical Province (BCMP); 3—The Proto-Circumpont Metallurgical Province (Proto-CMP); 4—the Circumpont Metallurgical Province (CMP); 5—the West Asian Metallurgical Province (WAsMP); 6—forest and forest-steppe cultures and communities outside the metallurgical provinces; 7—the Scythian World or the Early Iron Age. The locations of the dated monuments are shown on the corresponding maps (Figs. 9.3–9.6, 9.10, 9.12, 9.13, 9.15–9.17, 9.19, 9.21).

The central place in this series, as already mentioned above, is occupied by the materials of the four metallurgical provinces. However, it is more reasonable to talk about three, since the Proto-Circumpont province is, in fact, only an early, albeit very long-lasting phase of the Circumpont province, for which the Proto-CMP served as the initial, basic stage.

It is worth noting that one of the most important features of the studied within the framework of such huge systems as the Balkan-Carpathian and Circumpontic metallurgical provinces of settled agricultural and pastoral blocks of cultures was a very complex relationship between them, which in turn was reflected in the pictures of the absolute chronology of both blocks. While their interaction with the northern block of forest metal-bearing crops was much less expressive.

The diagrams cannot fail to show sharp differences in the quantitative characteristics of the demonstrated aggregates: from the minimum—in 6 and 13—age definitions (Fig. 9.2) up to extremely representative—up to 462 dates, respectively (Fig. 9.11, Yamnaya AO). The authors believe that a special mention here of very small aggregates seems absolutely necessary for a correct assessment of reality and reliability in determining the age of a particular archaeological complex or culture.

The distribution of the probability sums for each of the bases

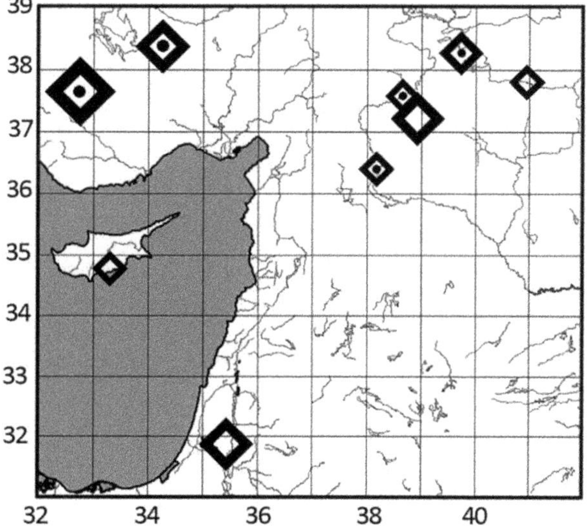

Figure 9.3. The area of pre-Ceramic Neolithic monuments dated to ^{14}C: without metal (rhombus), with native metal (rhombus with a dot inside).

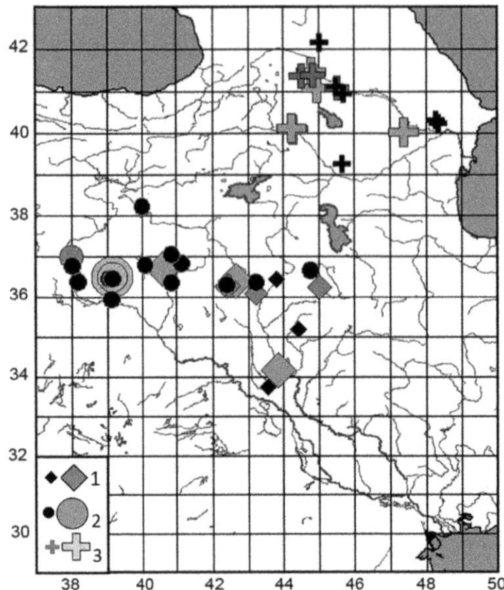

Figure 9.4. The area of monuments of cultures dated to ^{14}C of the Ceramic Neolithic period: 1—Hassuna-Samarra; 2—Khalaf; 3—Shulaveri-shomutepe.

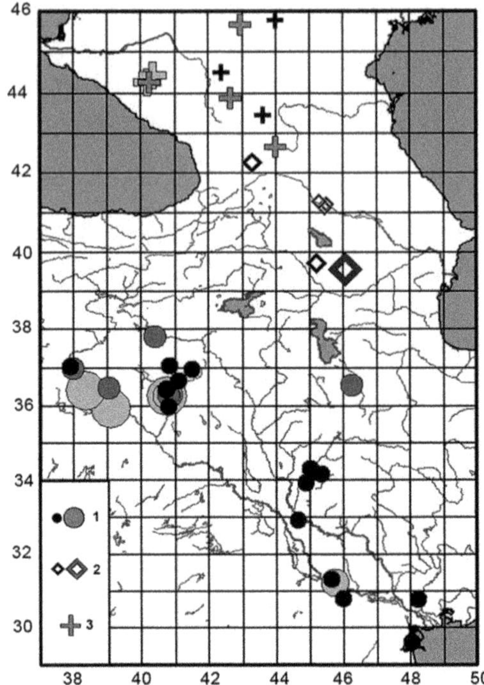

Figure 9.5. The area of ^{14}C-dated monuments of cultures of the Ceramic Neolithic period: 1—Ubaid/post-Ubaid; 2—South Caucasus (Leylatepe, Areni, etc.); 3—North Caucasus (KNZHK, Meshoko, Yaseneva Polyana, etc.).

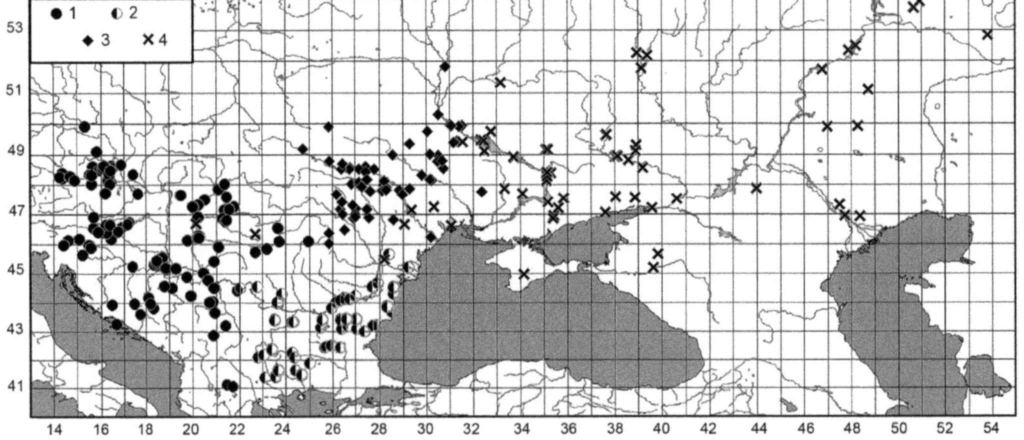

Figure 9.6. The area of ^{14}C-dated monuments of the main blocks of archaeological cultures and communities of the Balkan-Carpathian MP: 1—central block—Carpathian group; 2—central block—Balkan group; 3—Tripoli AK; 4—steppe AO.

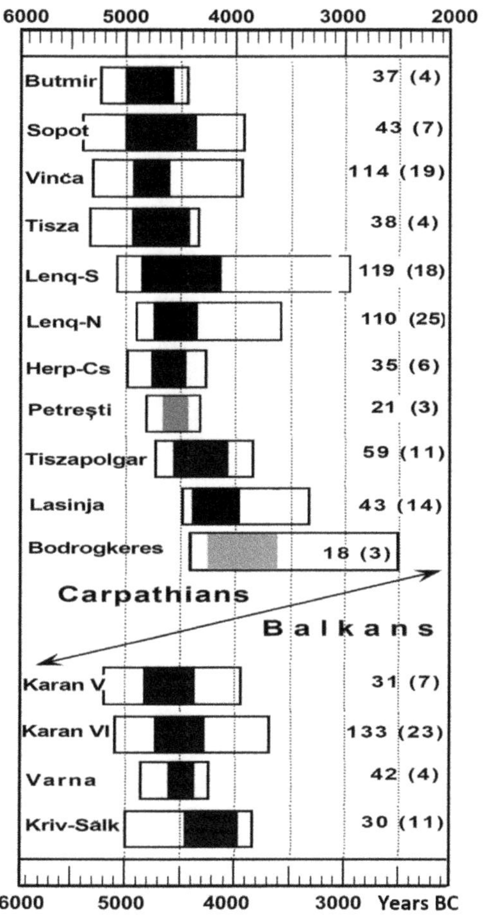

Figure 9.7. Chronological ranges of monuments of cultures of the Balkan-Carpathian MP Middle (top) and Lower (bottom) Podunavya.

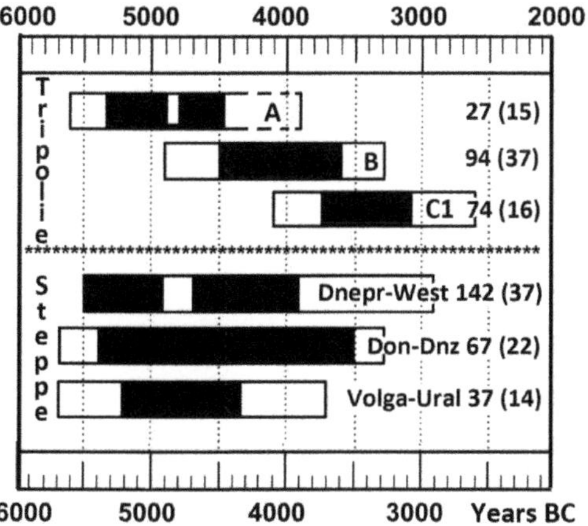

Figure 9.8. Chronological ranges of monuments of cultures of the Balkan-Carpathian MP: the Trypillya-Kukuteni block and in the zone of distribution of steppe AO (Dnieper-West, from the Dnieper to Transylvania).

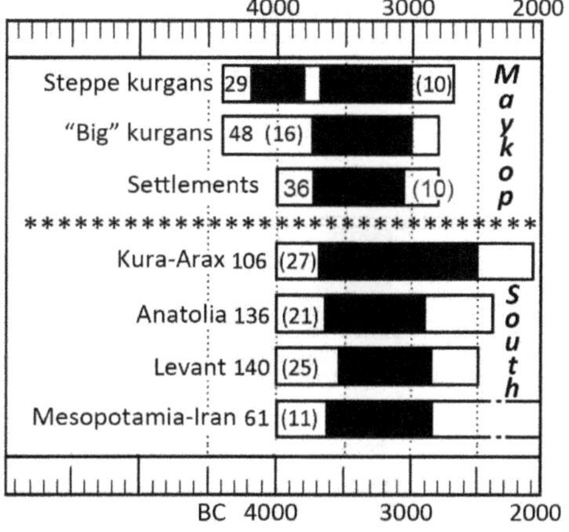

Figure 9.9. Chronological ranges of the main blocks of archaeological cultures and communities of the Proto-Circumpontian MP.

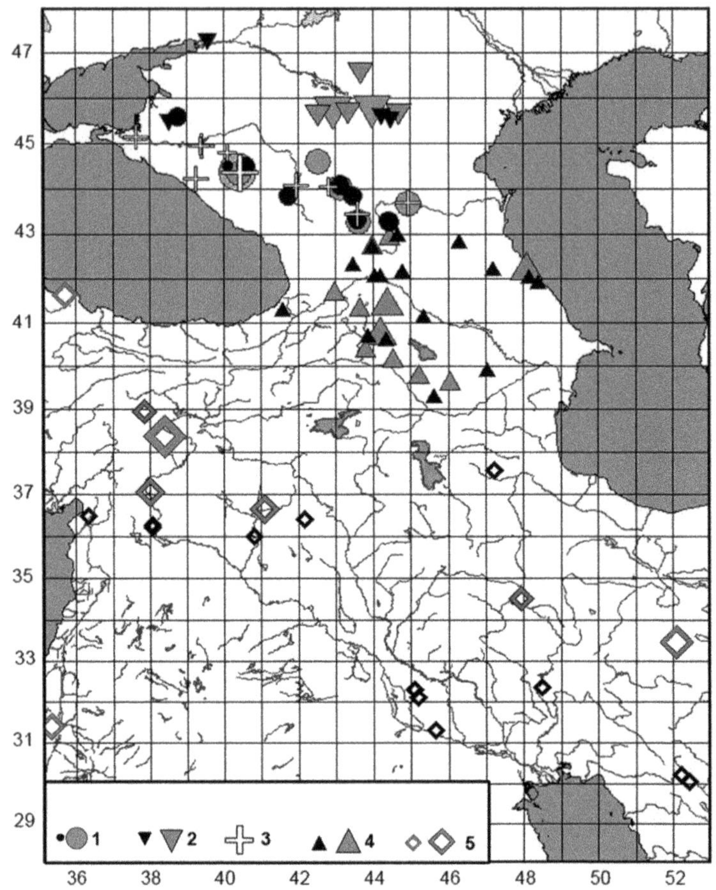

Figure 9.10. The area of monuments dated to [14]C Proto-Circumpont province. The North Caucasus and the Pre-Caucasian steppes: 1—the so-called "big mounds" of the Maykop culture, 2—the so-called "big mounds of steppe cultures," 3—settlements. South Caucasus: 4—monuments of the Kuro-Arak culture. Monuments of Eastern Anatolia, Mesopotamia, Western Iran: 5—so-called Late Uruk cultures.

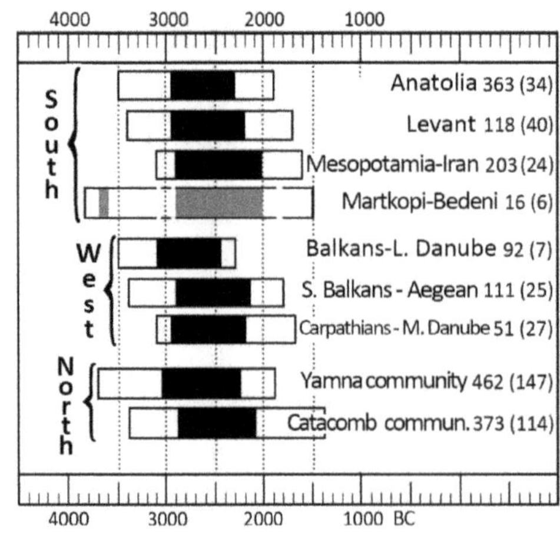

Figure 9.11. Chronological ranges of monuments of the main blocks of archaeological cultures and communities of the Circumpont MP.

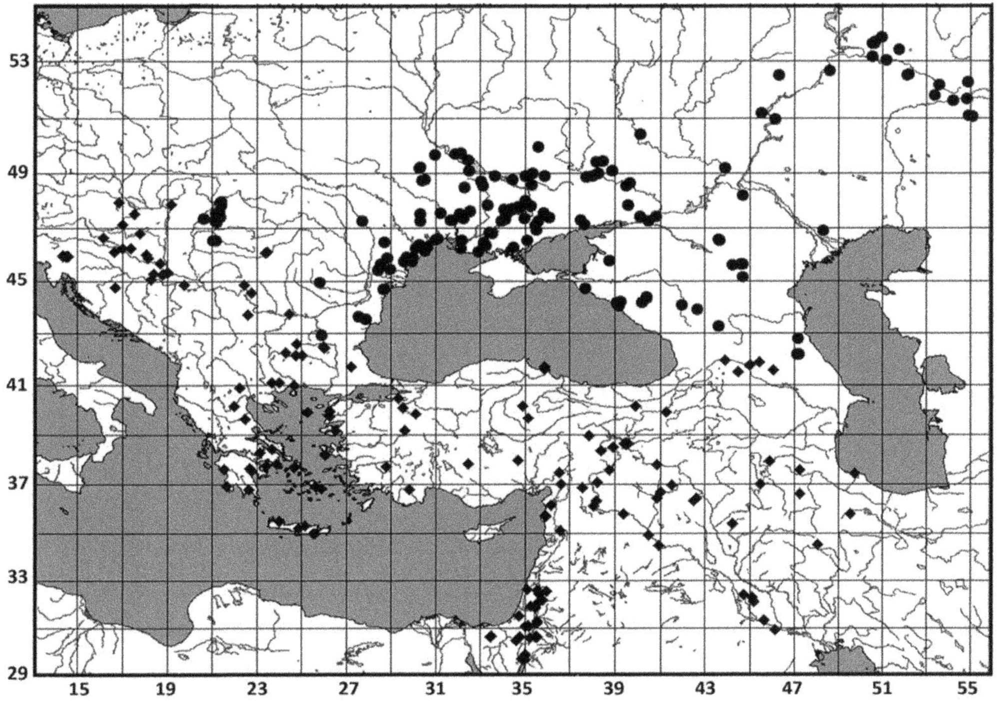

Figure 9.12. Circumpont MP. The areas of monuments dated to ^{14}C of southern settled agricultural cultures (rhombuses) and northern steppe pastoralists—Yamnaya AO (round icons).

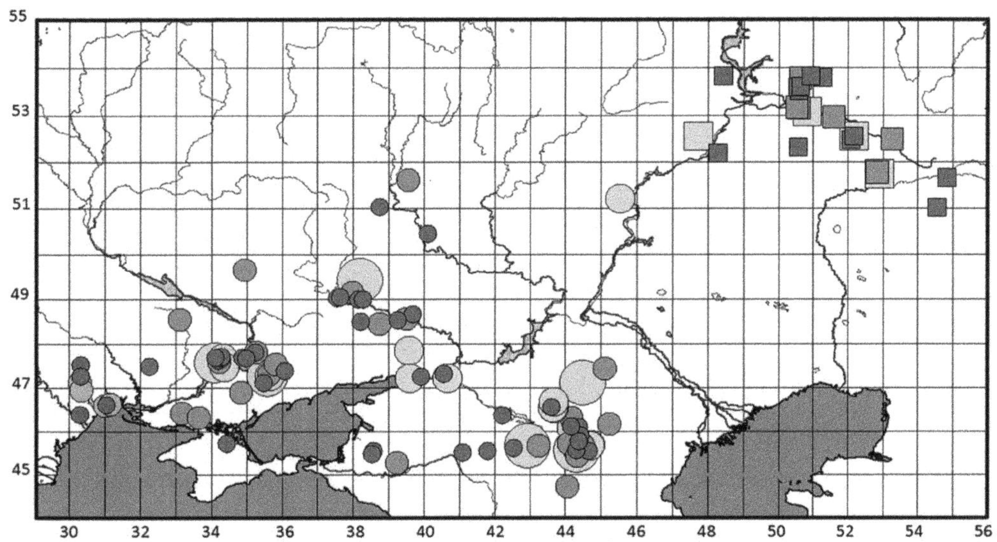

Figure 9.13. Circumpont MP. The location of the monuments of the northern steppe pastoralists dated to ^{14}C catacomb AO (round icons) and Poltavkinskaya AK (squares).

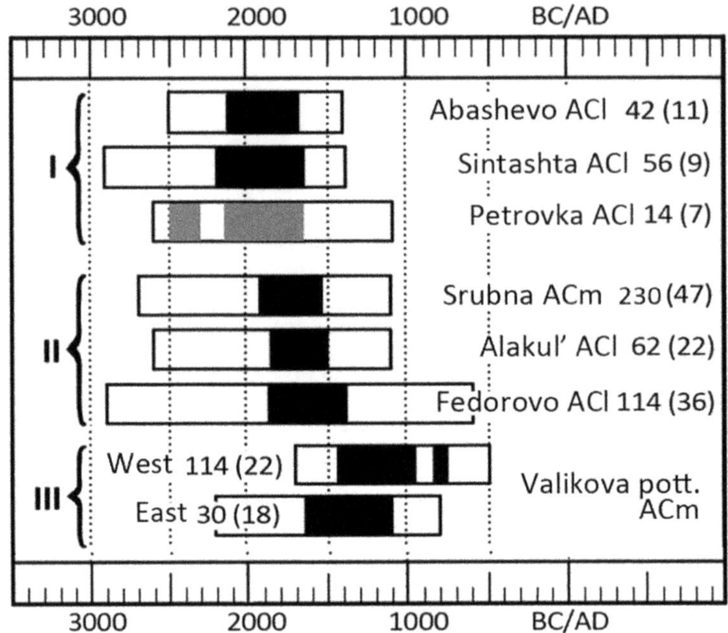

Figure 9.14. Chronological ranges of the AK and AO of the West Asian province. The figures I, II, and III indicate the main phases of the WAsMP.

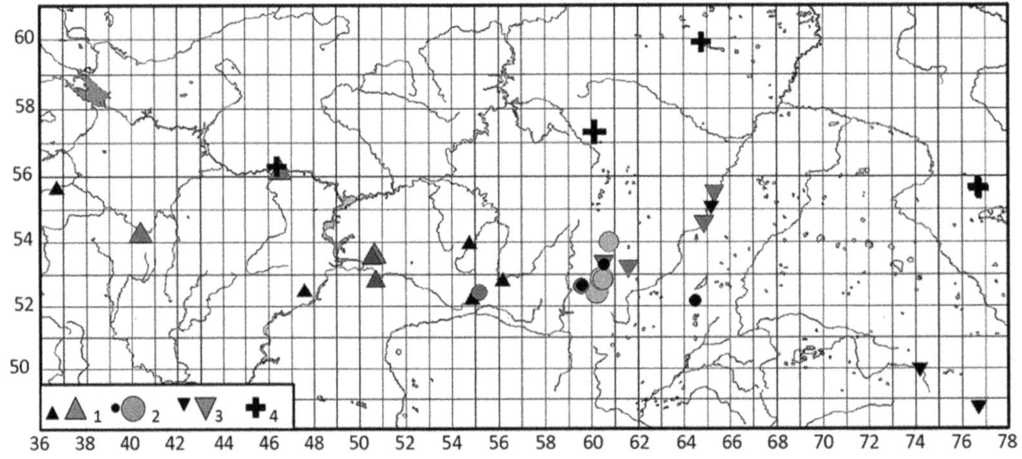

Figure 9.15. The early phase of the WAsMP. The location of the monuments dated by [14]C: 1—Abashevskaya AK; 2—Sintashtinskaya AK; 3—Petrovskaya AK; 4—Seiminsko-Turbinsky monuments that were not included in the system of the WAsMP.

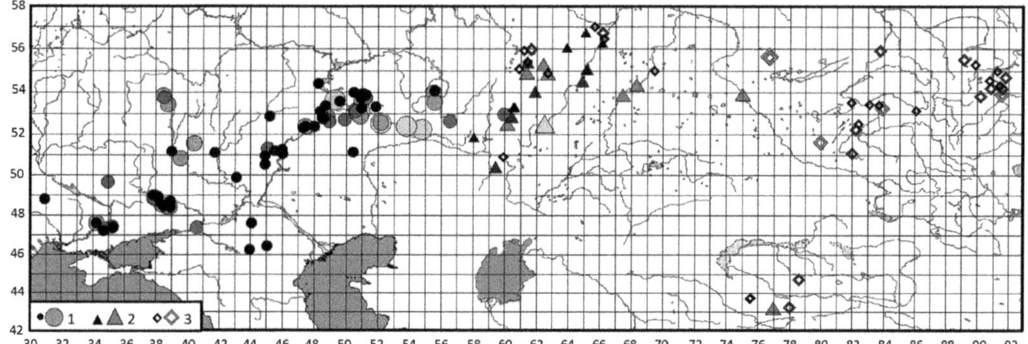

Figure 9.16. The middle phase of the WAsMP. The location of the monuments dated to ^{14}C:
1—srubnaya AO; 3—Alakulskaya AK; 4—Fedorovskaya AK.

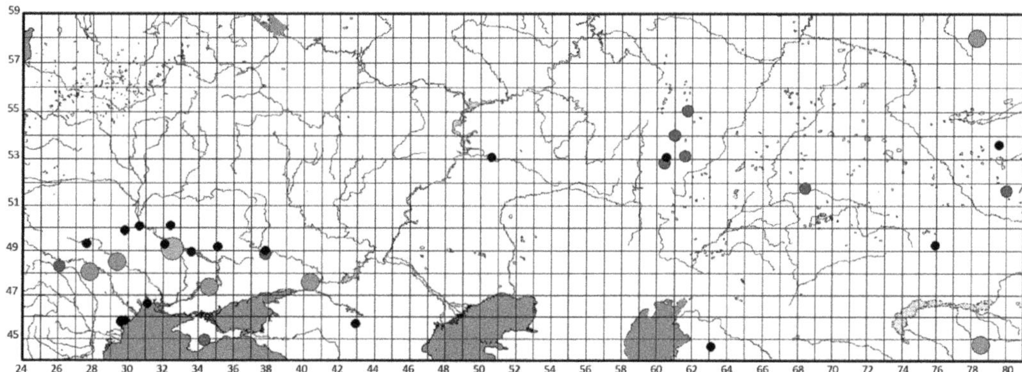

Figure 9.17. The late phase of the WAsMP. The location of the monuments of AO roller ceramics
dated to ^{14}C.

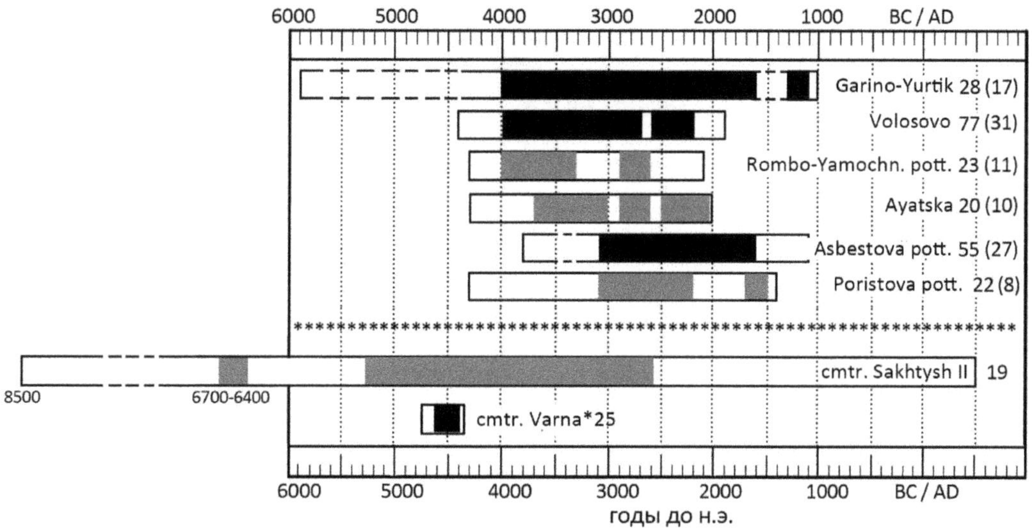

Figure 9.18. Chronological ranges of forest Neolithic AK, as well as the burial ground of
Volosovskaya AK, Sakhtysh II.

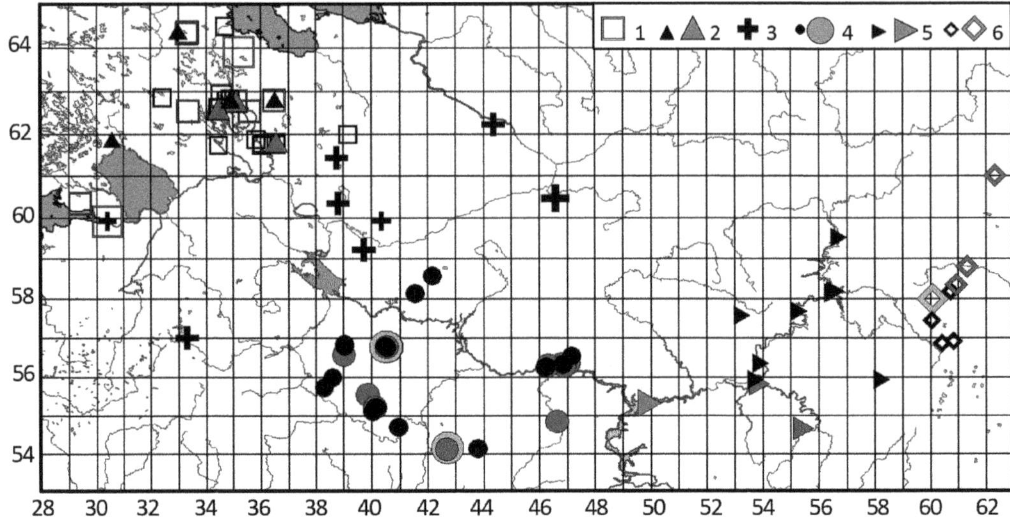

Figure 9.19. Areas of monuments of forest Neolithic AK dated to ¹⁴C: 1—asbestos ceramics; 2—rhombo-pit ceramics; 3—porous ceramics; 4—Volosovskaya; 5—Garinsko-Bor-Yurtikskaya; 6—Ayat (Trans-Ural).

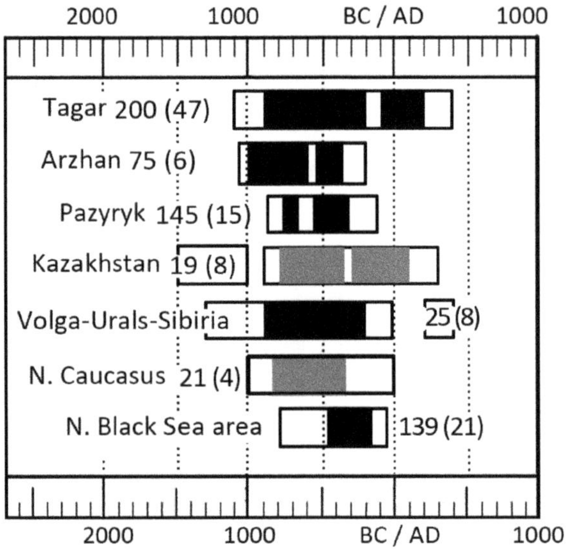

Figure 9.20. Chronological ranges of monuments of AK and AO "Scythian World" of Western Eurasia—from the Northern Black Sea region to the Sayan-Altai.

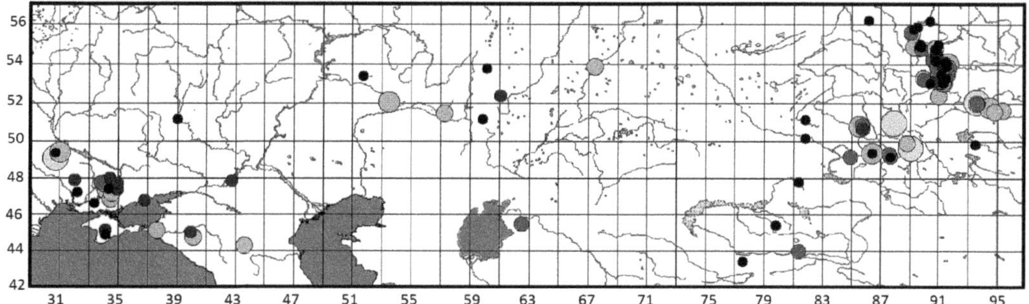

Figure 9.21. Areas of monuments of AK and AO "Scythian World" dated by ^{14}C in Western Eurasia.

is reflected in the diagrams as follows. The main rectangle, contoured with straight lines, means the range of the sum of the probabilities of determining the age of a particular population in two sigma, or 95.4%. Inside the main rectangle there is always one rectangle filled in either black or gray, meaning a range of absolute age probability sums of one sigma, or 68.2%. In the case of a single colored rectangle, it can be assumed that the distribution of probability sums within each analyzed population corresponds to a normal character or is close to normal.

Sometimes, inside the main large rectangle of two sigma, you can see two or even three small figures filled with black or gray. This already indicates the presence of anomalies in the probability distribution and their deviations from the normal nature. Therefore, in such cases, the probability of one sigma, or 68.2%, decays into two or even three constituent components.

Filling the inner small rectangle with gray color aims to draw attention to the insufficient (less than 25) number of dates for each of the aggregates. Thus, as it were, there is a signal of the relative unreliability of determining the age of aggregates of this kind—after all, a small number of analyses very often reflects on the unstable nature of the probability distribution. In all databases where the number of dates exceeded 25, small internal rectangles were filled in black.

Our research shows that when working with blocks of large databases or their aggregates, as well as when comparing them with each other, the most rational approach to estimating the age of a particular archaeological culture or complex can be considered a probability of one sigma. The probability of an age range of two sigma in many cases presents an extremely vague picture, which immediately reduces interest in such chronological definitions (see, for example, Fig. 9.20, Sakhtysh and some others).

Analytical procedures related to the calibration of all radiocarbon dating presented here, as well as chronological ranges for each of the 75 sets of dates, were carried out on the basis of the Research Laboratory for Archaeology and the History of Art, University of Oxford (Version 3.10 and later).

Radiocarbon dating and the processes of their detailed systematization make it possible to create broad cultural and historical canvases on which many unexpected vectors of distant and close interrelations of various archaeological communities appear. These studies continue the decisive, and sometimes even harsh breaking of traditional views on the dynamics and nature of the development of various societies and production centers of the Early Metal Era in Eurasia.

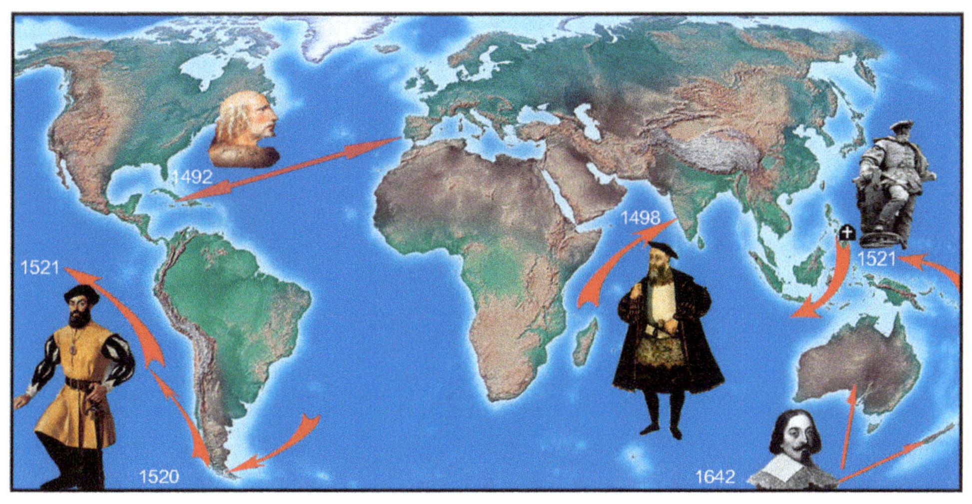

Part Two

T O THE HEIGHTS OF GLOBALISM

From Uncreated Realities to Created Ones

This book faces a difficulty, which is brought about by the author's desire to present readers, in this relatively small volume, with an extraordinarily broad view of the development of cultures and their associated ethnic groups on the vast continent of Eurasia. At the same time, we are also examining an equally grand temporal scope, and across this canvas, long and intricate branches of a considerable number of very complex problems have been wound and intertwined. The scope of this book, in other words, has prevented us from dwelling too long on details. Often, the author has limited himself to merely mentioning such topics, even if they are quite significant ones. In the first part of the book, we have tried to mitigate this difficulty by relying on generally accepted geological data that did not require special discussion; what we have covered was sufficiently determined by the crucial and foregrounded issues related to the earthly origins of the Eurasian world. This type of discussion has brought us to the bottom of the Pleistocene (associated with the emergence of *Homo*), up to and during which time the so-called geocreative realities clearly dominated; i.e., those created by the Earth/Nature.

The second part of this book is distinguished, first and foremost, by the fact that here the focus shifts to the *Homo*-creative or created realities, even as the geocreative (uncreated) realities are present but clearly recede into the background. This transition only became possible when addressing monuments of the subsequent geological epoch—the Holocene or post-Paleolithic. Our primary attention will be directed toward themes of ideology and worldview, which are also intertwined with the *Homo*-created creativity of ethnic groups, about which we will speak below.

The Holocene presents us with an astonishingly colorful fireworks display of cultures. And this, of course, could not go unnoticed, especially against the backdrop of the relative monotony/monochrome of the Paleolithic world, which spanned millions of years. Moreover, the multicoloredness and diversity of societies in the Holocene continually and universally increased, although in it did come in waves (Chernykh 2019, 103–110).

However, we are not only children of Eurasia but also children of the planet Earth, of which Eurasia is just a part (and its largest). Our planet is a sphere, or in Latin, *globus*, and hence the now-attractive terms such as globalism and globalization—used not only by political scientists but also by many others—will be discussed in the next chapter.

Chapter 10

Amidst the Relentless Currents of Globalization

10.1. Defining Globalism and Globalization

To comprehend the historical progression of globalization, it is crucial to first define the concepts of both globalism and globalization based on existing and available literature. Unfortunately, achieving a suitable definition proves challenging. We noted a similar phenomenon in Chapter 4, where we lamented the challenge of identifying scholarly consensus on the culture-religion dual concept. Luckily, the diversity in definitions within the double framework of globalism-globalization is considerably smaller, most likely because of the relatively short chronological period in which discussions on globalist issues have taken place.

The philosopher A. N. Chumakov argues that the term *globalism* is still relatively new—and frustratingly vague. Originating in the 1970s, the term gained substantial attention only in the late 1990s when specialists redirected their focus from global issues to "the understanding of the phenomenon of globalization itself" (Chumakov 2008, 7). Chumakov attempts to uncover the underpinnings of the term in all its associations, ultimately suggesting that

> the sphere of material production and spiritual activity, ecology and lifestyle, culture and politics—all this is now globalism's orbit, which . . . should be defined as an interdisciplinary field of scientific research aimed at identifying the essence of globalization processes, the reasons for their emergence, and their development trends. Additionally, the term *globalism* can be understood as the analysis of global problems and the methods through which we can confirm global problems' positive or negative consequences on humans and the biosphere. Furthermore, globalization is a centuries-old natural-historical process, of which global issues are a natural result. *Globalistics,* as a field of theory and practice, focuses on globalization and global problems (Chumakov 2008, 8, 15–16).

This definition, however, runs into the very same problem that Chumakov condemns in the first place—it is nebulous, unsatisfactory, and directed at specialists who have already devoted a fair share of attention to these issues. The Russian philosopher A. N. Zelenkov similarly expresses frustration with the term's nebulous and elusive character:

> The task of explicating the concept of globalization is extremely difficult. There is almost an infinite number of interpretations, and for decades, "globalization" has become one of the most popular used terms, not just within science, but in various cultural fields as well. . . . Today, a simple Google search for "globalization" yields an instant display of 28,700,000 links.

On the other hand, Brink Lindsey—the director of the Center for Trade Policy Studies of the Cato Institute in Washington D.C.—posits the definition a bit more clearly. In his 2006 lecture in Russia, he expressed that

> globalization is a vague and rather slippery word [that] means different things to different people. For many, globalization encompasses almost everything that is happening in today's world. . . . We are told that globalization is a very powerful phenomenon and, as such, can be the cause of everything that we see in the contemporary world. Following that logic, we can blame everything on globalization, especially the negative things. . . . However, I'm absolutely convinced that the movement toward globalization that we have seen over the past years is an example of a remarkable triumph of humanity.

Indeed, conversations regarding globalization and efforts to precisely grasp its meaning often seem limitless. Let us turn, for example, to another take on its definition by the renowned American historian Niall Ferguson, who has championed global imperialism as a net positive effect on humanity. In his book *Empire: How Britain Made the Modern World* (2003), he argues that

> it has become almost a commonplace that globalization today has much in common with the integration of the world economy in the decades before 1914. But what exactly does the overused word mean? Is it . . . an economically determined phenomenon, in which the free exchange of commodities and manufacturers tends to "unite mankind in the bonds of peace"? Or might free trade require a political framework within which to work? The Leftist opponents of globalization naturally regard it as no more than the latest manifestation of a damnably resilient international capitalism. By contrast, the modern consensus among liberal economists is that increasing economic openness raises living standards, even if there will always be some net losers as hitherto privileged or protected social groups are exposed to international competition. (Ferguson 2003, 15)

And, finally, we will conclude this introductory section with a brief but expressive definition by the historian David Christian (Fig. 10.1): "Globalization refers to the expansion of exchange networks, gradually encompassing the entire world since the year 1500 A.D" (Christian 2019, 416).

10.2. Globalism Supporters

To complicate the matter even further, those discussing globalist processes and their development are distinctly categorized into two opposing categories—those enthusiastically supportive and those vehemently against them. Let us begin with the positivists.

It is important to note that a significant portion of those participating in this discourse are emphatically in favor of globalization—the philosophers and policymakers mentioned in the previous section being among them. And whether we like it or not—and certainly,

Figure 10.1. David Christian, author of the topic-wise most significant book from his series "Origin Story," professor of Australian Macquarie University in Sydney.

a not-so-small minority loudly objects to it—globalization continues to expand and proliferate into every corner of industry, culture, and political systems. For a while now, the discussion of globalism has concentrated on a vital and pressing question—is it a force of good or evil? Globalization's opponents, for example, regard imperialism and empire (concepts so closely related to globalism) as the root source of evil, which manifests itself in cultural regression and traps its citizens in a constant state of struggle and anxiety.

The reality of the contemporary political landscape is that the United States (the most powerful and arguably destructive empire in the modern world) has shouldered a global burden, whether they admit it or not. Just like the British Empire before it, the American Empire professes to act in the name of freedom, even when its colonial interests are evident. In looking back at the height of imperialism, English politician and historian John Buchan remarked:

> I dreamed of a worldwide brotherhood based on a common race and faith, dedicated to serving the cause of peace, of Britain enriching the rest of the countries with its culture and traditions. . . . We thought we were laying the foundation of a world federation. "The Burden of the White Man" is now a meaningless phrase, whereas before it meant a new philosophy of politics and ethical norms, serious and definitely not shameful. (Buchan 1940, 125)

These well-intentioned imperialist dreams of unity and harmony, however, turned out to be utopian fantasies. And now, we turn to the negativists.

10.3. Against Globalism

Those critical of globalism—often referred to as anti-globalists—uphold the "altermondialism" worldview as an alternative to economic globalization which they consider as infringing upon human values such as environmental and climate protection, economic justice, and civil liberties. Altermondialsm (otherwise known as alternative globalization) gained traction after the 1999 World Trade Organization summit in Seattle, during which powerful opponents vocalized their distrust of the sweeping wave of globalization that they considered to be infiltrating politics, economics, and society. However, despite their mutual resistance to globalization, the speakers struggled to articulate a coherent set of arguments, thereby mirroring the ongoing struggle in reaching a consensus on the definition of globalism, as previously discussed.

When examining whether altermondialism presents a valid counterforce to globalization, communist academics Kefeli and Mozorov ponder the compatibility of the communist utopian idea of a universal, harmonious society with the concept of globalization. Both Marxist communism and globalization rest upon the idea of a new world order that eschews the shackles of social and economic oppression. However, Kefeli and Mozorov argue that the Marxist idea of a new world order was "intercepted by the ideologists of modern globalism in an 'American way'" (Kefeli and Mozorov 2006).

In responding to the major global problems of the late twentieth century, scholar Aurelio Pececci proposed alternative theories to address the new world order problem, notably alterglobalism, which emphasizes the role played by geocivilizations. Pececci identifies the Russian (Orthodox) geocivilization as an example, asserting its ability to chart its own developmental course alongside other geocivilizations. Pececci argues that Russia is exemplary because it is arguably the first nation to establish its own socio-economic structure. Furthermore, he argues that Russian geocivilization is unique in its self-sufficiency in raw materials and industry, and its ability to navigate global economic realities.

Philosopher and critic V. M. Mezhueva suggests that the socialist vision of society, in contrast to mere economic and political ideologies, could be characterized as culturological or, more precisely, humanistic. In the face of the global crisis of ideologies, the ideology aimed at salvaging Russia could rekindle the principles of socialist humanism and patriotism, with a focus on establishing a globally humanistic society. While this might seem oversimplified, Mezhuyeva argues that contemporary forms of global development span the spectrum from communism, representing genuine humanism, to American-style globalism. If communism strives for peace, labor, freedom, equality, and happiness for all peoples on Earth, can American-style globalism achieve the same? Kefeli and Morozov contend that the answer is most likely negative (ibid.).

We argue that Kefeli and Morozov's rejection of American-style globalization is compelling since the Marxist-Leninist perspective aligns more closely with the ideals of worldwide peace and fraternity based on shared values.

10.4. Globalism and Subglobalism

As evidenced by this chapter's discussion of the different facets of defining globalism, understanding what globalization is and what it entails is a seemingly limitless discussion that obfuscates more than it reveals. However, it is nevertheless important to define the phenomenon as we understand it, in order to guide the reader through the book's remaining arguments. Globalism and globalization are two separate albeit interrelated terms that are often used interchangeably, and it is therefore imperative to distinguish between the two. Globalism (or globalistics, as it is referred to in Russian scholarship) is the broader ideology or worldview that advocates for greater cooperation, integration, and interconnectedness among nations and peoples on a global scale. In other words, it is the holistic picture of the Homo sapien culture, which is intricately linked in its endeavors for the entirety of the globe within a specific and clearly defined period. Globalism upholds the importance of international cooperation, global governance, and addressing global challenges collectively and places ideas of human rights and environmental sustainability at its center.

Globalization, on the other hand, is the *process* characterized by increasing interconnectedness, interdependence, and integration of communities, societies, cultures, politics, and economies across the globe. It involves the movement of goods, services, capital, information, ideas, technology, and people across national borders. Each stage of this globalization is characterized by different developments and advancements, with each phase contributing to the overall refinement and complexity of the global integration process. Among these, one of the most significant factors depends on movement and the ability to influence one's neighbors. For example, one of the earliest examples of globalization among the Homo groups (*Homo habilis, Homo erectus, Homo neanderthalensis, Homo sapiens*) was their nomadic movement, often accompanied by herds of livestock.

The depth and scope of global studies fluctuate and are contingent upon the levels of interactions between Homo groups. Nevertheless, there is an overall tendency towards continuous expansion and growth, evident in the increasing prevalence of its distinct expressions. The dynamic of this trend, however, typically exhibits wave-like patterns, with the nature of these waves largely influenced by the activity of the globalization process itself. With the advancement of technology across the centuries, migration patterns changed from pedestrian to aquatic, with many forced to overcome thousands of kilometers sailing upon rough and choppy seas. Sailboats were then replaced with motor vessels, and those were then replaced with aviation. In our current age, technology has reached its current peak with the invention of computers and complex machines. What the future holds in terms of technological advancement remains a pertinent mystery.

In returning to the discussion of globalism and globalization, it is important to note that these terms will be included in this book insofar as a particular entity extends across the entirety of the Earth's surface, for example, across both the eastern and western hemispheres. However, there may be a notable exception for the largest continent, Eurasia. Therefore, we introduce yet another term—subglobalism—only if a given entity spans geographical areas beyond one of the four main sections of the Eurasian continent. This is particularly relevant when such entities traverse the boundary of delineating Eurasia along the Alpine-Himilayan geosyncline or the North-South diving line of the continent. The four principal parts of the continent—the Northern, Southern, Eastern, and Western—were discussed in Chapter 4 (see also Fig. 4.22).

10.5. The Great Migrations of Peoples and the Globalization of Planet Earth

The Great Migrations of Peoples often reveal the central narrative of widespread or remarkable rupture of cultural frameworks, in particular seemingly familiar hierarchies such as the divisions between strong and weak communities. Consequently, during such periods, the collapse of the previously perceived stable structure of this cultural fabric occurs. The weak cultures unable to adapt to relocation in new and unfamiliar geographical enclaves or those unable to overcome struggles with other ethnic groups often perish from these crises. In studying the migration patterns of early humans, we can determine significant reference points when periodizing the broader historical narratives. In some respects, such a chronological framework may even be more revealing than the traditional technological markers employed in archeology—stone, copper, bronze, and iron, for example. Notably, the coverage of cultures, as illustrated in the migration-based chronology, is often substantially broader than that of the technological timeline (see Fig. 10.2).

The end of the Paleolithic-Pleistocene era in 10,000 BCE coincided with the fifth Great Intercontinental Migration of Peoples, and it was during this time that the first wave of Eurasian migrants (Homo sapiens) first reached Australia, while a second wave ventured into America (Fig. 10.2, see also Chapter 2, Section 2.4). These significant transitions coincided with the arrival of the Holocene geological epoch—a post-glacial era that witnessed the emergence of vibrant post-Paleolithic cultures. In the Holocene, the uniformity of the Paleolithic cultures gradually gave way to a diverse array of civilizations, captivating enthusiasts of antiquity. It appears that the zenith of this striking diversity occurred around the transition to the New Age, when explorers encountered unfamiliar worlds across oceans. It is important to note that the chronological span between the two consecutive inter-continental migrations (the fifth and sixth) represents a substantial interval of at least fifteen thousand years—from the American migration through Beringia and fifty millennia to the Australian migration through the Torres Strait.

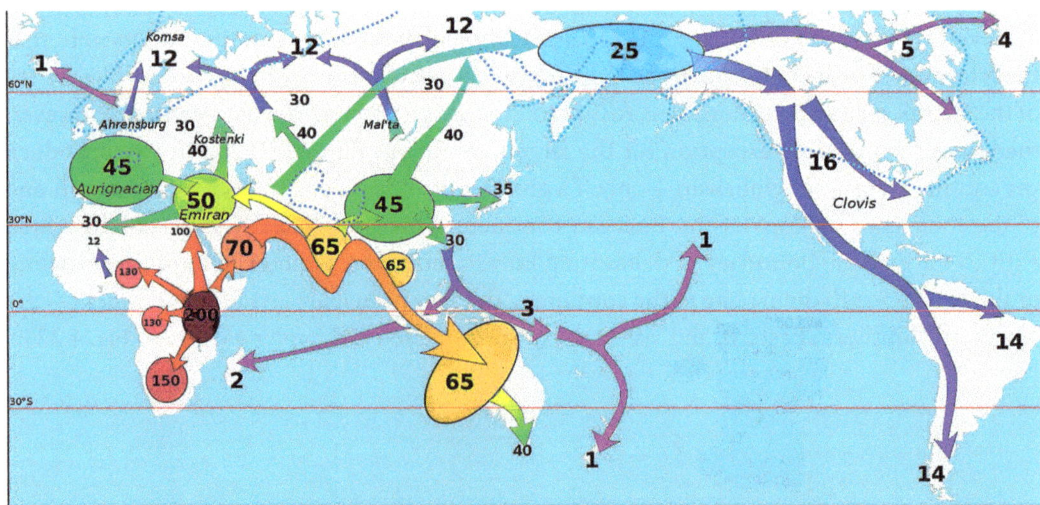

Figure 10.2. Ancient peoples migrations in the Palaeolithic period (Wikipedia commons, Fisheiro: Early Migrations mercator.svg).

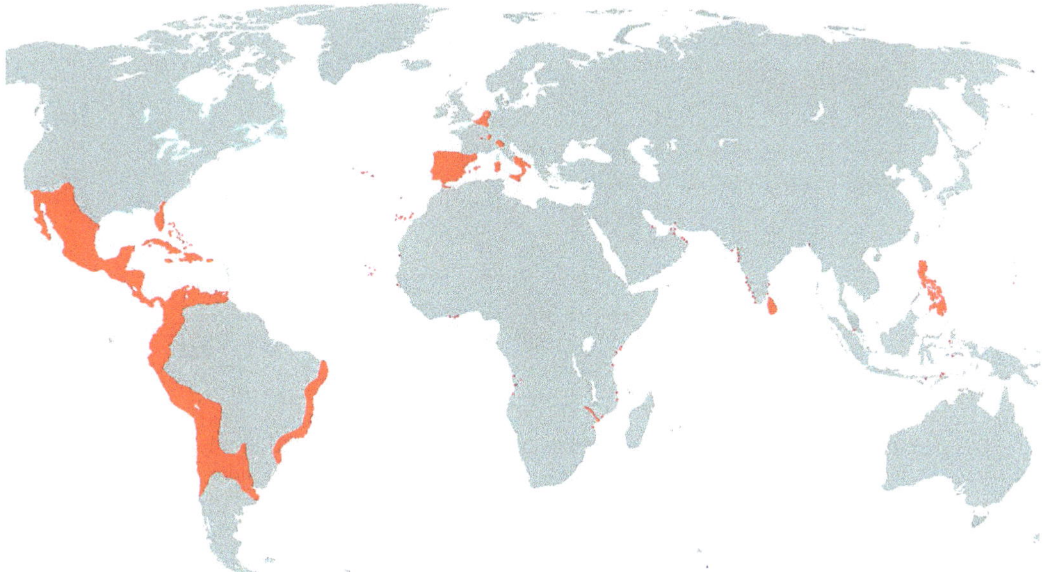

Figure 10.3. Spatial extent of the Spanish-Portuguese Empire during the so-called
Iberian Union 1580–1640 (Wikipedia).

Now we come to the focus of this chapter, which is the fourth to eighth centuries (300–700 CE), often referred to as the "millennium eve" of the New Age. This era witnessed the onset of the sixth Great Intramaterial Migration of Peoples, which was a voluntary migration for Eurasian populations but often coerced for certain African tribes subjected to enslavement. These global movements were instigated by the great geographical discoveries of the fifteenth to seventeenth centuries, with migration patterns persisting from the sixteenth to the twentieth century, extending effectively to the present day.

10.6. Global and Subglobal Formations

We will now endeavor to elucidate some of the significant divisions within the classification we have outlined. Firstly, it is pertinent to include early empires such as the Spanish-Portuguese Empire during the Iberian Union of 1580–1640 (see Fig. 10.3). Subsequently, the British Empire emerged as the most extensive global and sub-global formation (Figs. 10.4–10.6). Imperial entities like the Mongol Empire (Chingizov) and, to a certain extent, the Russian Empire, which succeeded it, can be categorized as sub-global formations. However, within Eurasia, their dominance was preceded by the Roman Empire. Moreover, among the sub-global formations, we may consider earlier colossal cultural pastoral communities like the Yamnaya and Scythian societies, which thrived across the expanses of the Eurasian Steppe Belt.

The classification we have outlined encompasses extensive metallurgical provinces, including the Circumpontine for the Early and Middle Bronze Ages, and the West Asian province for the Late Bronze Age. It appears that most of the ten formations mentioned can be categorized

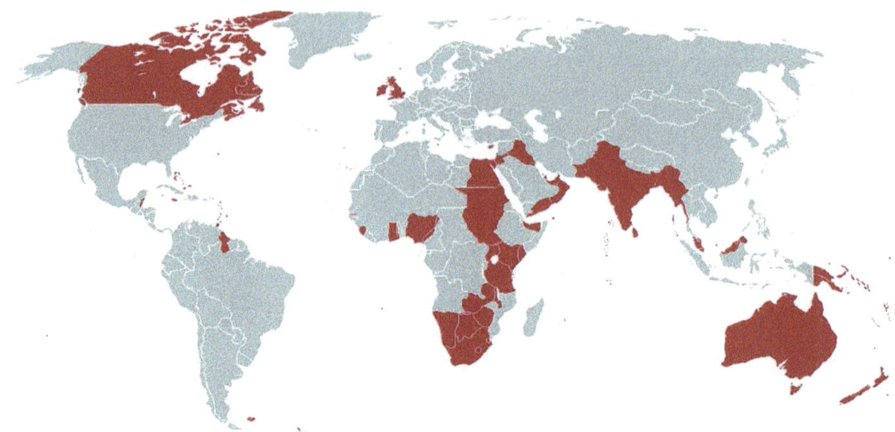

Figure 10.4. The territory of the British colonial Empire in the decades of maximum coverage—about twenty years of the twentieth century (Wikipedia).

Figure 10.5. Caricature of a prominent British colonizer—Cecil Rhodes, in the grotesque position of the ancient hero Rhodes Colossus. The author of this pictures is Edward Linleyc Sambourne, 1892.

Figure 10.6. This is what the fantasy looked like, in fact, the only one of the many reconstructions of the Colossus of Rhodes in ancient antiquity (Crispijn van de Passe the Younger, 1614).

similarly. These formations were extensively discussed in Chapters 8 and 9, supplemented with appropriate illustrative materials. However, detailed analyses of the main components were primarily presented in various works by the author, spanning different years (see, for example, Black 1972; 1977; 1978; 2009; 2013a, b; 2019; Chernykh 1992; 2017, among others).

While these books predominantly focused on the technological aspects of diverse cultures, the socio-cultural and ideological dimensions received comparatively less attention. This focus did not stem from a lack of interest in these topics but rather from the wealth of archaeological materials that necessitated such a choice. However, these subjects, despite their significant

126

interest, do not serve as the primary focus of this book, which deals with an entirely different set of issues. Therefore, the author opts to confine the discussion within the scope outlined here. Without question, each migration—whether global or sub-global—likely contributed to the expansion and refinement of migrants' understanding of the world surrounding *Homo* culture. While migrations often resulted in bloodshed and the demise of cultures and ethnic groups, they also facilitated significant advancements in knowledge about the vast world. These innovations, however, often came at a steep cost.

Chapter 11

The Millennial Eve of the New Age: Archaeology and History

However, the author should also outline the main directions of the proposed processes of detailing and systematization of formations. His preferences turned out to be mainly directed to the cultures of the northern half of the Eurasian continent, that is, to those that are distributed north of the great geofracture of the Alpine-Himalayan synline (see Chapters 1 and 3). This choice turned out to be due to the virtually constant—both in scientific and popular scientific historical and archaeological publications—preference in the characteristics of the southern Eurasian formations against the background of the northern ones. In this respect, the latter were almost constantly and sharply inferior to the southern ones. And this was especially evident on wide canvases in publications of this kind.

At the same time, among the blocks of northern *Homo* cultures, we will pay priority attention to those whose activity and effectiveness often led to noticeable, and sometimes even cardinal metamorphoses of the general Eurasian structural panel, and the expression of such fractures was not always manifested only in the northern half of the Eurasian continent.

11.1. Special Worlds of Eurasian Enclaves

By the end of the Pleistocene period, and by the beginning of the new period, that is, the Holocene, the general Eurasian structure of enclaves that we planned looked quite stable. However, the question also arises: is it possible to characterize the enclaves we have identified from a slightly different position—say, as if from an unusual geoethnic one? It seems that this phenomenon is quite real. In essence, enclaves are special worlds, which is quite definitely reflected not only in official/semi-official representations, but also in those assessments that we refer to as ordinary, everyday, quite everyday. In this series, of course, such worlds will be designated as European or the Chinese (Han) eastern world itself, which is many thousands of kilometers away from it. It is hardly possible to confuse with others the special world of Nomadic pastoralists, as well as the truly vast world of Forest hunters and fishermen adjacent to it from the north. Probably, we can hope that such a rather simplified bundle "world—enclave" will not cause any special objections.

But the question almost immediately follows from here: how, after our research, will it be necessary to correctly designate, say, the so-called special Eastern European world? After all, this world even officially stretches as if to the Urals—but where are its Western origins? And is it fair to correlate its eastern face with the mountainous strip of the Urals? Indeed, in reality, this kind of "world of Eastern Europe" corresponds much more fully to the western sub-enclave of the vast Ural forest-tundra enclave (see Section 6.4), but by no means Europe.

Similar issues are connected not only with the so-called Eastern Europe, but also with the vast expanses of forest-tundra enclaves up to Chukotka, Kamchatka and the coasts of the Pacific Ocean. And how should the titles of these North Eurasian enclaves sound? But let's leave it among the mysteries for now.

However, much greater perplexities and questions are, of course, caused by the common definition of the world of the Middle East or the Middle East. This world completely fits into the framework of two enclaves—the Iranian-Anatolian and the Arabian. But what does the East of Eurasia have to do with it? After all, it is protected from the true East—and together with the difficult heights of Tibet—by a special Indian world (Hindustan), and the Chinese world (Han) is still very far away. If we begin to focus on its man-made veil—that is, on the nature of *Homo* cultures with their anthropological, linguistic, ideological characteristics—then there will be no unity with the true East, and the sound of the term itself will become even more ridiculous.

Most likely, the appearance of this term and its use in a wide variety of spheres, both politics and science, was the result of burgeoning Eurocentrism, and with preferably more complicated detailing of the concept, Euro-Anglocentrism. The beginning of this process dates back to the middle of the nineteenth century. Moreover, in English-language literature, this is often linked to the Crimean War, when many in the West believed that Russia, in alliance with the Ottoman Empire, decided to defend (!) the Ottomans from European civilizations. After all, that empire seemed to be an obvious symbol of the East, always alien to Europe, but only the Near One. This also affected archaeology, and the start of the "Middle East boom" in archaeological science is often seen, for example, in a book published in 1902 under the very expressive title "The Nearer East" (Fig. 11.1) by the English archaeologist David Hogarth (D. G. Hogarth, 1862–1927). It is noteworthy that even the author himself began his work with the necessary explanation:

The Middle East is a term of **modern fashion** *[highlighted by me, E.Ch.] for the region that our grandfathers wanted to call simply the East. It is generally assumed that its territory coincides with those classical lands historically most interesting on the surface of the globe and which coincide with the eastern basin of the Mediterranean; but few could probably say where its limits should be and why. (Hogarth 1902, 1)*

We will confine ourselves to what has been said, although it is not at all difficult to assume, say, such a version. You can prove all your positions at any length and as convincingly as you like, but there is no doubt about this: both in the media and in political statements, Eastern Europe will remain Europe for a long time, although, of course, Eastern; and the Middle East will continue to be called the East, even the Middle East. All this will be somewhat similar to those endless discussions about the Urals as the divider of the essentially inseparable continent of Eurasia into two continents. But it is extremely difficult to explain this very peculiar phenomenon; the roots of this kind of mysterious adherents are hidden in the depths of the psychological depths of the *Homo* species.

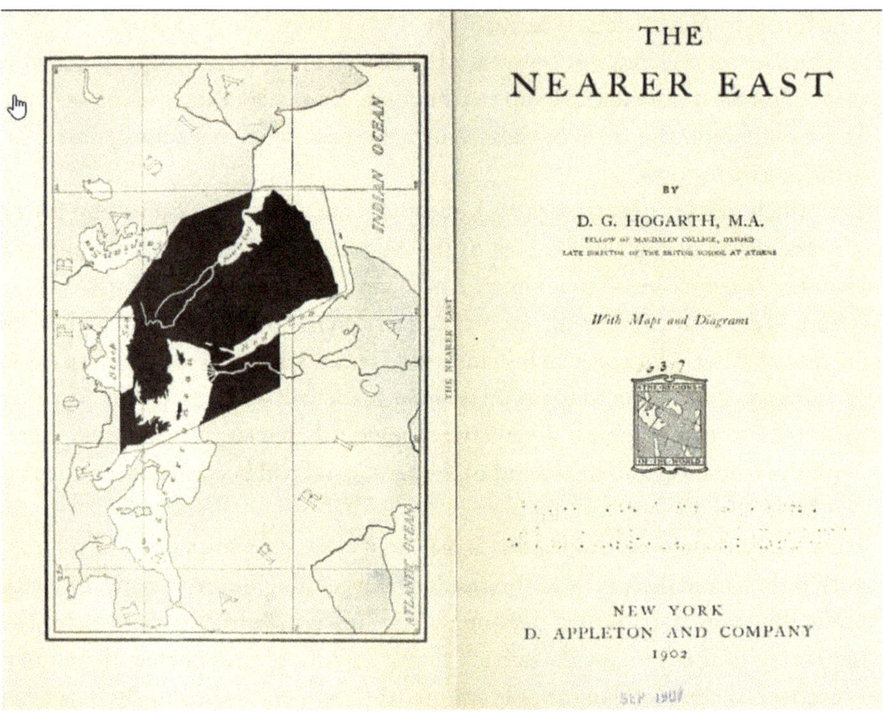

Figure 11.1. Flyleaf of David Hogarth's book "The Middle East," published in 1902 with a drawing of the Middle Eastern area (in the understanding of the author of the book).

11.2. Enclave Worlds in Incessant Struggles: The Steppe Belt Is in the Lead

Inter-enclave battles—they are also battles of worlds—are essentially endless, and even a brief enumeration of them within the framework of our work is absolutely unrealistic. We will focus on the obvious leader of these clashes: the western and eastern cultures of the double enclave of the Steppe Belt. In all likelihood, one of the most significant incentives for the leadership of this double was the so-called central or median geographical position of the Steppe Belt on the Eurasian continent (see Sections 3.4 and 6.4).[1] The double enclave itself, or, as we also call it, the block of two related enclaves is adjacent to all the other four enclave blocks; the only exception is the pair that we recognize as "alien" to the Eurasian world: Hindustan and Arabia. Another incentive was undoubtedly the domestication of the horse, the development of horsemanship and the new, rapid tactics of military attacks, so unknown to sedentary neighbors and often extremely unfortunate for them.

---

[1] The Steppe Belt has been discussed in several books (see, for example, Black 2009; 2013; 2017; 2019).

This situation manifested itself very early, already from the very start of the Early Metal Era. The Steppe Belt—its western enclave, almost up to the Southern Urals—turned out to be included in the system of the Balkan-Carpathian Metallurgical province, and this is the fifth millennium BCE. But at that time the western relatives of the Balkan cultures did not yet know metal. Then a millennium later, during the birth of the Proto-Circumpont Metallurgical Province, we were able to observe a paradoxical phenomenon: mining and metallurgical production is localized in the south—in Anatolia, in the South Caucasus, and the lion's share of metal, including gold and silver, is concentrated in the underground burials of the nobility in the North Caucasus and the adjacent steppe spaces. Then, when in the third millennium BCE the Proto-CMP rises to the status of the Circumpont metallurgical province covering vast expanses, the situation changes dramatically again. The localization of metal acquires a kind of normal character: the bulk of the metal taken into account turns out to be concentrated then already in the southern archaeological sites, where the metal was produced. Apparently, during that period, the military power of the communities of the Steppe Belt and the South was leveled, with the very likely appearance of their own cavalry in those cultures that are located south of the ridges of the Alpine-Himalayan geosyncline (see Sections 5.7 and 5.8).

11.3. The Western Wave of the Steppe Belt: The Scythian World

In the history of the rise of the leadership of nomadic, mobile pastoral cultures of the Steppe Belt, two obvious peaks of apogee are quite definitely outlined. For this reason, in the history of Eurasia, it was even possible to single out the iconic Millennia of the West and the East (Chernykh 2019, 135–156, 181–222). However, in this regard, it is possible to talk about millennia only conditionally. The first and earliest of these conditional millennia, called the Western one, refers to the first millennium BCE, and covers up to five centuries in it (from the eighth to the fourth/third). This millennium was closely connected with the world of the Scythians and ended with the formation of the cultures of the Sauromats/Sarmatians. The second—the Millennium of the East—coincides in its early centuries—the fourth, fifth and sixth—with the Great Eurasian Migration of Peoples, and the later ones reach the middle of the second millennium CE. That is, in this case, in reality, the millennium of the East really lasts about a thousand years. The gap between the two planned peaks of high success of the steppe peoples is approximately seven to eight centuries.

The Scythian world and the so-called Millennium of the West associated with it are, without a doubt and above all, an impressive vast layer of countless mounds. The most significant plan is represented here by the huge tombstones of the leaders (Fig. 11.2), under the mounds of which their burials are hidden, often accompanied by mass graves of killed horses. The graves of the nobility are saturated with a variety of metal products and, most often, weapons: iron, copper/bronze, or bimetallic. The graves often contain amazing riches of highly artistic gold and silver products. Most likely, it was in such a magnificent and brilliant guise that the Scythians intended to appear before the rulers of the otherworldly, extraterrestrial and, as they apparently believed, the eternal world. This, apparently, was the deep essence of their crown ideology.

Figure 11.2. Huge mounds of the Scythian world with stone and earth burial mounds in Xinjiang.

The Scythian nomads in the Iron Age managed to significantly expand the boundaries of their influences (possessions?) to the west (Fig. 11.3). However, the breakthrough of their opposite, eastern flank turned out to be incomparably more impressive. Bright Scythian complexes have already been studied 2.5 thousand kilometers east of the Dzungarian Gate—this border sign between the West and East of Eurasia. And the so-called Scythian world itself stretched an amazing strip of about 8.5 thousand km in length, and its general coverage turned out to be close to 8.0–8.5, but already millions of square kilometers.

Clear signs of the world of Scythian cultures appeared on the hilly plains of Ordos, outlined from the north by the bizarre meander of the Yellow River. A typical Scythian metal was found even further east, in Hebei Province—and this is already not far from the Pacific Bay of Bohai. And quite near its coast in the town of Linzi (Shandong province) there are various and numerous mounds. There under the embankment or mounds (?) of one of them in a very amazing length—215 meters!—the remains of more than 600 (!) intentionally killed horses were found in the trench pit (Fig. 11.4a, b). The impact of Scythian burial rituals in this case can hardly be disputed, although Chinese archaeologists believe that this amazing burial (grave No. 5) belongs to Jing-gong, the ruler of the kingdom of Qi from the Eastern Zhou block of states. Grave grave itself dates back to 490 BCE. In addition to horses, remains of dogs, pigs, etc. were found in the filling of the grave (Shandong Archaeological Research Institute1984; Linzi Qi Tombs 2007; Excavation of grave No. 5 of Eastern Zhou (in Chinese; Shandong Institute); Wu Weihua 2018 (in Chinese); Komissarov, Ulyanov 2010, 751).[2]

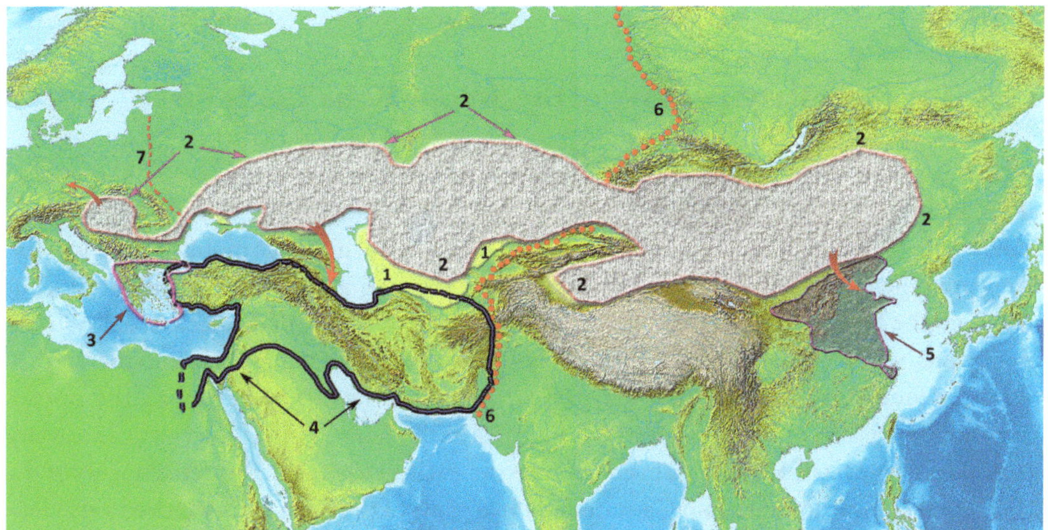

Figure 11.3. Schematic map of the main social formations against the background of the division of Eurasia. Symbols: 1—the southern borders of the Steppe Belt, almost completely covered by the Scythian World; 2—the Scythian World; 3—Greece and Macedonia; 4—the Achaemenid Kingdom; 5—Eastern Zhou; 6—the division of Eurasia into West and East; 7—the border of the European enclave.

---

[2] There have been numerous references to the remarkable and, in many ways, unique Grave No. 5 in Linzi. However, a thorough analysis of all the materials from this site, including the surface structures, has not been performed (at least to the knowledge of the author). Additionally, understanding how this complex relates to and correlates with other burial sites in this region, which is rich in such monuments, could be of significant importance.

Figure 11.4a. Linzi Burial Ground, province Shandong, the Qi kingdom of Eastern Zhou. Detail of grave No. 5 with burials of about six hundred dead horses; beginning of the fifth century BCE (Lu 2005, 233, Fig. 7.42).

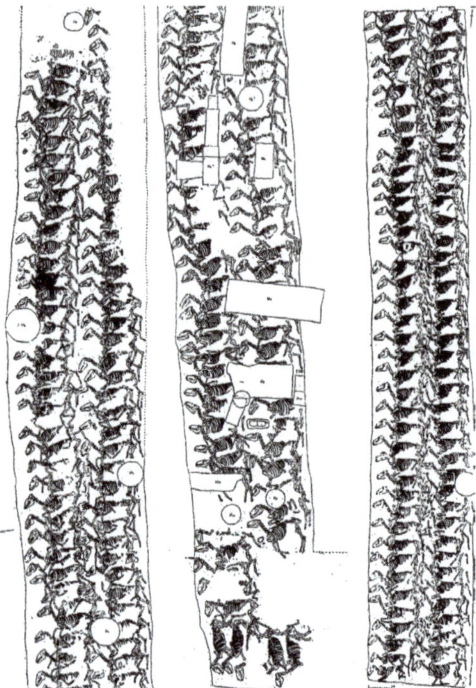

Figure 11.4b. Linzi Burial Ground, grave No. 5. The plan of the opened and cleared bones of horses (Excavation of grave No. 5, 1984; in Chinese).

11.4. The Early Eastern Wave of the Steppe Belt: The Huns

Now we have to make a long double jump—both temporal and spatial—by the middle of the first millennium, but already a new era, and now to the western half of Eurasia in the Millennium of the East that we are reconstructing. Once again, nomadic mobile formations are being promoted to the role of active stimulators of the ongoing shifts and breaks of cultures, but now from the East of Eurasia. Then cardinal changes were reflected in the vector of pendulum oscillations, which covered truly immeasurable spaces of the largest meganclave of the Earth.

By the fourth or fifth centuries, the relay of activity turns out to be among the eastern nomads, the earliest of whom were the horsemen of the Huns. And if the Scythian cavalry detachments appeared a millennium earlier near the Pacific coast, then the detachments of the Mongoloid Huns who broke into Gaul around the middle of the fifth century moved very close to the shores of the Atlantic.

In the offensive from East to West of the Eurasian continent, three main waves are distinguished, where the Huns (or Xiongnu in Chinese sounding) personified the first wave. It is also associated with the apogee of the Great Migration of Peoples in the first millennium of the new era. This period has become one of the most difficult and critical for the entire history of the continent. Then literally the entire map of the cultures of the continent, and especially its western half, turned out to be redrawn. Perhaps the most painful blows fell at that time to the seemingly indestructible Roman Empire.

Around the 370s, the Huns appeared in the west of the Steppe Belt, and by the beginning of the fifth century, their mobile detachments had already penetrated into the European enclave proper. They made an extremely painful impression on the local inhabitants. This people "surpasses all measure in their savagery. . . . They are ugly, similar to the eunuchs, have a monstrous and terrible appearance, so that they can be mistaken for two-legged animals. . . . Like mindless animals, they are in complete ignorance of what is honest, what is dishonest, they are not bound by respect for any religion or superstition, they are burning with a wild passion for gold." So the Greek Ammianus Marcellinus wrote about them in the fourth century (Marcellinus 31, 2–4).

The legendary figure of the Huns was, of course, their leader Attila (Fig. 11.5). "He was a man born into the world for the shock of peoples, the horror of all countries, who, by some unknown lot, made everyone tremble. . . . In appearance, short, with a broad chest, a large head and small eyes, with a sparse beard, touched with gray, with a flattened nose, with a disgusting color [of skin], he showed all the signs of his origin." This is the assessment of Attila by the Gothic historian Jordan (Jordan 2001).

And at the same time, in the European assessments of the leader of the Huns, there is also a bright, extremely curious and, it seems, the only truly paradoxical anomaly. In modern Hungary, a number of prominent historical and ethnographic communities venerate Attila as their first and glorious king, under whose leadership the Hungarians—they are the Huns—from the Asian depths broke into Europe up to the Danube and, having crushed many, revealed themselves to the world as such (Fig. 11.6a). Under the images of Attila, it is still possible to gather a fair number of people for the celebration in honor of this leader-vladyka (Fig. 11.6b), and even memorial steles are erected to him. Note, however, that anthropologists do not see any obvious reasons to deduce pure Caucasoid Hungarians from equally pure Mongoloid Huns.

11.5. The Second Eastern Wave: Turks and Turkic Khaganates

The second wave of the Millennium of the East was associated with the Turkic-speaking peoples and their rapidly increasing role in the history of Eurasia. Their destinies turn out to be absolutely unlike the previous ones—the Huns. And again, now, together with the Turks who appeared in the western half of Eurasia, we are plunging into an unimaginable tangle of various and constantly fighting with each other steppe pastoral peoples—after all, the Era of Great Migrations was not yet over.

The history of Turkic-speaking and basically steppe peoples is very often associated with Turkic associations called kaganates (Klyashtorny, Savinov 2005; Klyashtorny 2006). Moreover, in the scientific literature, for the most part, they do not even talk about the "Turkic khaganates," but about the "khaganate"—in the singular, although in fact there were several extensive formations of Turkic-speaking tribes. The chronological framework of a kind of "single kaganate" or a group of all Turkic-speaking kaganates is most often limited to a couple of hundred years—from 552 to 745.

Apparently, it is to the Turkic khaganates and their rapidly victorious rushes to the west that the Eurasian megastructure owes the formation of a special world of Turkic-speaking peoples in it (Fig. 11.7a). The assumption that the eastern half of the Steppe Belt served as the initial

Figure 11.5. Attila. On the left is the image of Attila from the "Hungarian Chronicle" by Wilhelm Dilich, seventeenth century. On the right: silver medal "Attila" issued by the National Bank of Kazakhstan in 2009 in the series "Great Generals" (Wikipedia).

Figure 11.6a. Image of Attila—"the glorious King of Hungary" in illustrations by Ferenc Lanzmar to the mausoleum of "The most powerful and glorious rulers of Hungary" 1664 (Wikipedia).

Figure 11.6b. Celebration in honor of Attila in Hungary on the "victorious" Kurultai field under a huge image of the leader of the Huns-Hungarians (Wikipedia).

area of this very remarkable linguistic world does not require, perhaps, special evidence. It is clear that the Turks were moving towards the "sunset" in the footsteps of their relatively recent predecessors—the "pioneers"—the Huns. It is quite difficult to say anything definite about the language of the Huns and its linguistic relationship with the pre-Turkic (Proto-Turkic) language.

The whole picture of the spread of Turkic-speaking peoples in Eurasia in principle has preserved itself up to the present. Perhaps another of the complex phenomena of this giant canvas is the spread of Turkic-speaking peoples to the far north, to the forest and even forest-tundra zones of Eastern Siberia, to Yakutia. The most acceptable hypothesis here seems to be that the original reason for the appearance of the peoples of this southern linguistic family in this area, so unusual for them, was the migration to the north of part of the tribes of the Eastern Turkic Khaganate, which suffered a heavy defeat from the Tang Empire. Therefore, the Turks moved to the north, among forest hunters and fishermen. But they did not part with horse breeding so expensive for their lives, which is why the shaggy and small-sized Yakut horse became so famous (Fig. 11.7b).

For Europeans, the Turkic-speaking Bulgars became one of the most noticeable among the peoples who broke into the Balkans. Their kingdom, which received the name of the First Bulgarian, even received recognition from Byzantium by 671, which made the Bulgars extremely "proud and began attacking camps and towns under Roman rule, and people were taken captive: why was the tsar forced to make peace with them, agreeing to pay an annual tribute to the shame of the Roman people. . . . It was strange to hear both far and near that subjugated all the peoples in the east, west, north and south as tributaries, now he himself had to yield to the despicable newly appeared people" (Theophan the Confessor, 679–680).

All the peoples of Turkic roots who inhabited and still inhabit the spaces from Xinjiang to the Caspian Sea (Kazakhs Kyrgyz, Uzbeks, Turkmens), the Volga basin (Tatars, Bashkirs), west of

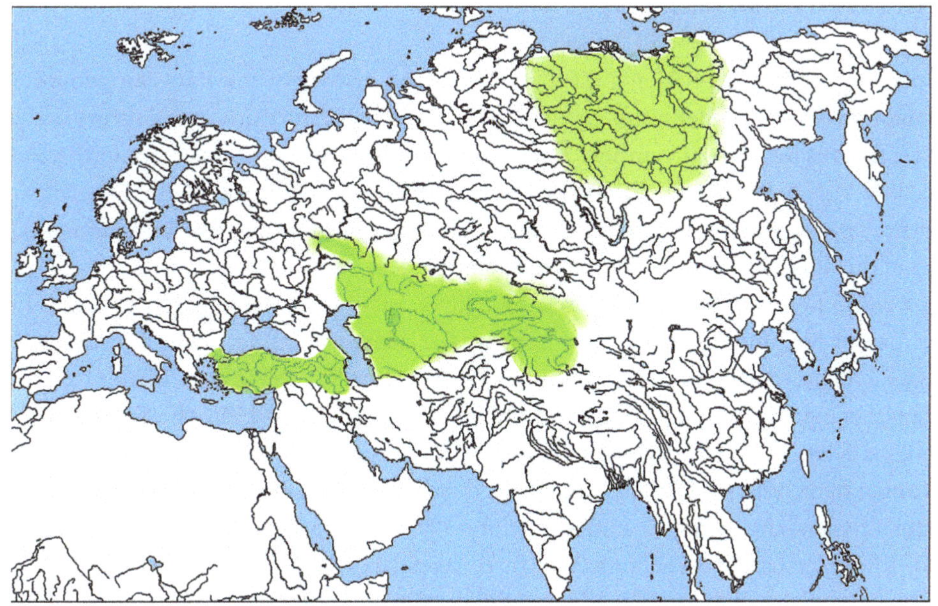

Figure 11.7a. Schematic map of the distribution in Eurasia of the main areas
of the speakers of the Turkic linguistic family.

Figure 11.7b. Short, thick-haired horses in Turkic Yakutia.

the Caspian Sea (Azerbaijanis), and, finally, Asia Minor (Turks)—turned out to be adherents of Islam. All of them still play a very significant role in the history of Eurasia.

11.6. The Third Eastern Wave: The Mongols and the Genghisid Empire

Apparently, of all the problems associated with the history of the Eurasian peoples in the first half of the second millennium, the Mongol conquests of the thirteenth century are among the most famous for many. Countless articles and books have been written about the Mongols of the time of Genghis Khan, about the tragic "three-hundred-year-old" Tatar-Mongol yoke. Therefore, here we will try to touch very succinctly only on the most important moments of the history of this third eastern and Westward-looking wave.

The central figure of this whole amazing epic is, of course, Genghis Khan (Fig. 11.8). In the history of his life, the year of the Leopard, or 1206 according to the Western calendar (Grousse 2007), was certainly the key one. Then, on Kurultai, the leaders of all Mongol tribes and clans, having erected a nine-bunched white banner, proclaimed him the Great Kaan—that is, Genghis Khan. Almost immediately, battles broke out with the increasing aggression of the Mongol detachments. Initially, the storm that broke out shook the East of the continent. But after 12 years, the Mongols appeared in the West of Eurasia, initially in the Central Asian zone of Khorezmshah's possessions. And three years later, detached from all such high recent benefits—material and spiritual—Khorezmshah says goodbye to his life virtually alone, and besides, among those affected by leprosy on some island of the Caspian Sea.

Then, five years before his death, Genghis Khan sends "Subeetai-Baatur on a campaign to the north, commanding to reach eleven countries and peoples: Kanlin, Kibchaut, Bachzhigit, Orosut, Macharat, Asut, Sasut, Serkesut, Keshimir, Bolar, Raral (Lalat), to cross the deep rivers Idil and Ayah, and also to walk to the city of Kivamen-kermen" (Secret legend, 262). Some names can be distinguished by consonance: Idyl is Itil or Volga, Kivamen-Kermen is Kiev . . . But the most curious thing is that the Genghisid warriors will certainly find all these places, they will take possession of some during Kaan's lifetime, and the rest after his death.

The Laurentian Chronicle also reported about these terrifying changes: "In the summer of 6731 (1223), the people appeared in Russia, and no one knows who they are, where they came from, what their language is, what tribe and what their faith is. Only God knows this. And they are called Tatars, and they have already captured many countries. . . . Many Russian princes have decided that if the Tatars come to our land, they will oppose. . . . And a great battle broke out, but the Russian princes were defeated. Many of them died, many were tortured to death." The chronicle was about the tragic battle of Kalka, but after this tragedy there was a break of 14 years. And then Genghis Khan's grandson Batu Khan (Batu) led the Mongol tumens to Kiev and other Russian lands.

In 1240, the Mongols stormed Kiev. Eighteen years later, Baghdad—and there, in the extreme southwest of their conquests, the Mongols reached the center of the great world of Islam. And in 1279, the Southern Song state, the last of the Chinese states, fell. However, back in 1271, the Mongol Yuan Dynasty was proclaimed. This was the apogee of the Mongol Empire of the Genghisids. Its spatial coverage was approximately 27–30 million square kilometers (Fig. 11.9). And all this was accomplished within a little more than seven decades.

Figure 11.8. Equestrian monument to Genghis Khan; the height of the complex without a pedestal is 40 m.; erected in 2008, 54 km south-east of the capital of Mongolia, Ulaanbaatar.

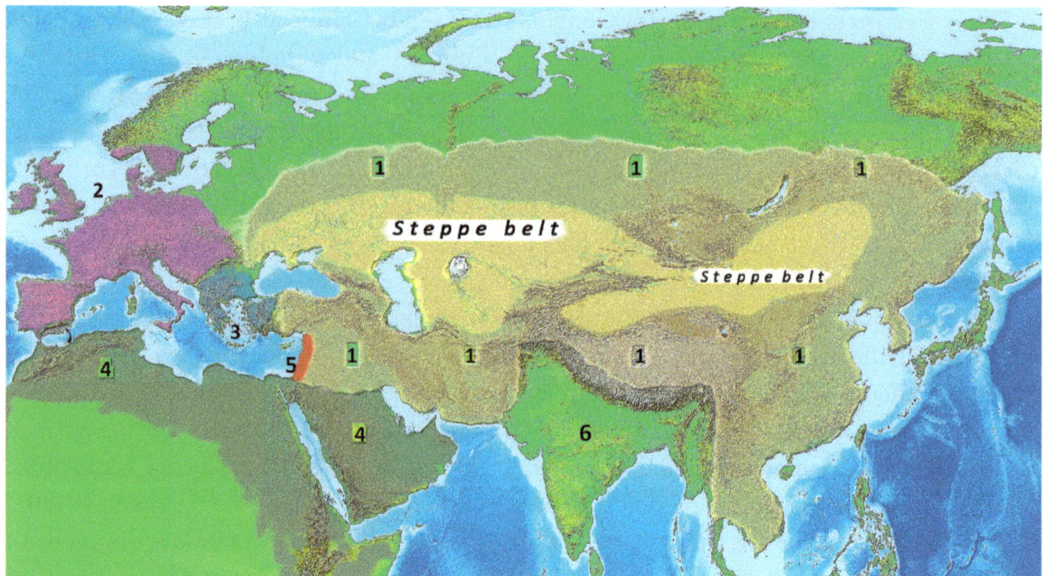

Figure 11.9. Schematic map of the main social formations in the Millennium of the East: 1—Mongol Empire, maximum spatial coverage; 2—Catholic states and principalities of Europe; 3—Byzantium; 4—Muslim states; 5—zone of active clashes between Crusaders and Muslim detachments; 6—Hindustan.

11.7. The Steppe World and the Phenomenon of the Great Wall of China

The European world, an enclave in the extreme West of Eurasia with its domain—the European mega-peninsula—could voluntarily or unwittingly avoid a tough confrontation with the raids of nomadic and semi-nomadic societies of the Steppe Belt. Perhaps the most striking example of this is Rome: initially republican, and then imperial. After all, having covered vast territories of the Mediterranean, including the coast of North Africa, as well as the western half of the Iranian-Anatolian enclave, Roman legions were not noticed in the zone of domination of nomadic cultures of the Steppe Belt (Fig. 11.10). The absolute contrast to this is quite obvious in the east of the mainland far from Rome, where all the endless battles were reflected both in written sources and in an absolutely unique visual and, perhaps, their brightest evidence, in the labyrinth of the Great Wall of China (Fig. 11.11).

The first documentation of the initial actions for its construction dates back to the kingdom of Chu—the largest of the block of cultures of the so-called Spring and Autumn period (Chunqiu), dating from the end of the eighth to fifth centuries BCE. Although we don't know much about these steps, it seems that they caused quite active imitations already in the next period of the so-called Fighting Kingdoms, when by the end of the third century BCE, the Qin kingdom, having brutally suppressed its rivals in this extremely unstable and dangerous "fighting" environment, managed to exalt itself to the rank of empire (in our understanding, of course). The general length of time it took to create the entire maze exceeded two thousand years. The final construction of the Great Wall fell to the builders of the Ming Dynasty (1368–1644), which was already very late in this series. They have built, restored and renovated almost 9,000 kilometers (!) of walls of a very different nature. And in the end, the general extent of

Figure 11.10. Maximum coverage of possessions The Roman Empire to the end of the reign of Emperor Trajan in 117.

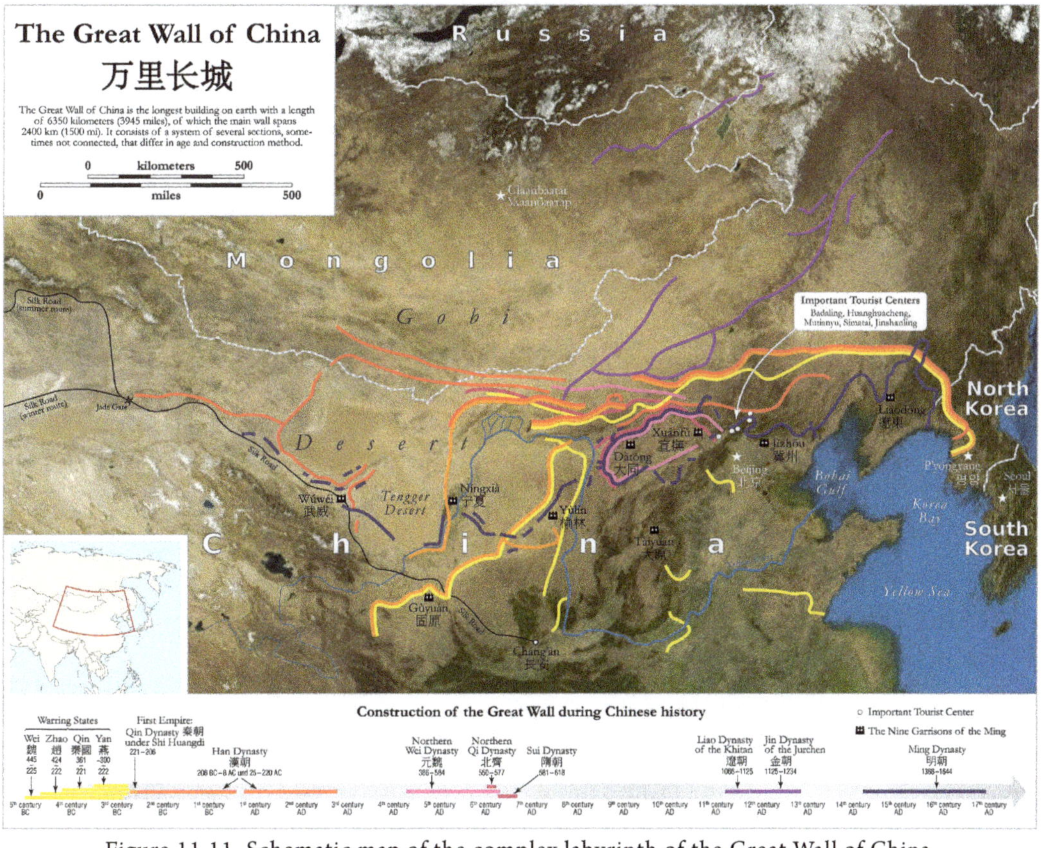

Figure 11.11. Schematic map of the complex labyrinth of the Great Wall of China.

this phenomenon, which is difficult to imagine on a scale, coupled with numerous and diverse branches (Fig. 11.12 a–c), exceeded 21 thousand kilometers! And since those distant times, it was the Great Wall of China that many began to consider one of the most vivid and expressive symbols of the great Eastern power, reliably isolating it from the other world.

a

b

c

Figure 11.12. Some sections of the Great Wall of China.

However, almost at all times, the effect of this unique structure remained incomprehensible and mysterious for many—after all, detachments of nomadic horsemen could, it is believed, easily overcome or somehow bypass the labyrinth of these endless walls. Here, for example, is the opinion of those who have studied this phenomenon for many years:

> The Great Wall simply did not work as an obstacle protecting China from barbarian raids. Since the walls were first built on the borders of China, they have given only a temporary advantage over daring raiders and looters. When Genghis Khan and his Mongol hordes began the conquest of China in the XIII century, the boundary walls proved to be an insignificant obstacle. The Great Wall did not protect the most active builders of the walls, the Ming Dynasty, from their most dangerous opponents, the Manchus of the northeast, who began to rule China as the Qing Dynasty from 1644. Conquerors could bypass pockets of strong resistance until they found weaknesses or breakthroughs, or—which required less effort—bribe Chinese officials to open the gates of the forts of the Great Wall. When the Manchus decided in 1644 to make a final dash to Beijing, a Chinese general who defected to them led them through the Great Wall. As Genghis Khan allegedly said, "the strength of the walls depends on the courage of those who defend them." (Lovell, 19)

Probably, it would not be bad to add to the words of Genghis Khan: "It also depends on the duty of honor!"

So should the costs of such fantastic labor of countless millions of workers be considered in vain? Here, for example, the preceding (before 1368, that is, the beginning of the Ming state, the Mongolian, but not the Chinese) Yuan dynasty of the conquerors of China has sunk into eternity. As if the fear of the eternal expectation of nomad aggression should have disappeared at the same time, but after the fall of the Ming Dynasty, one of the rulers of the Manchu dynasty that replaced it was not truly Chinese! The Qing Empire composed an expressive verse about the Great Wall:

> "They built it for ten thousand li, stretching it all the way to the sea,
> But all your expenses were in vain.
> You have exhausted the strength of your people.
> But when did the empire belong to you at all?"

Is it possible to define these immeasurable energy costs of Chinese cultures for the construction of a Great—many thousands of kilometers—wall for thousands of years as an obvious sign of the irrational in cultures. After all, all these unimaginable construction coverage, in fact, did not seem to achieve the main goals? It's hard to say. It is curious that the starting signal for this fantastic, though perhaps "not too successful" defensive structure was a reaction to the appearance of the Western Scythians here, although the main incentive for the construction of the defending Han Chinese was, of course, the deeds of the eastern nomads. Perhaps it was only in this most striking archaeological and architectural material phenomenon of Eurasia that a certain unity of the nomadic worlds-enclaves of the West and East, so dissimilar to each other within the Steppe Belt, was reflected.

Nevertheless, although the Chinese boundary walls historically have not been all that is attributed to the Great Wall, and although, at the core, modern flattering writings about the Great Wall hide much of China's confusing past, the wall and the ideology behind it should not be discounted, as if it were not related to history the thing. As a strategy that has survived more than two millennia, the Chinese border wall is a monumental metaphor for reading China and its history, for characterizing the culture and worldview that managed to charm and assimilate almost all neighbors and conquerors. (Lovell, 22)

However, to this day in China, the phenomenon of the Great Wall is proud and, presumably, quite rightly so.

In this connection, another extremely curious parallel also suggests itself. The ridges of the Alpine-Himalayan geosyncline seem to be completely suddenly and sharply deployed by some omnipotent cosmic forces to the south. This is obvious already to the east of Tibet, when thereby the Han remains undisguised from the malicious cunning of the warlike nomads of the Steppe Belt. And after all, the same AGG ridges successfully (at least before the invasion of the Genghisids) defended sedentary and more western southern farmers from rapid raids from the north. The question is quite natural: maybe there is a reason to see in the phenomenon of the Great Wall a peculiar—although, of course, very curious and already man-made—attempt to "correct the error of nature." For this purpose, they erected on the northern and western borders of the Han enclave a kind of unshakable and reliable for the defending heights, to the deep regret, the geosyncline ridges that suddenly turned to the south? After all, it was from the western and northern steppe heights that these virtually continuous waves of conquerors, so hated by local inhabitants, were coming. . . . Although, however, even the sky-high heights of the AGG could not serve as a barrier during the raids of nomadic Mongols-Genghisids in the thirteenth century, just as the unique Great Wall of China could not.

11.8. Waves of the West and East in Archaeological and Written Sources

Solving the problem of "docking" and reliable interconnection of paintings created on the basis of written documentation, on the one hand, and on the basis of archaeological materials on the other, is a very difficult process. Maybe the main problem here is rooted in their obvious dissonance. Written sources convince us, for example, that the spatial reach of the Great Mongol Empire was truly incredible—from the Carpathians in the West to the Pacific coasts in the East. But what does archaeology tell us about this? Either nothing, or something extremely inexpressive. Probably, we can repeat something close to this about the imaginary "empire" of the Huns, unknown to us from the heights of the archaeological discipline.

The complex issues of the conjugation of written and archaeological sources have a direct link to the so-called Mongolian syndrome. By this syndrome we mean a funeral rite that is not associated with the immersion of the remains of the deceased underground. Here, for example, is the observation of the famous traveler and mystic philosopher Nicholas Roerich in 1927–1928 in Tibet: "Burial is completely unknown to modern nomads of Tibet. They either expose the dead on the tops of mountains, or throw them into lakes and rivers, or follow the Tibetan custom of cutting the body into pieces and feeding the meat of vultures, which abound in the Tibetan highlands. The old literature of Tibet knows the custom of burying the body in 'stone boxes' made of stone slabs, but traces of such burials have not yet been found, although it is impossible to deny the possibility of their finding" (Roerich, 25). Even in our time, one can see such a funeral rite in Tibet, although not everyone is able to contemplate the process of the deceased's relocation to another world without a shudder (Fig. 11.13). Or, for example, how they buried, according to "Altan Tobchi," the ruler of the Universe:

> Genghis Khan, after the final defeat of the Tanguts, returned and ascended to heaven in the year of the Pig. . . . In the sixty-sixth year of his life, in the twenty-second year of his reign, in the year of the Red Pig, on the twelfth day of the seventh month, he became a tengri [deity]. The Argamaks were harnessed to a large cart, and the golden remains of the Lord of the Khagan were placed on it. Then, quite unexpectedly, the cart refused to move, and no one, no matter how hard they tried, could push it from its place. Then . . . the great people became worried, and the Sunni Kulugatey-bagatur, having created worship, said: "Although the merciful soul has left you, we will still return your ashes like precious jasper, show it to your Borte-khatun, deliver it to your people. . . . And the big cart moved, creaking. And all the people rejoiced! . . . They buried his clothes and stockings in this place and put a yurt over them. . . . His real corpse, as some say, was buried on Burkhan-Khaldun. Others say that they buried him on the northern slope of Altai Khan, or on the southern slope of Kantei Khan, or in an area called Yehe-Utek." (Lubsan Danzan, Chapter 13, 102–105)

Since then, endless expeditions and various kinds of amateurs have been unsuccessfully searching for the graves of Genghis Khan and Attila. But if only the emperor's "clothes and stockings" were buried separately, then should the seekers hope for the desired luck? The "Mongolian syndrome" is inherent in the cultures of almost all nomads of the eastern enclave of the Steppe Belt. That is why we become almost helpless in trying to answer a huge block of various questions concerning the Eastern nomads: What did they look like? What were their clothes like? What were their weapons? After all, everything was behind it: their anthropology, their connections, their idea of the world, the impact on the world around them. Therefore, we are entirely dependent here on written sources—but how often they are contradictory!

The peaks of dissonance—almost polar here—we must recognize the juxtaposition of the two worlds when referring to the nomads of the West, to the Scythian world. Their archaeological sources—and these are gigantic and innumerable golden mounds of the leaders—are so rich in a variety of contents that in the most diverse written documents, the vast scale of the deeds of

Figure 11.13. Tibet, the relocation of the deceased to another world with the help of vultures. The body of the deceased is cut up, and vultures are waiting nearby (Wikipedia; https://fishki.net/1924277).

these peoples and, above all, their rulers, can hardly be rejected or questioned. Otherwise, try to explain the way in which they managed to get all these treasures. After all, the buried leaders had to move with all these riches, with their retinue, with cavalry to the extraterrestrial world and look there according to the material ideas that prevailed in the Scythian world. And it is very difficult to say about the presence of something spiritual in these Scythian beliefs, which we put at the forefront when trying to understand the ideology of the Eastern nomads.

And one more thing. In the turbulent mixture of the Great Eurasian Migration of Peoples, among steppe ethnic groups moving hundreds and thousands of kilometers, it can be extremely difficult to correctly determine the belonging of a particular burial to a particular culture, to a particular ethnic group. This problem almost constantly and painfully worries, for example, researchers of the Huns in the west of the Steppe Belt (see, for example, Kovalevskaya 2005, 91–116).

* * *

We will conclude this chapter with one more remark. Most likely, an attentive reader may have noticed a strange gap: in the rather long series of Eurasian geoareas touched upon in this chapter, which served as the basis for the formation of a wide variety of worlds, for some apparently not entirely clear reason, one of the most remarkable geoareas of Hindustan and the associated world of Indian cultures suddenly fell out.

Yes, without a doubt, this world was extremely peculiar and did not quite resemble the neighboring worlds. But what is the reason for such a seemingly sudden detachment? The change is explained, however, quite simply. Precisely because the world of India—both in its original geohistory, as well as in closely related geoethnic and geopolitical pictures—requires a more detailed analysis. The author intends to present the results of this already in the next chapter.

Chapter 12

The Geoarea of Hindustan in Paradoxical Estimates

12.1. Geomorphological Isolation of Hindustan

It would be reasonable, perhaps, to begin this chapter with a brief repetition of those topics that we discussed in Chapter 5, namely, with a reminder of two southern geoareas alien to Eurasia—Arabia and Hindustan. And if the Arabian Peninsula is very closely connected with the northeast of Africa and the African/sub-Saharan littoral plate, then the situation with the Indian Peninsula looks completely different in this regard. The latter is the northern part of a very powerful and extensive, hidden under the waters of the Indian Ocean, the Australo-Hindustani tectonic plate, but continuing for many tens of millions of years—even to the present day—its steady progress in the north direction (Figs. 5.1:13, 5.4). And here is the experts' assessment of such a situation: "The subcontinent of Hindustan is the only fragment of Gondwana that turned out to be entirely in the Northern Hemisphere during its collapse and joined the so-called Northern Hemisphere about 40 million years ago. towards the new supercontinent of Eurasia. . . . The territory of the subcontinent began to experience an increasing uplift" (Hain 2001, 338, 348). Note, however, that geologists note this uplift to this day (see Chapter 3, Section 3.3, as well as Trifonov et al. 2012).

Consequently, a completely new stage in the geomorphological composition of the Eurasian continent has been presented since the edge of the last forty million years or so. Much later—that is, about one and a half or two million years ago—the earliest *Homo* appeared on the Hindustan peninsula from the African continent (see Chapter 2, Fig. 2.2). These were the forerunners of the most ancient cultures of the first intercontinental migration of mankind or the first wave— the oldest stage of the globalization of the planet Earth. Both the very beginning and the entire history of this geoarea up to the present day were connected with the cultures of this migration.

The isolation of the Hindustan peninsula from its neighbors was greatly facilitated by its geo-morphology: it was an alien, albeit very large fragment on the southern face of the giant body of the Eurasian Littoral Plate (Figs. 5.4, 5.6). The heights of Tibet and the Himalayas, unique in their expressiveness, served as its northern border. The western boundary of Hindustan was also marked by the impressive valley of the mighty Indus River with tributaries; the eastern boundary was already connected with the Ganges Valley (Fig. 12.1).

Figure 12.1. Hindustan and its land borders: north—Tibet with a monastery (painting by N. Roerich, 1933); west—Indus Valley, east—Ganges Valley (both aerial photography).

Figure 12.2a. Ganges Valley during a mass visit to the valley of the sacred river by pilgrims.

Figure 12.2b. Pilgrims in the Ganges Valley.

Figure 12.3a. Tibet, Lhasa: Potala Palace and Buddhist Monastery complex.

Figure 12.3b. Tibet: Taktsang-laghang Buddhist Monastery ("Tigress Nest").

Be sure to note: at least the northern and eastern sides of this peninsula soon acquired a distinctly pronounced character of truly sacred areas. For Hindus, the Ganges is still a sacred river, whose waters can purify people from all sins with the help of lightning and immersion in it (Figs. 12.2a, b). And a considerable number of monasteries are concentrated on the peaks of Tibet (Figs. 12.3a, b), belonging to different schools of Buddhist faith. However, the sacred places of Tibet attracted not only adherents of Buddhism, but, for example, such as the famous Russian artist, deeply in love with Tibet, and the mystic philosopher Nikolai Roerich (1874–1947); we placed his expressive picture at the top of our combined illustration (Fig. 12.1).

12.2. India/Hindustan in the View of Ctesias of Cnidus

Perhaps our voluntary or involuntary return to the topic of exceptional importance touched upon in the previous chapter is not without curiosity (see Section 11.8). It was about very significant differences in the assessment of various situations based on archaeological sources, on the one hand, and historical/written, on the other.

Here we will focus on historical documentation, and its hero will be Ctesias of Cnidus. The following words about this ancient historian, whose activity ended about two and a half thousand years ago, began his presentation by a prominent, modern, talented historian Nikolai Sergeevich Gorelov (1974–2008):

Figure 12.4. Ruins of the ancient city of Cnidus in the extreme southwest of Asia Minor—
the seat of the Greek historian Ctesias.

Just as the "Legend of the Indian Kingdom" was a medieval image of a distant country, the "Indica" of Ctesias of Cnidus became the first picture of India for the Greeks of antiquity. It cannot be said that India was not mentioned earlier than the 4th century BCE—excerpts of the writings of Hecataeus of Miletus and Scylacus from Coriander speak in favor of the fact that the Greeks knew about the country located in the east, but Ctesias was the first to really describe it. Who was this Ctesias of Cnidus?

At the same time, we note that Cnidus or Cnidos is an ancient city, a colony of Lacedaemonians/Spartans on the extreme southwestern cape of the peninsula of Asia Minor (Fig. 12.4).

According to fragmentary information, for seventeen years he labored at the court of the Persian kings Darius II and Artaxerxes Mnemon, and then returned to his homeland and compiled a fundamental work, "The Peach" (they say this work numbered twenty-three books), in addition, Ctesias undertook a description of India, and also acted as authors of several treatises. The profession of a doctor in the family is Ktesiya . . . it was hereditary . . . But in addition to a successful medical career, Ctesias also had a remarkable gift as a writer. Descendants willingly read his books, but did not feel much confidence in the information reported there. (Gorelov 2004, 185)

So, what did Ctesias tell readers about India in his "Indica"?

About the peoples

Indians outnumber all other people combined [Figs. 12.5, 12.6]. No other people live behind the Indians. . . . Indians are supremely righteous people. . . . In an uninhabited area of the country there is a sacred piece of land named after the Sun and Moon . . . . Thirty-five days a year, the sun cools down there, giving people the opportunity to celebrate in his honor and return home unharmed . . .

In the middle of India there are dark people called pygmies who speak the same language as other Indians. They are very small; the tallest of them are two cubits tall, while most are one and a half cubits tall. Their hair is knee-length or even longer, and their beards are bigger than everyone else's. When the beard grows, they do not wear any other clothes, but let their hair fall much below the knees, while the beards hang down to the very feet. After the hair covers the whole body, they are girded and covered with them instead of clothes. Their genitals are so big that they touch their ankles, and thick. They are also snub-nosed and ugly in appearance . . .

There are also kinocephalians who do not build houses and live in caves. They hunt animals with a bow and arrow and are so agile that they catch them on the move. Women wash only once a month during their period and at no other time. Men don't wash . . .

Figure 12.5. Pictorial reconstruction of various types of *Homo* according to Ctesias.

Figure 12.6. Pictorial reconstruction (~1260) of the archer and his victim (sciopod) according to Ctesias.

Men and women have tails hanging down from behind, like dogs, but they are longer and fluffy. Kinocephals copulate with their women on four paws, like dogs, to get closer in a different way [is considered] a disgrace. They are righteous and live longer than humans, for one hundred and sixty-two years, and some of them for two hundred years . . .

Indians never suffer from headaches or toothaches or eye diseases. They don't have boils in their mouths or suppurations on their bodies. Most of them live up to one hundred and twenty, one hundred and thirty and one hundred and fifty, and some even up to two hundred years.

About nature

In India, the river flows through the plain and through the mountains, on which the so-called Indian reed grows, so thick that two people can hardly grasp its stem. In height, it is equal to the mast of the ship. There are even larger specimens. Among this reed there are stems both male and female. Men have no core, and they are very strong; women have a core . . .

There is a lake (eight hundred stages in the perimeter), on the surface of which, when the wind is not blowing, oil is arranged. Then the pygmies sail across the lake in small boats and collect it in troughs. . . . There is also a river with honey flowing from the rocks.

About the weather

It's very hot in India.... The sun shining there is ten times more than in other countries, and many die from the heat.... There is no thunder, lightning, or rain in India, but winds and hurricanes are frequent, which demolish everything in their path. In most places in India, it is cool in the morning and unbearably hot in the second half of the day.

About animals

The animal is a manticore: its face is like a human one [Fig. 12.7]. It's the size of a lion, and the color is red, like cinnabar. He has three rows of teeth, ears like a human. The tail of a scorpion, at its end there is a sting larger than an elbow. There are lateral stingers on both sides of the tail, and, like a scorpion, there is a sting on the head. So, whoever comes near him, he kills with his deadly poison ...

Sheep and goats in India are more than donkeys. Basically they will give birth to four or six cubs at a time. They have long tails; therefore, they are cut off from the females so that they can copulate.

Wild donkeys, the size of horses and even bigger, live in India. They have a white body, and a red head, while their eyes are blue. On their foreheads they have a horn, one cubit long [Fig. 12.8].

There is a worm in the Indus River that looks like those that live on fig trees, it is about seven cubits long; its thickness is such that a ten-year-old boy can hardly grasp this worm. He has a tooth on his upper and lower jaw, he catches prey with them.... They catch it with a large hook on an iron chain, tying a kid or lamb to it. After that,

Figure 12.7. Manticore, according to Ctesias (image from the British Library).

Figure 12.8. Unicorn. The image on the wall of the castle in Edinburgh, Scotland.

the worm is suspended for thirty days, having previously substituted vessels. During this period, so much oil flows out of it that it could fill ten attic bowls. At the end of thirty days, the worm is thrown away, and the oil is delivered to the king of the Indians, and only to him, because no one else is allowed to take a drop. This oil burns everything it hits, whether it's a tree or an animal. And it is impossible to calm this fire with anything, only with a large amount of clay, and thick.

And now, I hope, it will also be permissible for the author of the book to offer his commentary to the text of Ctesias of Cnidus.

It is extremely doubtful that today there will be a person who will believe—at least at a minimum—the impressive story of Ctesias about this amazing piece—a fragment of the southern littoral plate—of our Earth. And it's not even that Ctesias himself did not have a chance to visit India, because his narrative was formed only on how diligently he listened, how he memorized and recorded the tales of those Persians who allegedly saw it all with their own eyes. For the author of this book, the paintings created by Ctesias became an obvious sign of that imaginary world that once seemed, without a doubt, true and absolutely real. Apparently, the "Indica" very prominently reflected the fact that in various versions it will "decorate" the world of *Homo* for many hundreds of years, and we will talk about this below.

Ctesias created his "Indica" in about 400–397 BCE. Seven decades later, that is, in 327, Alexander the Great approached with his victorious army to conquer India to its western border: the Indu River. It is possible that the great conqueror of considerable areas of South

Asia (and this, excluding his native parts of the Balkans, the entire vast Iranian-Anatolian geoarea) was quite familiar with the writings of Ctesias, especially since he himself would take part in the creation of fascinating fairy tales a little later about this world. And if he was familiar with the works of Ctesias, did this not serve as an incentive for his famous campaign? In any case, the researcher of this era, Nikolai Gorelov, whose name we have just mentioned, believed:

> If Ctesias of Cnidus had not written down this story, it is still unknown whether Alexander the Great would have decided to go to conquer the Indian kings. The imagination of the famous commander was captured by stories very similar to fiction, which confirmed the simple and clear idea that nothing is impossible in the world. (Gorelov, 186)

12.3. Letters of Alexander the Great to Mother Olympias and Aristotle

Alexander, the tsar, sends greetings to Olympias, mother, and Aristotle, my teacher. A lot of time has already passed, oh, my mother [Fig. 12.9], since your motherly love didn't know anything new about me. . . . As I told you before, I defeated Darius in three battles; as a result of his defeat, I became the master of all Persia. . . . Then, having gathered everyone, I set out on my way to Egypt, which has a large number of cities and extensive possessions, and arrived in Judea. The inhabitants of this country show worship for the living God, who inspired me to make good conditions for them. I turned to him with all my heart. I did the people a favor, and abolished all existing taxes, and I even gave them a good share of the loot taken from the Persians. They proclaimed me king, the ruler of the world, and then through their lands I arrived in Egypt after a certain number of days. I have not spent much time here to subjugate this whole country to my domination. Upon entering their capital, the Egyptians also greeted me as a king, the ruler of the world. Based on the answer of the oracle of the country, I gave my name to a city in Egypt [Alexandria]. I rebuilt it completely, starting from the foundation, and ending with a lot of columns and statues. I only expressed contempt for their gods, which are not gods . . .

After that, I decided to reach the edge of the earth. This decision was immediately executed as soon as it was made. When we toured parts of the earth located under the sun, we met several terrible and impassable places. Thirty days later we crossed these difficult places; we arrived on the plain all together; we found the savages, and they gave themselves up to flight. Then, penetrating further, we discovered the Pillars of Hercules and the palace of Semiramis.

Here we rested for several days; moreover, we postponed the hike; we met people who have six arms and six legs. After we put them to flight, we continued our journey to a place on the seashore. After making a stop there, we saw the waves produced by the sea crayfish, which grabbed the dead horse, and returned to the sea. Soon a lot of sea monsters attacked us, from a species that did not smoke much at us, we captured one crayfish. The flame of the fire that we built saved us. We left this place, and arrived at another, also on the seashore, where we saw an island. I prepared the boat and went there. I found people who spoke the same language as us, who were wise, and naked, like those who came out of the womb of their mothers. After leaving this place, we walked for several days, and met people who are six-legged and three-eyed, then dog-headed people; we put a lot of work to put them to flight. Finally, we found ourselves on a vast plain; in the middle is an abyss. I made a bridge over which an entire army passed.

However, perhaps the most expressive and significant comprehension of the great commander Alexander the Great certainly looked like his "discovery" of the sources of the Egyptian Nile in India (!):

Alexander was prepared on the banks of the Hydaspes [this is a tributary of the Indus] many thirty rowing vessels, hemioles, many vessels convenient for transporting horses and for ferrying troops along the river, and he decided to sail down the Hydaspes to the Great Sea. Even earlier, when he saw crocodiles in the Indus (this is the only river except the Nile where they live), and on the banks of Akesin such beans as the Egyptian land grows, and heard that Akesin flows into the Indus, he decided that he had found the sources of the Nile; The Nile originates somewhere here in the land of the Indus, flows through a huge desert, loses the name of the Indus here and, having entered a populated country, receives from the local Ethiopians and Egyptians the name of the Nile, or, as Homer established,

Figure 12.9. Olympiad, mother of Alexander the Great (gold medallion from Aboukir, Egypt) and Alexander himself (Hellenistic bust).

the name of Egypt of the same name with the country. Under this name, the Nile flows into the Inner [Mediterranean] Sea. In letters to the Olympics, among various information about the Indian land, Alexander writes that, in his opinion, he found the sources of the Nile, but he gives little evidence on such an important issue, and even those are weak. Having gathered more accurate information about the Indus from the locals, he learned that the Hydaspes flows into the Great Sea with two mouths and has nothing to do with the Egyptian land. Then he had to give up what he had written to his mother about the Nile (Arrian, book six, 1).

Alas! It is a pity, of course, that I had to give up such an amazing "discovery." Meeting with six-legged and three-eyed characters who can immediately express themselves in Greek, of course, also seems important, but it really is worth little against the background of the source of the African Nile revealed in India! And besides all the pronounced phantasm of his letters, it's also not at all easy to explain.

Perhaps we should turn again to the "Indica" of Ctesias. Of course, one can even suspect that the thinking of this relatively little-known ancient historian was somehow traumatized when he composed his own work based on rumors about India. But one can't think that way about the great Alexander the Great? After all, he was in India and saw everything with his own eyes. And this is exactly what he wrote to his mother, Queen Olympias, as well as to the famous ancient thinker and philosopher, his teacher Aristotle. And his personal chronicler Callisthenes, who accompanied Macedonian in his glorious campaigns, also seemed to confirm all this.

12.4. Significant Changes in the Tales of Unknown Lands

However, not everyone rushed to confirm the contrived by Ctesias and allegedly already truly known by Alexander the Great. About four hundred years after Alexander's Indian battles, at the beginning of the second century CE, the ancient Greek historian and geographer Arrian Flavius (c. 86–160), who held a number of senior positions in the Roman Empire in the second century, created the book "Anabasis" (Fig. 12.10) about the campaigns of Alexander the Great and, in particular, about Indian, where he wrote:

> Alexander crossed this Indus with his army at dawn and entered the land of the Indus. In this work, I do not write anything about the laws by which they live, nor about the strange animals that live in this country, nor about the fish and monsters that are found in the Indus, the Hydaspes, the Ganges and other Indian rivers; I do not write about the ants that mine gold for the Indus, nor about the vultures that guard him. All these are stories created more for entertainment than for the purpose of truthfully describing reality, as well as other ridiculous fables about the Indies, which no one will investigate or refute. Alexander and those who fought with him refuted many, except for those that they invented themselves. (Arrian, book five, 4)

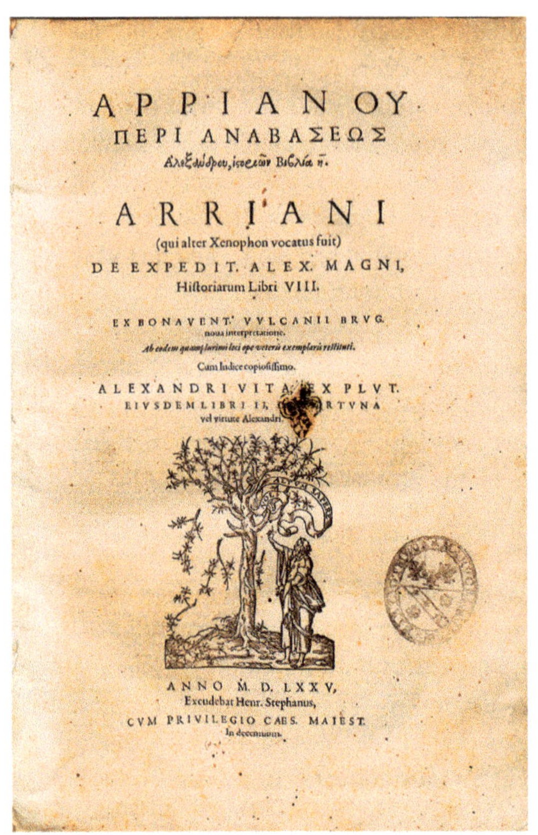

Figure 12.10. Arrian Flavius, the "Anabasis of Alexander the Great" created by him; publication of the title cover of the book in 1575.

Here, of course, Arrian's expression "what they invented themselves" cannot fail to catch the eye. Nothing else comes to mind, except what we were just talking about: discussing the fascinating stories of the Greek about unknown miracles.

However, we should probably move chronologically closer to our time and come across, say, the works of Hermas Sozomen (ca. 400–450)—a Byzantine lawyer, an early Christian writer-historian, and author of "Church History." His attention is no longer at all occupied by the six-legged and three-headed persons of *Homo*, he absolutely does not care where the sources of the Nile are hidden. He is much more concerned about the problems of the victory of Christianity over the disastrous errors and malicious heresy of this world: after all, it is in the canons of Christ's teaching that the absolute of truth rests. And the adherents of Christianity are provided with a truly happy future life in the extraterrestrial world—however, only with very strict adherence to the prescriptions of this teaching. And the young boy Frumenty became a guide to these saints:

We know that around the same time, the inhabitants of the inner, so-called India, adopted Christianity. The preaching of [Saint] Bartholomew did not reach them; they were converted to faith in Christ by Frumentius, who was their priest and preacher of divine teaching. . . . [For example, like many other Greek sages.] Democritus of Kos visited a lot of cities, climates, countries and peoples and, as he himself says, spent about eighty years on the other side. In addition to these, some other wise men of Greece, ancient and new, did the same. Imitating their example, a certain philosopher from Tyre of Phoenicia, Meropius, arrived in India. He was accompanied by two youths, Frumentius and Edessius, both his relatives, whom he taught sciences and guided with wise instructions. Having surveyed everything that was there in the Indian land, he was already on his way back on a ship sailing to Egypt . . . [It happened that when the ship, for lack of drinking water or other edibles, entered one pier.], the native Indians attacked him and killed everyone, not excluding Meropius himself. But the youths, taking pity on their youth, were taken alive and brought to their king. The tsar made the youngest of them a cupbearer, and the eldest, Frumenti, the caretaker of his house and the administrator of his property, because he found in him prudence and the ability to make such orders. . . . They wanted to return to Tyre to their relatives, but since the king's son was still very young, his mother asked both young men to stay for a while and take over the management of the kingdom until her son reached manhood. Respecting the request of the queen, they began to manage the affairs of the kingdom and rule over the Indians. Frumentius, perhaps prompted by divine visions, or, with the help of God, began to find out by himself whether there were no Christians in India, or whether there were any Romans among foreign merchants. Carefully searching for such people, he invited them to his house, received them kindly and favorably, urged them to come together for prayer and to form meetings according to the custom of the Romans, and exhorted them to constantly serve God, for which he built prayer houses. When did the king's son reach adulthood; then they asked for leave from him and the queen, who reluctantly let them go, and after parting with them amicably, they arrived in the Roman Empire. . . . Frumentius, postponing his return to Phoenicia for some time, arrived in Alexandria: for it seemed to him indecent to prefer the fatherland and kinship to the care of divine affairs. There, having met with the primate of the Alexandrian Church, Athanasius, he told him about the affairs of the Indians and explained that they needed a bishop who would take care of the Christians there. Athanasius, having called together the native priests, consulted with them about this and ordained Frumentius himself as a bishop of the Indian country, as a most worthy and capable man to spread the faith. . . . Therefore Frumentius returned to the Indians again and, they say, performed his sacred duties with such glory that all who knew him were surprised at him and glorified him no less than the Apostles. Moreover, God himself glorified him, having performed through him many wonderful healings, signs and wonders. (Sozomen, Book 2, Chapter 24)

Essentially the same version about the two youths with Frumentius was also adhered to by Socrates Scholastic (Book 1, Chapter 19). However, it is assumed that this Socrates (ca. 380– 439 [?]) created his "Church History" earlier than Hermas Sozomen, and the latter simply copied a lot from him, albeit with some corrections.

So supposedly a completely new life began for the inhabitants of India, who had not known the high truth until then. The boy—the blessed Frumentiy from above—received, as if at once, the right to lead the whole state, and not only spiritual, but even administrative. And we see how the pole of narrators' tales is sharply shifting—from descriptions of crazy freaks and unknown monsters to pictures of an absolutely different direction and character—to understanding the world through the prism of the dogmas of religious faith. And, according to Sozomen, it is as if almost all, and maybe even all, the inhabitants of India have seen the light and have aspired to Christianity. This is how a completely new life allegedly arose, began and flourished among the inhabitants of India who had not known the absolute truth until then.

12.5. At the Origins of the Idea of the Chimera of India

Now, of course, we should pay tribute to Ctesias of Cnidus, even if not a very significant ancient historian. After all, it is to him that India owes its reputation as a fabulous "wonderland." And we have to literally wade through the real wilds of all kinds of his fictions. This is how the Greeks imagined India when Alexander the Great began his campaign. More sober minds did not trust much in the tales of Ctesias. And Alexander the Great himself also did not believe in the incredible size of India. However, maybe all or at least most of the educated Greeks shared the idea of India as a very different country from the world around it, where you can meet a lot of mysterious and unprecedented. It is not surprising, therefore, that Alexander also really wanted to lift this mysterious veil.

Here, for example, the ancient philosopher Arrian Flavius, who was not very keen on phantasms, expressed his doubts quite unequivocally. One might have thought that his position would indicate the gradual rejection of the thinkers of that time, at least from the insane abundance of fabulous phenomena created by the authors either on unverified rumors, or if they wanted to please readers with such coveted miracles and all sorts of unknown things. But, alas! The chimera of India has shown its amazing stability and, in any case, it has been observed for almost two millennia.

But more than that: the chimera of India almost constantly and inextricably took part in one or another of a wide variety of general Eurasian actions—sometimes very unpleasant and even stained with thick blood. Sometimes it seemed that this country was becoming, as it were, a kind of center of the world—if not in the global interpretation of this concept, then at least in the general Eurasian one. And it also presents us with a considerable mystery related to the problems of the dynamics of the addition of the megastructure of the Eurasian world.

Chapter 13

A Dive into the Fabulous Worlds of Chimeras and Phantasms

About one and a half thousand years have passed since the tale of Ctesias of Cnidus, and Christian Eurasian communities have plunged into a new myth that tells about the existence of a truly vast empire, headed not by a crown and hereditary ruler, the emperor, but by a priest-presbyter John (Figs. 13.1, 13.2). The legend of Prester John spread quite quickly and extremely widely in many regions of Southern Eurasia from the middle of the twelfth century among almost all European and crusading peoples. The legend has been developing and multiplying in its variations for about four centuries. Even in the Russian testimonies, which are very remote from us by the depth of their age, the legend was reflected as a "Legend about the Indian Kingdom," the ruler of which was "Tsar Pop Ivan," although the Russians did not participate in the crusades.

The legendary element of the legend of the blessed kingdom was most clearly reflected in the dream of a world filled with all the benefits, of a priest-king who stood up to protect Christians from the hordes of the wicked. Most often it is believed that the first news about Prester John was presented in the chronicle of Otto Freisinger in 1145, and after that it was reflected in many other chronicles. And since 1165, a letter from Prester John, King of India, to the Emperor

Figure 13.1. Image of the figure of Prester John, https://www.todocoleccion.net/arte-grabados/grabado-coloreado-preste-juan-emperador-abisinios-c-1740).

Figure 13.2. Presbyter John's proposed geoarea in Northeast Africa (Wikipedia).

of Byzantium, the "ruler of the Romans" Manuel I Komnenos, has been widely distributed in Europe (Figs. 13.3, 13.4). The letter tells us that the kingdom of Nestorian Christians still exists and it is, without the slightest doubt, the largest and most powerful in the world.

Several hundred copies of the letter have been preserved, and these letters have been translated into several languages, including even Hebrew. During the Second Crusade, the Crusader knights were convinced that Prester John would provide powerful support to the Crusaders and help them win Palestine from the infidel Muslims.

However, the true authorship of the letter of Prester John remained unknown. It is even possible that the number of authors was considerable, and they did not exist next to each other, and not at the same time. However, they were all united by a fundamental similarity in opinions—or essentially in their dream!—about the greatness of the faith to which they devoted their whole lives. The text of Presbyter John's letter certainly deserves, even with small abbreviations, an introduction to the readers of our book, and therefore we will make a brief virtual visit to the kingdom—or even to the empire—of "His Majesty Presbyter John."

13.1. Prester John to the Ruler of the Romans

1. Presbyter John, the power and valor of God and our Lord Jesus Christ, the king of kings and the ruler of those who command Manuel, the ruler of the Romans, sends in response a wish of health and assurance of his good will.

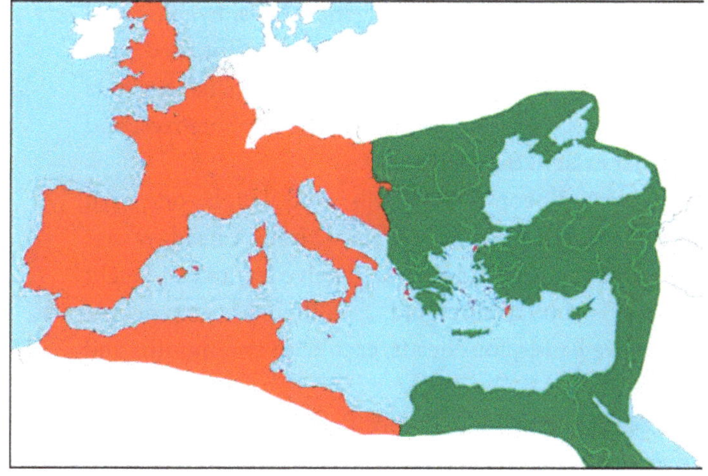

Figure 13.3. The Roman Empire after the collapse at the end of the fourth century into two parts: the western, the Roman proper, and the eastern-Byzantine.

Figure 13.4. A gold coin of Manuel I Komnena (Wikipedia).

2. They have announced to us that you honor our greatness and our power is known to you. Through the apocrisiary, we learned that you want to send some funny and funny things to our court . . .

7. If you wish to come under our domination, then at our court you will gain even greater dignity and height of position and will be able to enjoy abundance, and if you wish to return, you will leave with rich gifts . . .

9. If you want to hear how great our power is and to what limits our power extends, then know: I, Presbyter John, am the lord of the lords, and none of the kings on this earth can compare with me in wealth, valor and strength. Seventy-two kings are my subjects.

12. In the three Indies we rule [Fig. 13.5], and our possessions extend from the inner India, where the body of the holy Apostle Thomas rests, through the desert and at sunrise and returns along the Great Descent in the Babylonian Desert to the very Tower of Babel.

13. Seventy-two regions serve us, and only a few of them are Christian, each has its own king, and they are all our subjects.

14. Elephants, two-humped and one-humped camels, hippos, crocodiles, metagalinarias, giraffes, finserts, panthers, wild donkeys, white and red lions,

polar bears, white thrushes, mute cicadas, griffins, tigers, lamias, hyenas, wild bulls, sagittarians, wild people will be born and live in our country, horned men, fauns, satyrs and women of the same breed, pygmies, cynocephalians, giants four hundred cubits high, one-eyed, cyclopes and birds called phoenixes, and almost all kinds of animals that exist in the world . . .

Wild horses, wild donkeys, horned people, wild bulls, wild people, one-eyed people, people with eyes behind and in front, people without a head, whose eyes and mouth are located on the chest, and they are twelve feet tall and eight feet wide, and their skin color is similar pure gold. And there are also people with twelve legs, six arms, twelve hands, four heads, each with two mouths and three eyes [Figs. 13.6, 13.7].

There are women in our country with huge bodies and a beard up to their breasts, a flat head, and they dress in skins and are beautiful hunters by nature, and for hunting they tame wild animals instead of dogs—and, so, they go to a lion with a lion, and to a bear with a bear, to a deer with a deer, and so on same with the others . . .

In some of our other provinces there are ants the size of puppies, and they have six legs and wings like sea locusts, their teeth inside their mouths—the ones they chew—are larger than dogs, and outside their fangs are longer than those of forest boars, with these fangs they kill people and animals. And after killing them, they immediately devour them . . .

These same ants are underground from sunset to the third hour of the day and all night they extract the purest gold and bring it to the surface. From the third hour of the day until sunset, they are on the surface and feed at this time. And

Figure 13.5. The figure of the presbyter sitting on the throne against the background of his possessions in Africa (https://history .fandom.com/ru/wiki/).

Figure 13.6. "Monstrous tribes" from the book Rabanus Maurus (Rabanus Maurus) "About the Universe." Raban Moor, German theologian, writer, poet (about 780–856).

Figure 13.7. "The people of Divya found by Tsar Alexander the Great." The inscription on the sheet and engraved on copper pictures (Russian folk pictures, 12, 13).

then they go underground to dig for gold. . . . During the day, while the ants are on the surface, none of the people dare to leave the shelters outside, because the ants are ferocious and cruel . . .

15. We have other peoples who devour human flesh and miscarriages and who are not at all afraid of death. And if one of them dies, then his tribesmen greedily pounce on the corpse, saying: "Eating human meat, we perform a sacrament."

17. These very peoples, as well as many other tribes, were imprisoned by the young Alexander the Great, King of Macedon, among the highest mountains in the northern countries. At will, we take them out against our enemies and let them devour them, so that neither people nor animals remain.

18. But when they have destroyed our enemies, we will lead them back to their homes. That's why we accompany them, because if they return without us, they devour all the people and animals they meet on their way.

19. These wicked tribes before the end of the ages in the time of the Antichrist will come from the four corners of the world, surround all the fortresses of the saints and the great city of Rome, which we are going to give to our firstborn, together with all Italy, and with all Germany, and with both Gauls, with England, Britain and Scotland, and we will give him Spain, as well as all countries up to the Arctic Sea.

20. These wicked nations are as numerous as the sand on the seashore, and truly no tribe and no kingdom will be able to resist them.

21. In our country honey flows and milk abounds . . .

22. A river called the Indus flows through one of our provinces, where the pagans live. This river originates in Paradise and its bends cover the entire area . . .

25. In this region there are many trees, and it abounds with large snakes, which have two heads, and on them horns like sheep, and their eyes shine like large lamps . . . But when the pepper ripens, people gather from all the surrounding provinces, and they bring with them chaff, straw and dead wood, which they cover the grove with, and when a strong wind blows, they light a fire inside and outside the grove so that not a single snake can crawl out of it. And so the snakes die from the unbearable heat, except for those who hide in burrows.

26. And as soon as the flames subside, men and women, large and small, with two-pronged pitchforks enter the grove and take out all the roasted snakes, piling them in huge heaps, like threshed straw. And these, when they dry well, the Indian sages, having crushed at the same time with some medicinal herbs, add to the medicine according to their understanding. And this very remedy surpasses all healing drugs, so that even infertile ones are able to give birth, and if we speak in general and briefly, it cures all diseases . . .

28. If someone, after fasting, eats three times from this source, then from that day on he will not suffer from any disease and throughout his life will look like he is thirty-two years old . . .

30. If someone carries such a pebble in his pocket, then the light will always be with him, for no matter how small the pebble is, but if you get it, the more you look at it, the more the light flares up. If a proper spell is cast on him, he makes a person invisible, removes hatred, establishes consent, expels evil . . .

They know neither hatred nor envy, live in peace, do not quarrel with each other over property. No one stands out among them, except those whom we send to collect our taxes. Every year they load fifty elephants and fifty hippos with the purest balsam and the same amount with precious stones and pure gold. For these lands are rich in precious stones and red gold . . .

People who eat the bread of heaven live for five hundred years. And truly, every hundred years they are rejuvenated and completely renewed, having drunk three times from a spring that gushes from under a tree standing on that island. And this is water . . . it relieves them of old age, and with its help they are freed from the burden of years and look like thirty or forty years old. At the end of five hundred years they die, and in accordance with the custom of that people they are not buried, but are transferred to the island mentioned above and placed on trees growing there, the crown of which is very thick and never falls off . . . The dead bodies there do not smolder and do not turn into dust, but remain as if alive and will remain as one of the prophets predicted, until the arrival of the Antichrist. . . . In the time of the Antichrist, so that the Word of the Lord, who said to Adam, would be fulfilled: "You are dust and you will return to dust," the earth will open up to an incredible depth and swallow them up. And, having accepted them, it will close, becoming the same as before . . .

Towards the north, at the very edge of the world, lies an area belonging to us called the Cave of Dragons. This place is impassable and ferocious in ferocity, as well as harsh, the deepest depth and pitted with caves, and also abounds in secret passages. Innumerable thousands of terrible dragons live there, which the inhabitants of the surrounding areas guard with great care, so that none of the Indian spellcasters or those who arrived from any other place tried to anger the dragons. It is customary for Indian rulers to have dragons at weddings and other feasts, and without dragons they consider the celebration a failure. And just as ordinary shepherds ride foals, subdue and tame, train and tame, give them names, put a bridle and saddle on them and ride wherever they want, in the same way these people who guard and restrain dragons—the bosses over dragons—with their spells and potions, these very dragons are subdued, tamed, trained, and they tame them, give them names, put a bridle and a saddle on them and ride when they want. . . . This dragon nation sends to Our Majesty every year as a tribute a hundred people, the rulers of the dragons, and a hundred dragons have already been subdued, so that they are like sheep among people and play with people like dogs wagging their tails. These very people with dragons serve as our messengers, and when it pleases our grace, we send them along with dragons flying through the air to all countries of the world, wanting to learn about everything new that is happening . . .

38. Next to the desert among uninhabited mountains, a certain stream flows under the earth, access to it opens not otherwise than quite unexpectedly . . .

41. Behind the stone river, ten tribes of Israel live, which, although they have their own king, but they are our slaves and subjects of Our Majesty . . . [Fig. 13. 8].

44. Gold and silver, precious stones, elephants, one-humped and two-humped camels, as well as dogs, Our Majesty has abundance . . .

Figure 13.8. The servile members of the ten slave tribes of Israel carry tribute to the ruler of the Empire.

Figure 13.9. The Second Crusade at the gates of Constantinople, when Christian warriors sincerely believed in the help of Prester John in battles with Muslims (Miniature by Jean Fouquet from the "Great French Chronicles").

46. There are no thieves and robbers among us, there is no place for adultery and greed here. There is no separation between us. Our people have abundance in every wealth. Each of us has several horses, several houses. It is well known that no one can compare with us either in wealth or in the number of tribes . . .

47. When we go to war, we command to carry in front of us, instead of banners, thirteen large and high crosses made of silver and precious stones, and each of them is on a chariot, and each chariot is followed by ten thousand horsemen and one hundred thousand foot soldiers, not counting those who look after the utensils, chariots, clothing and food [Fig. 13.9].

50. All the riches that exist in the world surpass and overshadow our greatness . . .

51. No one among us lies and cannot lie, because as soon as he begins to lie, he immediately dies.

52. We all follow the truth and love each other. Adultery does not live among us. No vice rules us . . .

55. Amazons are women who have their own queen, and they live on an island that stretches in all directions for a thousand miles and is surrounded on all sides by a certain river without beginning and end, like a ring without a stone. The width of this river is one thousand five hundred and sixty-five stages. There are fish in it, pleasant to the taste, which are quite easy to catch. There are also other fish, the size of large shells, and they have four conveniently located legs, a long neck, a small head, pointed ears and a large tail. . . . They are naturally so

tame, as if they were raised by man, and in running they are as fast as the sea winds, and they voluntarily lie down in pairs on the shore—male and female, waiting to be taken.... The men of the women mentioned do not live with them and do not dare to come to them—unless they want to die on the spot—but they live on the outer bank of the river mentioned. For if any man sets foot on this island, he dies on the same day ...

64. We have the most beautiful women, but they do not come to us except four times a year and for the sake of childbirth, and so, having been cleansed from their uncleanness thanks to us—just like Bathsheba from David—each returns to the house.

65. Once a day, our courtyard has a meal. Thirty thousand people are present at our meal every day, not counting those who arrive and depart ...

67. In front of the gates of our palace, near the place where the competition is held, there is a mirror of amazing height, and one hundred and twenty-five steps lead to it.

72. The mirror is guarded by twelve thousand armed warriors both day and night, so that it does not accidentally crack or break ...

73. During each meal, seven kings, sixty-two dukes and three hundred and sixty-five counts serve us alternately at our meal, not counting those who have been sent to our court on various business.

74. Every day, twelve archbishops eat on our right hand, twenty bishops on our left, not counting the patriarch of St. Thomas the Apostle, the protopapes of Samarkand and the arch-protopapes of the city ...

77. Once he was told in a dream: "Build a palace for your son, who will be born to you and will be the king of the kings of the earth and the lord of the lords of the whole earth."

78. And such grace will descend on this palace.... No one there will suffer from hunger or disease, and none of those who will be inside the palace will die on the day they came. And if he suffers from terrible hunger and is near death, then if he enters the palace and stays there for some time, he will leave there in good health, and if he tastes a hundred dishes there, he will become so healthy, as if he had never suffered from any diseases in his life at all.

81. And if someone from this source, having fasted three times for three years, three months, three weeks and three hours, eats three times within three hours, so that not before the hour itself and not after the hour, but during, that is, between the beginning and the end of each of these hours, then he will not die and will look like a young man before three hundred years, three months, three weeks, three days and three hours have passed ...

96. Next to this palace is our glass chapel, not made with hands—the greatest among all the miracles—because there was nothing on this place, and on the first day of our life it appeared where it is still today for the glory and decoration of our name ...

99. We cannot tell you enough about our glory and power now. But when you come to us, you will tell us whether we are truly the lord of the lords of the whole earth. However, know at least a little: in one direction our possessions extend in length for four months of travel, and no one knows how far our possessions extend in the other direction.

100. And if you can count the stars in the sky and the grains of sand on the seashore, then you will measure our possessions and our power.

***

Given in the city of our Bibrik on March 18 on the 51st the year of our birth. About confirmation: everything that seems incredible and what is told above, a certain cardinal, named Stefan, stated truthfully and reliably, and he told it to everyone.

***

The book, or history, of Prester John, which was translated from Greek into Latin by Christian, the Archbishop of Mainz, has been completed. This Christian was installed by Archbishop Conrad. This Manuel ruled in Greece from 1144 to 1180 CE.

13.2. On the Homeland of Prester John

This question is almost primordial. And perhaps the reader will be surprised when he learns that in the discussion on this story, the author preferred to turn to Jewish sources. After all, in fact, the vast majority of the most diverse—positive or negative—reactions to this myth belonged to researchers and commentators, adherents of various Christian trends.

Below we will try to explain the reason for such a strange preference, and for this we will pay attention mainly to the article published in 1995 with the catchy title "Prester John: Fiction and History" or in Russian sounding "Prester John: Fiction and History." The author of this article was a prominent Israeli scientist, writer and religious Zionist activist Meir Bar-Ilan. However, he began his article with a mention of the Jewish book of Ben-Sira, which was published

> …in 1519 in Constantinople, and the appendix to it contains "a copy of the letter that Priest Juan sent to the pope in Rome": The main theme of [this letter] is as follows: A long time ago in a very remote country there lived a king who was not only a great king, but also a Christian priest. The name of this king was Prester John, and he ruled 72 countries. His land was rich in silver and gold, and many wonderful creatures lived there. This king wrote letters to several popes in Rome, telling them that he was a faithful Christian and was familiar with all kinds of unknown animals, such as: men with horns on their foreheads and three eyes, women who fought on horseback, men who lived 200 years, unicorns, etc.

This legend, like many others, can be interpreted, although not without difficulty. Indeed, early scholars who investigated this subject proved that this legend has a historical core, and it is possible to distinguish between fiction and history. However, it will not be easy to clarify the whole story, and the purpose of this article is only to partially advance the discussion of this historical legend . . . (Bar-Ilan 1995, 292)

It is not easy to separate fiction from real history in this legend. Therefore, only three topics will be discussed here: 1) the geoarea of Prester John's being, 2) the connection between his letters and the novel about Alexander [the Great], and 3) the origin and distribution of his Hebrew letters in Europe.

13.3. Is Prester John from Africa or India?

It is quite clear that we are not going to analyze in detail the arguments of the disputants on all three topics delineated by Bar-Ilan. Let us confine ourselves to the main conclusions of the Israeli researcher mainly on the first of these topics, that is: "Where is the true homeland of Prester John: India or Ethiopia?"

Of course, in India, Bar-Ilan answers:

> If this identification has been vague so far, it seems that the connection of the letters with the "Novel about Alexander [the Macedonian]" together with other Indian problems negates the entire probability of finding Prester John in Africa. Indeed, it is time to find out how the confusion arose between India and Africa as the land of Prester John.

But there is no reason for us to find out the ways of this confusion. It is much more interesting to understand the reasons for the complete rejection of Africa as the supposed center of the "presbyterian empire." After all, it was there that such a center was placed on the maps by supporters of the African homeland of "His Majesty" (see, for example, Figs. 13.3 and 13.5). After all, the northeast of Africa—and especially the area where Abasia or Abyssinia, aka Ethiopia—is located very close to the Holy Land. This, without a doubt, the true and most important shrine among adherents of Judaism. In addition, the text itself from the letter of the presbyter to the Roman ruler Manuel contains a very unpleasant note for the Jews: "Beyond the stone river there are ten tribes of Israel, who, although they have their own king, are our slaves . . ."

Therefore, the further the original center of the presbyter is from the African Abasia, the more calm and reliable everything will seem. In any case, this is much more like the reason for the detachment of Africa/Ethiopia from the Presbyter John's supposed homeland by some persons. But there is nothing about this, except for the statement given here about the slavery of the ten Jewish tribes, or at least I could not find them. And India, including its special role in the early and often fantastic ideas about Eurasian history, was discussed in the previous chapter (Section 12.5). Indeed, at least her virtual role was such, and we will see this more than once below.

Chapter 14

The Counter-Examination in Two Millennia

A counter-examination is usually understood as a long-term polemic or dispute going back centuries or even millennia, or disputes on some large but differently understood and not unambiguously interpreted topic. We will try to touch on such a topic or problem in this chapter, although in the history of *Homo* it is far from the most tremulous and exciting for millions of people. After all, the entire history of mankind, or at least that part of global history that is connected with the life of the species *Homo sapiens*, is saturated with counter-views— and incomprehensibly diverse and diverse—in essence. And one more thing should probably be said without fail: the roots of almost all disputes of this kind are completely or at least to a considerable extent found in the sphere that we consider ideological. Many of the notable people in this kind of discussion are remembered, for example, by the fact that they can expand almost limitlessly on the topic. And it is they who generate results that often appear colored by thick flows of blood.

The counter-examination that has attracted our attention is far, however, from fights of such a tragic nature—and we will be glad of it. We will talk about the diversity in the assessments of the life that surrounded our very or rather distant ancestors: about life, on the one hand, real, and on the other—virtual, largely invented by people themselves. And which of the sides of this pair looks more blissful and attractive for a person? It should be taken into account that both of these sides appear in the psyche of *Homo* almost always mutually and closely intertwined. It seems to me very doubtful that we will be able to meet a person whom all medical and special services recognize as mentally absolutely normal, but who would not add absolutely anything from the world to his real life in something fabulous. But only the fractions of these facets can vary very much. And besides, they are very dependent on the epoch in which a person's daily life takes place, and on his secondary environment—in these cases, the dependence of the psyche of each individual is extremely great.

Below we will start with the facet of a person's understanding of his being, which reflected virtual life, in some ways densely colored with fairy-tale pictures.

14.1. From Ctesias and Alexander the Great to Christopher Columbus

In Chapter 12, two persons at once—Ctesias of Cnidus and the great reformer of the then world of southern Eurasia, Alexander the Great—captured, in fact, the foreground of our book. And reading into the text of Ctesias, I—the author of this book—have retained the desire to understand: what is the reason when a not-so-famous ancient Greek medical historian offers all this magnificent fiction? I have almost no doubt that he was aware that these were, of course, his inventions, although at the same time he often tried to refer to the opinions of his interlocutors, as if preparing a specific alibi for himself: "After all, I only convey what I have heard from people who are false and faithful." Fictional, however, and Ctesias himself probably really wanted to see and embarrass! Maybe such desires were born from the fact that everyday life offered him monotonous and boring repetition? And by that fictional world, he probably destroyed or at least diluted the monotony of this boredom.

This can be assumed for Ctesias, but Alexander the Great? After all, his life was hardly similar to the Ctesian monotony we assumed. And then how to understand his very little witty inventions about six-armed, six-legged and three-eyed humanoid creatures? The commander invented this intentionally. I am sure of it; after all, there have never been such anywhere. Is it possible that the inventions were needed, say, in order to show his distant "mother" and the famous teacher, the philosopher Aristotle, in those years, that real life outside their environment is completely different, and is saturated with very interesting and unusual? Perhaps.

And then the ancient figures were replaced by early Christian figures. In the story of Hermas Sozomen (ca. 400–450), a Byzantine lawyer and author of "Church History," a simple boy Frumenti—in the name of Christ and Christian truths—pretty soon, together with his younger brother Edessius, became almost the ruler of the Indian kingdom, surprising the Indian aborigines with his Christian wisdom. And Blessed Athanasius said, for example, to Frumentius: Who among us better than you can disperse the darkness of paganism and bring there the dawn of the divine gospel? The "accession" of Frumenti, and, as it seems, not only ideological, was accomplished at the request of (!) the tsarina mother; after all, her son, the tsarevich, was very young at that time. But when her boy grew up, both alien youths Frumenti and Edessius, with the permission of their mother and the king who had reached maturity, were released from their posts.

So fantastic tales multiplied, and without them, probably, life even for writers of such narratives seemed boring and boring in reality.

The next and, perhaps, much more striking document in all the problems of the fairy-tale, virtual world that interests us was the "Letter of Prester John to the ruler of the Romans." We have dealt with it in sufficient detail in the preceding chapter, and therefore we will limit ourselves here to only brief comments. It seems almost certain that both by the nature and by the very assertive style of the presentation of the views of his unknown author of the letter, this impression was clearly superior to the legend of Ctesias. However, the crowning feature in this tale was, of course, the myth of the vast and most powerful empire/kingdom of Prester John, proclaimed with all the almost irresistible energy. At the same time, it was clear that the original, head center of this empire, without any doubt, nested in the same legendary country of India.

The most remarkable thing was also the fact that the myth that appeared to the peoples was able to conquer a great many people. And the confidence in this made itself felt in various areas of Eurasia for at least three or four subsequent centuries, and not only in the southern half of the continent. It seems that a considerable number of diverse, and sometimes even curious actions and situations were connected with the myth of the presbyter-emperor. In our book, preference was given to only one, but very remarkable curiosity. But this situation turned out to be associated with one of the most iconic figures in our common history, Christopher Columbus.

Perhaps there is not even much point in reminding readers of the well-known truth that the great naval commander literally burned with the desire to reach India, but crossing the ocean from west to east, and not from east to west. And he achieved this by discovering, however, as everyone also knows, a New World instead of India. However, he remained completely confident that this particular continent is, of course, the true India. It turned out that he also believed that it was here that the real possessions of Presbyter John were located; after all, India was in fact the fatherland of the presbyter in the legendary and vast country of his rule.

But the author still wants to give the floor here to the Russian writer—the famous Yakov Mikhailovich Svet (1911–1987). After all, Y. M. Svet (Fig. 14.1), according to the author of the book, is the best of those Russians who were closely interested in the deeds of the great navigators who discovered new, unknown lands on our planet, and all this was also described with talent. So, Christopher Columbus's flotilla of three sailboats, among which there was a vessel named "Ninya" (Fig. 14.2), reached some islands unknown to navigators in the western hemisphere of our planet.

On Thursday, June 12, 1494, the command was given to whistle everyone up. The crew of the "Ninya" lined up on the deck, and the notary of the flotilla, Don Fernand Perez de Luna, read out an amazing document. It was an act by which it was solemnly and formally assured that Juan's land was not an island, but a continent and "the beginning of the Indies and their end, so that, following this land, you can reach Spain by land," and the final paragraph of the act was formulated as follows: "and anyone who ever dares to assert contrary to what is listed here, let him pay a fine in the amount of ten thousand maravedis, and let his tongue be cut off, and if a groom [a sailor of the second article] or a person of equal status says this, he will receive a hundred lashes and will be deprived of his tongue."

Then all the sailors of the "Ninya" confirmed under oath that they agreed with this act and affixed their signatures. And by the end of the day, signatures were collected on two other ships, and the notary Fernand Perez de Luna put the act in a casket that was locked with three keys.

Very many biographers of the Admiral indignantly described the events of June 12, 1494. They accused the head of the expedition of deliberate deception—they say, he knew perfectly well that Cuba was an island, but went to great lengths to convince the royal couple that this land was part of the Asian continent. Undoubtedly, these authors would be right if we were talking about any navigator of the era of Cook, Kruzenshtern and Bellingshausen. But the

Figure 14.1. Svet Yakov Mikhailovich (1911–1987), one of the most prominent Russian writers and translators of special literature, devoted to the study of not only outstanding world-class navigators, but also those countries and organizations that provided the organization of such trips.

Figure 14.2. The sailboat "Ninya" which became the lead ship of the Christopher Columbus flotilla.

Admiral lived, sailed and thought in the XV century. There can be little doubt that he himself was unshakably convinced of his rightness. After all, it was not for nothing that two years later he had heated arguments with Bernaldes, proving to him that it was possible to pass along the shores of Cuba to the Red Sea. And the Admiral's opinion was sincerely shared by many of his companions . . .

Act . . . oath . . . There is no doubt that the solemn gathering on the deck of the "Ninya" causes bewilderment among people of the XX century. But it was not for nothing that the Admiral spent seven years wandering through the offices of the Castilian kingdom, he perfectly studied the bureaucratic mores of the crown departments, he knew that in the eyes of their highnesses, a paper signed by witnesses, notarized in all its form, is worth a hundred times more than a verbal message or a simple report on a voyage to the "Asian" shores.

Ten thousand maravedi fine, curtailed tongues, whips. . . . All this sounds wild to the modern ear, but the final paragraph of the act was written into this document in 1494. And to the Admiral's credit, it should be noted that these terrible punishments remained only empty threats. Michele Cuneo cited one curious case: Abbot Lucena, an inquisitive "passenger" of the expedition, openly stated that, in his opinion, Cuba was not a mainland, but an island, and the Admiral did not consider it possible to deprive him of his language. Lucena, upon her return to Isabella, was strictly ordered not to leave Hispaniola. He thus got off with a kind of "subscription not to leave."[1]

However, it is likely that something even more interesting in this seemingly curious curiosity looked different: apparently, Columbus was definitely sure that the land he discovered was not only India, but also the Land of Prester John.

Perhaps even more curious and significant for world history was the fact that following the discovery of the New World by Christopher Columbus with the subsequent change of the name of this continent to America, the quite obvious start of the sixth Great Intermaterial Migration of Peoples sounded. Then the continent of America—this failed India—was opened. With this really Great Migration, that seemingly inescapable, albeit sluggish-sounding, two-thousand-year-old counter-examination ended, expressed in battles—albeit not very hot—between unshakable supporters of virtual fairy-tale life against those who did not accept it and preferred to notice mostly signs of the harsh reality of everyday life. Myths about six-armed, headless little men, about sheep the size of an elephant and about all sorts of other things have ceased to captivate the broad masses of immigrants. For them, a new world opened up across the ocean and, of course, a different one, completely unlike the fairy tales of Ctesias and his followers. Strange discussions seem to have subsided, or maybe even evaporated altogether. However, it is unlikely, because they are eternal, but the undulating ones will then flare up, then go out.

But, unfortunately, so far we have not even remembered the positions of supporters of a completely different view of this harsh existence.

14.2. From Flavius Arrian to the Sixth Great Migration

In Chapter 12 we touched briefly on the figure of the late Antique philosopher Flavius Arrian, who very decisively renounced the fairy-tale plots of Ctesias and his followers. However, it cannot be said that his forcefully expressed renunciation became an example for others. Just "realists" wrote about their impressions of the world around them, but often without an energetic fuse of the struggle with "storytellers." Even at times they could refer to their texts, but that's all. There was no direct dialogue here, and it seems that the participants of this kind of discussion did not set it in their tasks.

Perhaps, as part of the discussion of this topic, it will be quite interesting to compare the texts of some "storytellers" and "realists," moreover, who composed their reflections at a relatively close time. The famous Marco Polo will speak from the "storytellers" in our book, and the "realists" will be represented by Guillaume or William de Rubruk and Plano Carpini, who are significantly less known to readers.

The Venetian merchant and traveler Marco Polo (1254–1324), in fact, forever decorated his name with the book "Il millione." Her title in various versions of translation into Russian does not sound like "A million," but either as "The Book of Wonders of the World" or "A book about the diversity of the world." At the same time, it is very significant that the well-known standard

---

[1] Svet Ya. *Columbus* (Moscow: Molodaia Gvardiia, 1973), 196.

for the indispensable content of "storytellers" is extremely and, above all, successfully presented to the reader by the splendidly colored cover of the Russian edition of this book (Fig. 14.3a).

Let us now turn to what are perhaps the most interesting issues of this work. But at the same time, I must say that for me the whole book of Marco Polo under any of its titles—the Book of Wonders of the World or a Book about the Diversity of the World—is still a collection of fairy tales, stories, and various (sometimes curious, and often quite primitive) fudges. Here, for example, one of the most memorable pages in his essay was literally covered with golden plates of the Japanese island of Zipungu/Chipungu/Chipingu:

> The island is very large: the inhabitants are white, beautiful and courteous; they are idolaters, independent, do not obey anyone. I will tell you, they have a great abundance of gold: there is an extremely large amount of it here, and they do not export it from here; no merchants come from the mainland, and no one comes here, that's why, as I told you, they have a lot of gold. I will now describe to you the outlandish palace of the local king. To tell the truth, the palace here is large and covered with pure gold, just as our houses and churches are covered with lead. It costs a lot—and not to count! The floors in the chambers, and there are many of them here, are also covered with pure gold two fingers thick; and everything in the palace, and the halls, and the windows are covered with

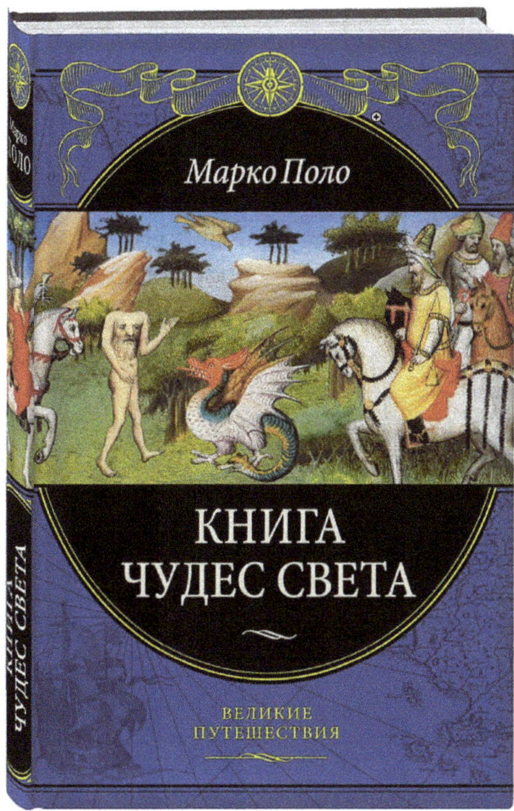

Figure 14.3a. The cover of a book on Marco Polo published by a Russian publishing house.

Figure 14.3b. The cover of a book about Marco Polo, published in 1874 in England, depicts a prison with bars on the window and Marco himself dictating to his cellmate the text of his book.

177

gold ornaments. This palace, I tell you, is immeasurable wealth, and it will be a miracle if someone says what it is worth. There is an abundance of pearls here; it is pink and very beautiful, round, large; it is as expensive as white. They also have other precious stones. A rich island, and its riches cannot be counted . . .

The fantasy of the riches of this island surpasses even the ruler of that world, the great Khan Kublai (Kublai Khan), who is almost wrapped in golden beauties, and recall that Marco Polo could supposedly see all this personally.

Or about something else, much more understandable to my readers. We are faced with the figure of the once quite real Mongol conqueror Genghis Khan in his battles with the mythical Tsar Pop Ivan: the ruler of that world, who was either the heir of the equally mythical and already-known Presbyter John, or independent of him. And here it is:

In 1201, he defeated his rival Zhamukha. The Tatars chose a king for themselves, and he was called Genghis Khan in their own way, he was a brave, intelligent and daring man; when, I tell you, they chose him as king, Tatars from all over the world that were scattered in foreign countries came to him and recognized him as their sovereign. This Genghis Khan ruled the country well. . . . It's even amazing what a lot of Tatars have gathered here. Genghis Khan saw that he had a lot of people, armed him with bows and other their weapons and went to fight foreign countries. They conquered eight regions; they did no harm to the people, they did not take anything away from him, but only took him away with them to conquer other people, [and] they conquered a lot of people. And the people see that the government is good, the king is merciful, and followed him willingly. Genghis Khan recruited so many people that they roam all over the world, but decided to conquer more land.

However, then Genghis Khan decided to marry the daughter of Tsar Pop-Ivan, but

the priest Ivan heard that Genghis Khan was wooing his daughter, and he was angry. "What is the shamelessness of Genghis Khan!" he began to say. "He's wooing my daughter! Or does he not know that he is my servant and slave?! Go back to him and tell him, I will burn my daughter, but I will not marry him; tell him from me that he should be executed by death as a traitor and traitor to his sovereign!" Genghis Khan heard the shameful abuse that Priest Ivan was punishing him, his heart swelled and almost burst in his stomach; he was, I tell you, a man of authority . . .

And that was the beginning of the outbreak of bloody battles between Genghis Khan and Priest Ivan:

Two days later, both sides armed themselves and fought fiercely; it was the most evil fight ever seen; there were many troubles for both sides, and finally Genghis Khan won. And then the priest Ivan was killed. From that day Genghis Khan went to conquer the world. He reigned, I will tell you, for another six years from that battle and conquered many fortresses and countries; and at the end of six years he went to the fortress of Kangi. . . . Genghis Khan died of natural causes during a campaign on Tangut, in 1227, and an arrow hit him in the knee here; he died from that wound. It's a pity, he was a brave and smart man.

Perhaps the most remarkable among the assessments of the deeds of Marco Polo was the non-recognition of his work by a fair group of experts on the Middle Ages. Since the first publication of the book, some have been extremely skeptical about it.

Some even in the Middle Ages considered this book to be just a novel or a fable because of the sharp difference between its descriptions of a developed civilization in China and other early stories just mentioned by Guillaume de Rubruk, Plano Carpini and others. And in much later centuries, there were also doubts about Marco Polo's narrative about his travels in China, for example, due to the fact that the Great Wall of China, for example, was not even mentioned in the book, and this was considered simply impossible for visitors like Polo. And here the difficulties with identifying the many geographical names that Marco used also surfaced.

The commentators of Marco Polo's Book tried with the greatest effort to establish at least an approximate route of the senior Polos from the Volga River to the headquarters of the great Khan. Very few reference points for the compilation of this route were given by the stingy, concise manner of presentation of the narrator of the Prologue. Commentators could only operate with indirect evidence scattered in the Book. They also used the writings of Arab geographers, descriptions of Catholic missionaries and ambassadors who traveled to the headquarters of the Tatar khans in the thirteenth century, and other historical materials.

Probably, also for this reason, many people are asking questions: did Marko actually visit the places he mentioned in his itinerary, and did he misappropriate the reports of his father and uncle or other travelers? Some even doubted whether he was in China at all (Fig. 14.3b). That's how, for example, even the title of a book by a major British expert on Chinese history, Frances Wood, was formulated as "Was Marco Polo in China?" But all this was accompanied, and is still accompanied, by tough disagreements about the realism of this person. For example, even the monuments erected to the famous Venetian traveler in the Mongolian capital Ulaanbaatar (Fig. 14.4) or in the park of the Chinese city of Hangzhou (Fig. 14.5) seem to say that his travels are real.

Let us recall that he wrote his book—and this is his main work—in the prison of Genoa, with which Venice was almost in constant hostility. Why, Marco Polo did not write, but dictated the text to a cellmate and the writer of chivalric novels Rusticano or Rusticello. At the same time, since the beginning of his travels, which is 1271, he completed the dictation of the text in the year 1299, that is, 28 years later! Marko had not kept any records before, otherwise they would have been discussed somehow. They were just very free memories of countless and diverse kingdoms, principalities, localities, meetings with great or just significant personalities of the Eurasian lands. Or perhaps not memories, but something completely different; say, fantasies.

Figure 14.4. Marco Polo Monument in the center of Mongolia's capital Ulaanbaatar.

Figure 14.5. Marco Polo Monument in Hangzhou City Park, China.

For this reason, many of the shortcomings of the book, Marco's supporters were not against blaming the shortcomings of this volunteer secretary. And these are the very remarkable words that Marko begins his book with:

> And I'll tell you one more thing: since the Lord God created the forefather Adam with his own hands, and until now there has not been such a Christian, or a pagan, or a Tatar, or an Indian, or any other person from other nations who would find out and know about parts of the world and about great curiosities as precisely as Marco has found out and knows. And he therefore said to himself: it would not be good if all those great curiosities that he himself had seen or heard the truth about were not recorded so that other people who had not seen or heard it could learn from such a book. I'll tell you another thing: for twenty-six years he collected information in different parts of the world, and in 1298. from R. X., sitting in a dungeon in Genoa, he forced Rustican of Pisa, who was imprisoned with him, to write down all this.

Actually, we note that from a psychological point of view, similar-sounding statements can probably be defined as super-narcissism:

Narcissistic personality disorder characterized by a belief in one's own unique-
ness, special position, superiority over other people, grandiosity; an overesti-
mated opinion of one's talents and achievements; preoccupation with fantasies
about one's successes. (Wikipedia)

And here are some more words with which Marco Polo concludes his book:

> But I believe that it was God's will that we return and the people could learn
> about the things that exist in the world. Because, according to what we said in
> the introduction at the beginning of the book, there has never been a person,
> be he a Christian, Saracen, Tatar or pagan, who has traveled the world as much
> as the noble and famous citizen of the city of Venice, Mr. Marco, the son of Mr.
> Nicholas Polo.

Or, to put it another way, he only confirms his super-narcissism. And thus, as it were, he puts
into the hands of his harsh critics an additional and extremely weighty argument for the rightness of their doubts about the true nature of his writings.

14.3. The Works of the "Realists" Plano Carpini and Wilhelm de Rubruk against the Background of the Book "The Storyteller," by Marco Polo

Most likely, it would be more correct to start this section with an assessment of Marco Polo's
"fairy tales" against the background of Carpini and Rubruk, since his work took place at a later
time in relation to these "realists." Perhaps he was even familiar with their writings, but did not
share their style and views. But it is possible to compare it this way: the author of the book
hopes very much that in any case the result will turn out to be quite similar.
Let us start with de Rubruk. These are the words with which he begins his "report" to King
Louis IX:

> To the most excellent and most Christian sovereign, Louis, the glorious King of
> the Franks, Brother William de Rubruk, the smallest in the order of the Friars
> Minor.

And about the king's command to the least of the minorites before his journey:

> "He will go to the land of foreign peoples, experience the good and the bad in
> everything" . . . I have done this deed, my Lord King, but oh, if only as a wise
> man, and not as a fool, for many do what a wise man does, but not wisely, but
> more stupidly; I am afraid that I belong to their number. . . . I, not daring to resist
> the vow of obedience, composed as I could and knew how, asking forgiveness
> from your incomparable meekness for both superfluous and insufficient, or for

what was said not quite clever, but rather stupid, since I am not a smart enough person and am not used to composing such long narratives.

This is what Rubruk wrote in the introduction to his report, and it is impossible, of course, not to notice how essentially polar—both in style and in the whole "sound"—these relatively brief words of Rubruk differ from the super-narcissism of Marco Polo's introductory ideas about his own person. However, it seems as if one can assume that Rubruk wrote to King Louis IX, and so the court style of letters certainly demanded. But it is unlikely: after all, Marco, who was in an irresistible narcissism, counted on the attention of the entire World of that time, including kings and even emperors:

> Sovereigns and emperors, kings, dukes and marquises, counts, knights and citizens, and anyone who wants to know about different peoples, about the diversity of the countries of the world, take this book and make it read to yourself; you will find here extraordinary all sorts of curiosities and various stories about Great Armenia, about Persia, about the Tatars, about India and many other countries; all this our book will tell clearly in order.

So the letters to the supreme rulers, despite the obligatory court style, necessarily highlighted somehow the character traits of his writer. The same was probably reflected in the letters of the "realists."

In this chapter, we will briefly present the deeds of two characters, in whose writings and reports we will find important details for us. All these figures went on difficult journeys back then—even before Marco Polo. John del Plano Carpini lived for about 70 years (ca. 1182–1252) and, it is believed, was the first of the Europeans to go to explore the completely unfamiliar and vast empire of the Mongols. The task for this journey was expressed in 1244 by Pope Innocent IV (Fig. 14.6). Prior to that, Plano Carpini had been in a rather high position of the so-called custode of the Franciscan Order in Saxony. He went to Mongolia accompanied by another monk, Benedict the Pole, and through the Czech Republic, Poland, Kiev, the lower reaches of the Don and Volga, Khorezm, Semirechye, the depression of Lake Alakol. So he and his companion reached the area of the location of the main headquarters of the Mongols in the upper reaches of the Orkhon River. In 1246, Carpini visited Sarai, where he met Batu, then a nomadic headquarters near Karakorum, where he attended a reception at the newly elected great Khan Guyuk, and in 1247 he returned safely to Rome. His main "report" on the accomplished journey is undoubtedly the book Historia Mongalorum quos nos Tartaros appellamus (The History of the Mongols, which we call Tatars). The book mainly contains systematized ethnographic information about the nomadic and warlike people of the Eurasian Steppe Belt, completely unfamiliar to Europeans.

Guillaume/William de Rubruk (c. 1220–1293), a Flemish monk and also a Franciscan, went to Mongolia already on the highest commission of King Louis IX in 1253–1255 (Fig. 14.7). His report on the work at the behest of the monarch was "Itinerarium fratris Willielmi de Rubruquis de ordine fratrum Minorum, Galli, Anno gratia 1253. ad partes Orientales," and in Russian, much more briefly, "Journey to the Eastern countries." Apparently, the most

Figure 14.6. Pope Innocent IV sends Dominicans and Franciscans to the Tatars (in black robes are Franciscans).

Figure 14.7. The upper part of the image: de Rubruk hands King Louis IX a report on a trip to the Mongols/Tatars; the lower part: de Rubruk with his partner (Illustration from the book in Cambridge, Corpus Christi; William of Rubruck Corpus Christi MS 066A, 67).

Figure 14.8. The capital of Mongolia Karakorum; palace the great Khan Ogedei and the silver tree/fountain, built by master Wilhelm Boucher (reconstruction).

important center of his survey of unknown Mongolia was the capital of this vast steppe empire: Karakorum (Fig. 14.8).

The works of Carpini and Rubruk differed sharply from the book of Marco Polo in their real grounding. Therefore, my intended reader should immediately and without hesitation willingly trust the authorship of the "realists" and what and how they describe everything that happened to them. Some borrowings or echoes of phantasms can also be encountered in their texts, but such anomalies are very easy to separate from the realities that form the basis of their reports. The beginning of the envoys' visit was almost always associated with the insistent demands of local ministers for gifts specifically for them, but for those whom they serve, other, of course, much more valuable gifts are needed.

Such rather vile situations, as the envoys themselves assessed, were reflected in the writings of both Plano Carpini and de Rubruk. But we will limit ourselves here to just some examples that quite clearly reflect the atmosphere into which Western envoys had to plunge, and persons of fairly high rank. First, upon arrival at a certain place, about which the ambassador had not really heard, it was necessary to find representatives of those services that were supposed to take you to the rulers. And initially we will turn to an example from the narrative of Rubruk:

> On the advice of merchants, I brought with me from Constantinople, as gifts to the chief chiefs, fruits, nutmeg wine and delicious biscuits (biscoctum), so that the way was more accessible to me, since they do not look at anyone with a merciful eye if he comes empty-handed. (Chapter 1)

But all this turned out to be not easy at all:

> It was our guide who wanted me to come to every boss with a gift, but there were not enough funds for this. For every day there were eight of us who ate bread, not counting those who came by chance, who all wanted to eat with us. (Chapter 15)

However, it has also been more difficult:

> So, when the servant sprinkles in this way on the four sides of the world, he returns to the house; and two servants with two bowls and as many dishes stand ready to take the drink to the master and his wife, who is sitting on the bed next to him, but higher. And if the master has a lot of wives, then the one with whom he sleeps at night sits next to him during the day, and everyone else should come to that house on that day, and there is a meeting there on that day, and the gifts brought are put into the treasuries of this lady. (Chapter 3)

But Plano Carpini also has similar descriptions of troubles with gifts:

> Moreover, both princes and other persons, both noble and non-noble, beg for a lot of gifts from them, and if they do not receive them, they value the

ambassadors low—moreover, they consider them as if for nothing; and if the ambassadors are sent by great people, they do not want to take it's a modest gift from them, but they say: "You come from a great man, and you give so little?" As a result, they do not consider it worthy to take, and if the ambassadors want to do their business well, then they should be given more. . . . Therefore, we also had to give away most of the things that pious people gave us for food, when we needed gifts. And it should also be known that everything is so in the emperor's hand that no one dares to say, "this is mine or his," but everything belongs to the emperor, that is, property, pack animals and people, and on this occasion the emperor's decree has even appeared recently. (§3.2)

Or somewhat different the situation:

> We were told that if we took the horses we had to Tartary, they could all die, because there were deep snows, and they did not know how to dig grass under the snow with their hooves, like the horses of the Tatars, and it was impossible to find them anything else to eat, because the Tatars did not have any straw, no hay, no fodder. Therefore, after the meeting, we decided to leave them there with two servants who were supposed to guard them. Consequently, we had to give gifts to the thousandth in order to earn his mercy in order to get ourselves carts and escorts. (§ I.V)

> Note that the situation with "gifts" has been perfectly preserved up to the present day, only now all these actions are called bribes, and the situation as a whole is corruption.

Now it will be interesting to get acquainted with a much more strange situation, when the ambassadors and their generally quite small retinue very often suffered from hunger, although it was as if the hosts who received them had to provide the envoys with good food. Again, let's continue the example with the text from Plano Carpini:

> And when we returned, we stayed for a few days and again returned to him [the local ruler]; together with him we stayed safely for a month, amid such hunger and thirst that we could barely live, since the food given out for four was barely enough for one, and we could not find anything to buy, since the market was very far away. (§2.8)

But for de Rubruk, the theme of a hungry stay literally flooded with gold—following, of course, the assurances of Marco Polo—made the East look even more disgusting:

> We were given only cow's milk for food, very sour and smelly. We were running out of wine; the horses had muddied the water so that it was not fit to drink. If we didn't have the crackers that we had, and the grace of God, we probably would have died of hunger. (Chapter 13)

On the eve of the Assumption, our acolyte arrived at the court of Sartach, and the next day the Nestorian priests, dressed in our robes, were in front of Sartach. Meanwhile, we were taken to another owner, who had to take care of the room, food and horses for us. But since we didn't have anything to give him, he did everything badly. And we rode with Batu, descending near the Volga, for 5 weeks. Sometimes my friend felt so hungry that he told me almost with tears: "It seems to me that I will never get to eat." The market always follows the Batu courtyard, but this bazaar was so far away from us that we could not go there. For we had to walk, for lack of horses. (Chapter 22)

And on the way between him and his father, we felt a strong fear: it was the Russians, Hungarians and Alans, their slaves (Tatars?), whose number they have is very large, gather at once for 20 or 30 people, run out at night with quivers and bows and kill anyone they catch at night . . . .Our guide was very afraid of such a meeting. During this journey, we would have starved to death if we had not taken some crackers with us. (Chapter 20)

Now try to compare the texts of the "Book of Miracles . . ." Marco Polo and these two Western envoys, the realists. The result of the comparison may probably please supporters of a negative assessment of Marco's work, but upset those who so zealously defended his exceptional truthfulness. Is it possible to imagine that Marco in his Book of Miracles would have screamed about the lack of food and the hunger of the messengers? Or about the demands of worthy gifts not only to rulers, but also to servants of lower ranks, on whom petitioners always depend. This problem is essentially eternal, up to the present day. And in the messengers, even very prominent ones, the servants of the local authorities identify very quickly and certainly real petitioners, from whom you can definitely get something.

In this discussion we have touched upon on the problem of "counter-examination"—a very sluggish dispute, and lasting for two millennia—requires, of course, completion. Probably the most striking evidence of the achievements of both the "storytellers"—that is, Marco Polo, and the "realists" Plano Carpini and Guillaume de Rubruk—are maps on which their routes are outlined in sufficient detail. True, their paths are not mapped by them, but according to the descriptions provided by them. The "realists" own the maps in Figs. 14.9 and 14.10; and the "storyteller" Marco presented his route in Fig. 14.11.

The differences even in the maps of Plano Carpini and de Rubruk are quite obvious, although in general their routes should most likely be recognized as fundamentally similar. Certain differences are due, of course, to the different length of the routes—after all, de Rubruk's journey turned out to be longer. Otherwise, the general outline of their movements seems to be quite similar; apparently, their descriptions also coincided in many details.

You get a completely different impression from the Marco Polo card. And if the northern (land) half of his proposed route can somehow cause an assessment at least somewhat positive, then sailing across the oceans—the Pacific and Indian, when our hero managed to circumnavigate the whole of Eurasia on some imaginary ship from the south, is another matter. Moreover, this story can claim Marco Polo's champion place among his countless chimeras. Here we can

Figure 14.9. Plano Carpini and the map of his journey.

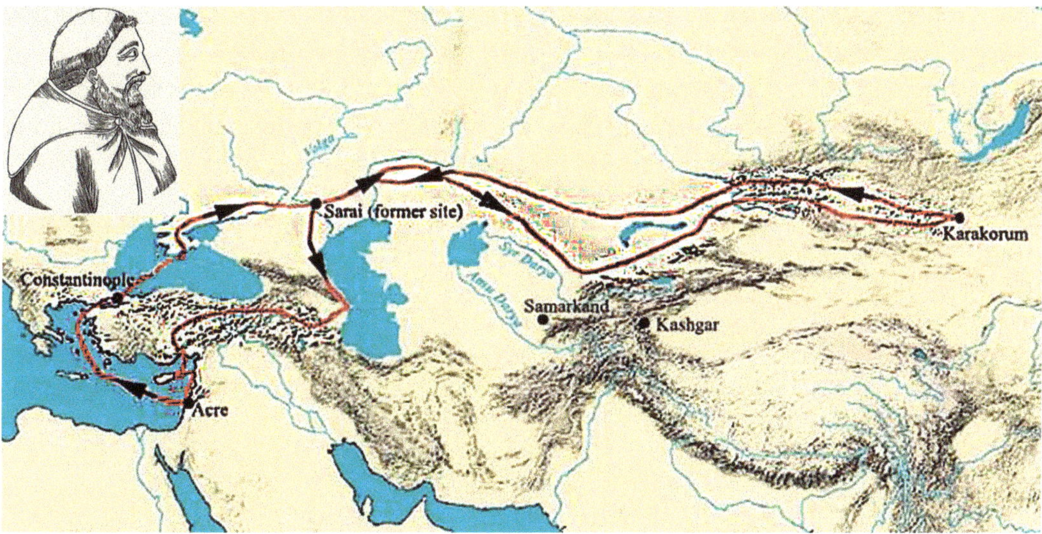

Figure 14.10. Guillaume/William de Rubruk and the map of his journey.

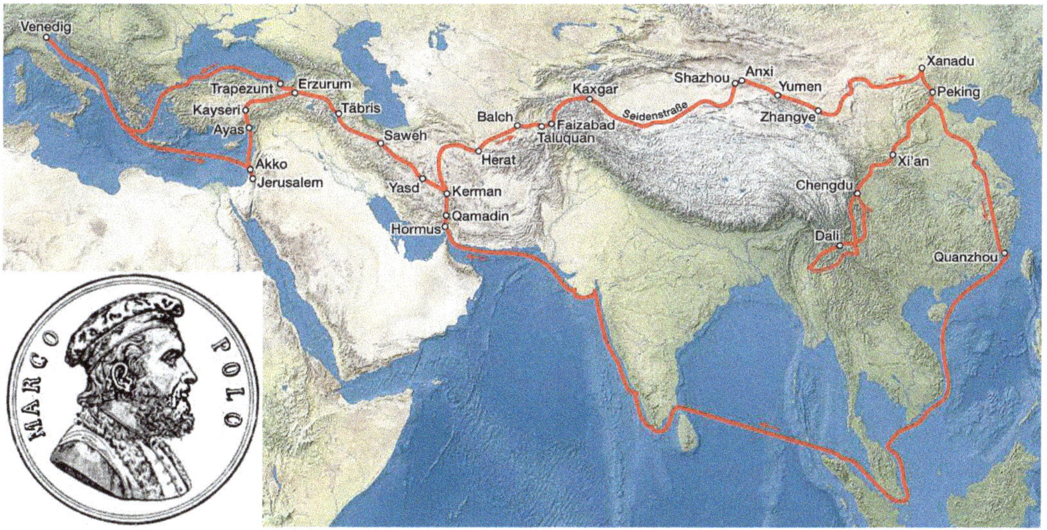

Figure 14.11. Marco Polo and the map of the journeys reproduced by him on paper.

recall once again how I. P. Magidovich, one of the largest Russian experts on the problems of ancient and medieval navigation, characterized even the northern part of the proposed route of Polo:

> The commentators of Marco Polo's Book tried with the greatest effort to estab-
> lish at least an approximate route of the senior Polos from the Volga River to
> the headquarters of the great Khan. . . . Commentators could only operate with
> indirect evidence scattered in the Book.

For this reason, it seems to me that we should put an end to this discussion, which may be quite strange and provoked only by the author of the book. In addition—and quite understandably—we do not have even the most modest participation of the disputants; even the later traveler, Marco Polo, did not react to the achievements of his predecessors.

The author invites readers of this book from a two-thousand-year-old controversy to rise in the next chapter to much brighter heights, where contrasts are incomparably more significant and fateful, and which the author finds eternal and all-encompassing.

Chapter 15

Us and Them: The Eternal and All-Encompassing Dichotomy

Let us embark on an exploration of the perpetual and all-encompassing dichotomy inherent in our daily life—the dual opposition of *us* versus *them*. Primarily, *us* serves as a benchmark against all things measured, juxtaposed with *them* in constant comparison. This dynamic manifests whether we consider our immediate neighbors or distant nations, various groups delineated by belief system or socioeconomic status, or even by racial or social origins. Furthermore, an individual's affiliation with a particular group can be incredibly fluid, akin to sliding or leaping between different groups.

Furthermore, the catalog of these opposing duplicates can be extensive, a challenge encountered by those who seek to construct their foundations upon psychological frameworks. Nonetheless, we endeavor to illuminate the structure of this concept, albeit with a certain degree of uncertainty regarding its success. Our journey commences with an examination of the foundational works within the field of psychology.

The human psyche is social, says Boris F. Porshnev in his book "Social Psychology and History," because it is largely conditioned by the socio-historical environment. Porshnev passionately defends his ideas about key concepts in social psychology, rooted in the Marxist-Leninist perspective on society's structure. A focal point of his argument is the psychological dynamic of "us and them," which he explores extensively in the book's second chapter. This dynamic, he argues, has driven human interactions since the dawn of civilization, shaping much of our history.

While not in complete agreement with Porshnev's views, the author intends to build upon his framework, particularly focusing on the "us and them" dynamic in Chapters 5 and 6. However, before delving into those chapters, let's first revisit the foundational works of psychology.

15.1. Carl Jung and the Role of the Subconscious

When we consider the concept of *us*, we understand it as those who are native or closely related, evoking a sense of familiarity and proximity, while *they* are considered foreign, perhaps even malevolent and dangerous. These habitual cognitive patterns most likely stem from the mechanisms of our subconscious which underlay our thought processes. The eminent Swiss psychoanalyst Carl Gustav Jung (Fig. 15.1) notably upholds the importance of the subconscious.

Those who deny the existence of the subconscious state that our current knowledge of the psyche is exhaustive. And such an opinion is definitely as false as the assumption that we know absolutely everything about the universe. Our psyche is a part of the world around us, and its mystery is also boundless. Therefore, we cannot define either one or the other. We can only say that we believe in their existence, and describe their functioning as far as possible. (Jung, 9)[1]

Jung's most important contribution to the understanding of psychological processes is the subconscious, which appears, according to Freeman, as "not just as a place of honor for repressed desires (like Freud's), but as a whole world—as alive and real as the individual's consciousness, the world of his mind, and even infinitely wider and richer than the latter" (Freeman, 3). Symbols are the language of the subconscious, while dreams are its means of communication. In Jung's formulation, the subconscious is the "great guide, friend, and advisor of consciousness." (ibid)

Figure 15.1. Psychologist Carl Gustav Jung (1975–1961); he was very tall, which is why the author preferred to choose his photo even without colleagues.

15.2. Unlimited Self-Admiration, or Social Narcissism

We have already encountered profound self-admiration in the case of Marco Polo in the previous chapter. However, our focus now shifts towards examples of group narcissism rather than individual manifestations in order to show the dangers of the *us* versus *them* dichotomy and how it plays out in the historical process.

---

[1] The author has already tried to look under the veil of this complex topic. In his book *"Homo Cultures: The Key Facets,"* and in the chapter "Homo Cultures through the prism of psychological and philosophical constructions" all the issues that we will discuss below have already been touched upon to one degree or another.

Let us begin by examining the National Socialist anthem dedicated to their martyr Horst Wessel (Fig. 15.2), a poet and active member of the Nazi Party who was murdered by a Communist assailant in 1930. His death served as a poignant opportunity for the Nazis to elevate him to the status of a revered figure in Hitler's Third Reich. The anthem goes as follows:

Die Fahne hoch! Die Reihen fest geschlossen!
SA marschiert mit ruhig festem Schritt.
Kamraden, die Rotfront und Reaktion erschossen,
Marschieren im Geist in unseren Reihen mit.

[Raise the flag high! The ranks tightly closed!
SA marches with steady and firm step.
Comrades, shot down by the Red Front and reaction,
March in spirit within our ranks.]

The revelations of Alfred Rosenberg (1892–1946) are also particularly noteworthy. Rosenberg was a Baltic-German Nazi theorist and a prominent racial supremacist ideologue who was responsible for promoting the notion of Aryan superiority. In his 1930 book *Der Mythus des 20. Jahrhunderts* (The Myth of the 20th Century) (Fig. 15.3), which garnered significant national praise, Rosenberg argued that

history and the task of the future no longer mean a struggle of class against class [as interpreted by Marxists] or a conflict between one church dogma and another, but a settlement between blood and blood, race and race, People and People. And this means: the struggle of spiritual values against each other. (Der Mythus, Introduction)

Thus racial history is the natural history and mysticism of the soul at the same time; and the history of the religion of blood, on the contrary, is the great world history of the rise and fall of peoples, their heroes and thinkers, their inventors and artists. (Der Mythus, 5)

Today, a new faith is awakening—the Myth of blood; the belief that protecting blood is also protecting the divine nature of man as a whole. This is a faith shining with the brightest knowledge that the Nordic blood represents the mystery that defeated and replaced the old sacraments. (Der Mythus, 65)

Rosenberg's racist ideas were powerful precisely because he accentuated the divide between *us* and *them,* and ultimately helped shape the ardent distrust and condemnation of the Other, which in his formulation, was anyone outside of the "superior" northern Aryan race. He continues:

Figure 15.2. A leaflet with a reproduction of a photo of Horst Wessel and with the text of the song "Die Fahne hoch," almost sacred for Hitler's Third Reich, against the background of a Nazi demonstration.

Figure 15.3. Hitler's leader of the Third Reich and his ideologist Alfred Rosenberg (1892–1946), executed in 1946, and the cover of Rosenberg's main work: the book "The Myth of the 20th century."

> It has always been an idea of honor, spiritual and mental. But the idea of honor, as well as its bodily representatives, was involved in the war of soul and spirit against values represented by alien races or the product of racial chaos. (Der Mythus, 84)

The concept of individual supremacy within the collective *we*, as articulated in Rosenberg's writings, exemplifies what can be termed the "narcissism syndrome." This syndrome reflects a longstanding pattern that has persisted throughout the course of human civilization.

15.3. The Dangers of Narcissus Syndrome

The narcissist syndrome as revealed in Rosenberg's writings appears to underscore an enduring and pervasive quality that suggests its eternal nature and inherent aspect of humanity's existence. The Hitler regime, for example, capitalized on this idea on a social and individual level and constructed a nightmarish edifice upon this foundation. However, they were by no means the first to engage in such reprehensible practices. The belief in the superiority of the Aryan race and its special role in shaping world history captivated an unprecedented number of hearts and minds, suggesting a potential for this pattern to perpetuate endlessly.

To compare the assessments of group achievements and setbacks invariably unfold against the backdrop of a pervasive phenomenon known as narcissism or the Narcissus syndrome. This syndrome represents another enduring and universal archetype within human culture at its most expansive. It is therefore crucial to dive into its specifics and manifestations in order to mitigate any potential discrepancies, beginning with an examination of its pioneers and main thinkers.

While Sigmund Freud (1856–1939; Fig. 15.4) is credited with introducing the term "narcissism" into our cultural and conceptual framework, it was the German psychologist Paul Näcke (1851–1913; Fig. 15.5) who was its true pioneer. In his lectures "Libido Theory and Narcissism" (1914) compiled in the *Das Ich und das Es* (often translated as *I and It* or *The Ego and the Id*, 1923), Freud referenced Näcke in conjunction with the term narcissism, acknowledging his significant contribution to the study of psychopathology: "The name for this placement of libido—narcissism—we borrowed from the described P. Näcke (1899) perversions in which an adult individual gives his own body all the tenderness usually shown to an extraneous sexual object."

While Freud embraced the idea of narcissism and attempted to expand its scope, his focus remained largely within the confines of sexology and libido, not extending to the individual or social aspects of this disorder. However, his direction gained widespread attention from psychologists Paul Näcke was a compelling figure, albeit less well-known in scientific circles. Born in St. Petersburg to a German father and French mother, he spent his early childhood and adult life in Germany, emerging as a polyglot and talented student in the fields of medicine, philosophy, history, archeology, and architecture. Näcke infamously advocated for the castration and sterilization of "degenerate criminals and degenerates in general" and held views regarding the unsuitability of integrating Black individuals into American society, with notable attention paid to the manifestations of narcissism—both individual and collective—by the renowned German psychologist Erich Fromm (Fig. 15.6). Many regarded Fromm as Freud's successor and a notable figure in the realm of neo-Freudianism. Fromm himself, however, rejected this characterization despite acknowledging Freud's important contribution to the field of psychology.

Fromm's assessment of narcissism underscores its pursuit of satisfaction and social approval through a collective ideology of superiority and inferiority, often leading to divisive consequences. The narcissistic obsession with superiority can manifest across various cultures and racial divides, with the group designated as *us* holding an unconditional right to determine and evaluate those categorized as *them*. This right is entrenched in the narcissism of the collective that considers itself unassailable, often exemplified by figures such as priests and shamans who assume the role of mediators of a higher power, dictating ideological assessments without room for dissent. Fromm's insights into this phenomenon allow us to study this specific dichotomy of *us/we* versus *them/they* and how it is intimately intertwined with the concept of narcissism. This binary framework extends beyond simple cultural distinctions to encompass broader divisions among human races.

Demonstrating a group's superiority through military victories has been a historically reliable method of consolidating power and garnering respect, as exemplified by Titus Livy's book *History of Rome* which glorifies military conquest as a hallmark of greatness throughout history:

> It is excusable in antiquity, interfering with the human with the divine, to magnify the beginnings of cities; and if it is permissible for some people to sanctify

their origin and elevate it to the gods, then the military glory of the Roman people is such that, if he called Mars himself his ancestor and the father of his ancestor, the human tribes will bear it with the same submission with which demolish the power of Rome. (Titus Livy, vol. 1)

Figure 15.4. The famous psychologist Sigmund Freud (1856–1939).

Figure 15.5. German psychologist Paul Näcke (1851–1913).

Figure 15.6. German psychologist Erich Fromm (1900–1980).

However, victories as a representation of greatness are not always as straightforward as they seem. Instead, it is often easier and more reliable to assert superiority through a multitude of material symbols such as gold artifacts, monumental architecture, opulent burials, extravagant attire, and so on. This allows for the *us/we* group to separate themselves from the *them/they* group in displaying overt shows of power.

But here is another case: if the *we* group does not have representative material evidence of greatness, or such materials look rather poor, how can one justify one's own greatness? It is precisely in such cases that, instead of material greatness, the sphere of spiritual comprehension often comes to the fore. And here again we have vivid examples from the pre-great-war life of Hitler's Germany, and the Soviet Union of Joseph Stalin's time with the famous "March of the enthusiasts":

> *And here it is, this country:*
> *Prosper, country of heroes,*
> *Country of dreamers, country of scientists!*
>
> *And in our amazing country:*
> *Whether you bend down at a machine,*
> *Whether you cut into a rock —*
> *The wonderful dream, not yet clear,*
> *Already calls you forward.*
>
> *And therefore there can be no doubt that*
> *There are no obstacles for us,*
> *Neither on sea nor on dry land,*
> *Neither ice nor clouds scare us.*
> *The flame of our soul, the banner of our country,*
> *We will carry through worlds and ages.*
>
> (1940, music by Isaak Dunaevskii, lyrics by A. D'Aktil')

True, it seems to me that such loud words are unlikely to frighten even close and comparatively weak neighbors. Perhaps such words will in some way contribute to the growth of spiritual strength in the ranks of the native *we* category, but nothing more.

Finally, we should probably turn to the last question in this series: to the definition of the category "*they*": who are they? For Hitler's Germany, we have already touched on this question, and the answer here was quite unambiguous: *they* are the Jews. This is how the main ideologist of Hitler's fascism, Alfred Rosenberg, defined this situation. And nearly all of the actions of the Hitler team—especially during the war years—fully confirmed this answer.

For the Russian communists of Stalin's time, the answer to a similar question—who are "they"?—would have been more complicated. We can, of course, rely on the "March of the Enthusiasts" and on the absolutely unambiguous words of V. Lebedev-Kumach in his 1937 song "Beat the Drum" calling for courageous battles: "Whoever is not with us is a coward and an enemy!"

It seems that all who are not with us are classified as "they," and in this case they constitute a countless, hostile, and all-encompassing world cohort. But is this really so? Hardly. In any event, the case in communist Russia was not very similar to Germany of the 30s–40s. And the fact that both authors of "March of the Enthusiasts" (Isaak Dunaevskii and A. D'Aktil', the latter born as Noson Frenkel) are Jewish shows the validity of this doubt.

Nonetheless, a society or culture's confidence in their exclusivity serves as a necessary or highly desirable psychological foundation, but it does not negate the challenges posed by issues of racial hierarchy.

The dual opposition of *us/we* versus *them/they* pervades numerous works throughout history, from Freud's *Das Ich und das Es* to B. F. Porshnev's *Social Psychology and History* (1970). Porshnev's example of this binary opposition traces the theme of evolution and highlights shifts from individualistic to collective identities. While parallels exist in both these works' titles, their contents diverge significantly. Poshnev's book which was published during the era of obligatory adherence to Marxist-Leninist ideology underscores the socio-political dimensions of group dynamics and ideological conformity.

Chapter 16

Great Geographical Discoveries, the Great Migration, and the Rise of Slavery

16.1. Romanus Pontifex of Pope Nicholas V

If one were to write a book about the history of slavery, it would be fitting to cite the formidable bull "Romanus Pontifex" written by Pope Nicholas V (Fig. 16.1) on January 8, 1454, as well as a slightly earlier one bull "Dum Diversas". Although verbose for a chapter introduction, his words exemplify the greed and domination inherent in subjugating the will of another person:

> We are carefully weighing . . . conceded to King [of Portugal] Alfonso and his successors . . . [the right] to attack, seek, capture, conquer, subdue all Saracens, pagans and other enemies of [Christ], wherever they may be . . . and to reduce their people to eternal slavery and to appropriate to themselves, to themselves and to their successors the right of a Kingdom, a duchy, a county, a principality, possessions, possessions and property.

At first glance, this excerpt might suggest the profoundly problematic view that slavery originated from divine decree. However, it is common knowledge that slavery was a common practice before this proclamation, tracing back to ancient times that predate even the Roman Empire. However, the matter is clear-cut. Consider the "Sermon on the Mount," delivered by Jesus Christ, and its concluding words:

> You have heard that it is said, "Love your neighbor and hate your enemy." But I say to you: love your enemies, bless those who curse you, do good to those who hate you, and pray for those who offend you and persecute you, so that you may be sons of your Heavenly Father, for He commands His sun to rise over the evil and the good, and sends rain on the righteous and the unrighteous. For if you love those who love you, what is your reward? Don't publicans do the same? And if you greet only your brothers, what are you doing in particular? Don't the pagans do the same? So be perfect, just as your heavenly Father is perfect . . .

This sermon holds paramount significance in Christianity and is often considered a testament to Christian life and the Kingdom of God. Spanish Bishop Bartolome de Las Casas (1484–1566) (Fig. 16.2a, b), for example, embodied these concepts when he proudly opposed the atrocities of the Spanish colonization of the Americas and particularly the actions of conquistador Francisco Hernández de Córdoba (1467–1517). His courageous writings, such as those published in his 1552 book (Figs. 16.3, 16.4a–c), condemn the violence and oppression inflicted upon indigenous peoples and align with the values espoused in Christ's sermon. In his words:

> However, while he [Hernandez] was about to go to Spain, God decided to move him to a better world so that he would give an account of other, even greater offenses that he himself had inflicted on the Indians of Cuba; after all, they served him, and he sucked blood from them and equipped his ships with this blood to capture innocent people who lived peacefully in their lands; but his most grievous sin, for which, no doubt, he held a special answer before God's judgment, and for which, by the way, he was asked before his death, were those ungodly deeds that he committed in the lands of Yucatan, for he killed and plunged into hell fire the souls of many Indians; and after all if he had left foreign lands, since the owners of the land demanded it, then many sins would have been forgiven him. In fact, did Francisco Hernandez sow peace, and goodness, and mercy, and justice, and friendliness in that newly discovered land of Yucatan? Did he come there as a kind and welcome neighbor? What should these Indians think of us and what respect could they have for our Christian religion if people who called themselves Christians caused them so much harm and trouble just because they did not want to tolerate aliens in their lands, considering them suspicious and dangerous people, and they had every reason to think that this visit would bring them nothing but misfortune, because it happened everywhere the Spaniards appeared. So, our righteous friend Francisco Hernandez, who in piety was quite equal to most other Spaniards, died. (Book 3, Chapter 98, 103)

The mystery underlying this quote is why Pope Nicholas V issued such a rigid and contradictory decree, particularly one so at odds with Jesus Christ's teachings. Speculation on this matter is challenging and parallels with the atrocities committed during the military-sanctioned Crusade campaigns. Pope Nicholas V's bull may have been influenced by the failures of subsequent crusade-like campaigns (like those against the Hussites (1420–1434) or the campaigns against the Ottomans that ended in the Crusaders' defeat in 1444). These actions influenced a shift in focus towards maritime exploration, as exemplified by Portugal's burgeoning naval ambitions.

Inevitably, colonizers regarded themselves as *us* and the colonized as *them*, a theme explored in the previous chapter. Las Casas, however, with his nuanced view of power, often challenged those in power. This reversal of roles—where *we* became *them*—presents a chilling paradigm. However, it underscores the complex interplay of power dynamics and ideological justifications throughout history.

Figure 16.1. Pope Nicholas V (leftward) and Portuguese King Afonso V (1432–1481).

Figure 16.2a. Spanish Bishop Bartolome de las Casas (1484–1566).

Figure 16.2b. Monument to Bartolome de las Casas in Seville; Indian figures are depicted to the right of the bishop, and to the left is the conquistador.

16.2. Amerigo Vespucci and the Great Geographical Discoveries

Major discoveries of the magnitude made by Amerigo Vespucci (1421–1520) typically occurred within the approximate framework of a century and a half. The beginning of this era can be attributed to Christopher Columbus (1451–1506) who stumbled upon the Americas in his quest to reach India. The conclusion of this period of "acquisition" of Australia and New Zealand by Europeans led by Dutch explorer Abel Tasman (1603–1659) in 1642. Although

Figure 16.3. The cover of the first edition of the 1552 book by the Spanish bishop Bartolome de las Casas: "The shortest message about the destruction of the Indies"; in the current Russian translation, the title of the book is "The History of the Indies," which is hardly true.

Figure 16.4a. Colonialists cut off the hands of Indians. Illustration from the book of Bishop de las Casas.

Figure 16.4b. Execution of the leader of an Indian tribe for refusing to convert to the Christian faith. Illustration from the book of Bishop de las Casas.

Figure 16.4c. Mass execution of Indians by colonizers. The gallows even got the name: "In memory of Jesus Christ and his 12 apostles"; there are 12 hanged people. Illustration from the book of Bishop de las Casas.

European discoveries extended beyond the timeframe provided, we will provide a brief overview of the primary regions on our planet where pivotal events of human history have occurred. These events were not only connected to the four main continents but also to significant islands within the Atlantic, Pacific, and Indian Oceans. While some Arctic Oceans, such as the Svalbard archipelago, could be mentioned, they pale in comparison to the importance of the southern oceanic regions during the time of the "Great Discoveries" (Figs. 16.5–16.7).

Figure 16.5. Christopher Columbus's flotilla of three ships: "Santa Maria," "Ninya," and "Pinta" on the way.

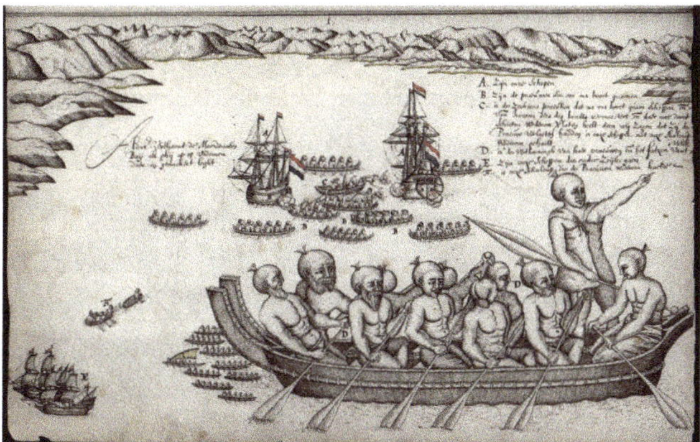

Figure 16.6. Maori canoes and ships of Abel Tasman in the Bay of Murderers (New Zealand, drawing of the participant of the voyage of Isaac Gilsemans, 1642).

Figure 16.7. Amerigo Vespucci (Richard Cavendish. History Today, 54, issue 3, 2004).

It is important to mention the naming of America after Amerigo Vespucci, who played a significant role in this process (Fig. 16.7). However, the naming of the continent was largely influenced by other factors, notably by the renowned German cartographer Martin Waldseemuler (1470–1520). Austrian writer Stefan Zweig (Fig. 16.9) describes these events in his book *Amerigo: A Story About a Historical Error* (2021). Scholars often express satisfaction and contempt in discussing Vespucci and his connection to the naming of the American continent, most likely because many believe that it should have been named after Columbus who "discovered" it first. However, it was Vespucci who was the first to identify a "new world" outside of Europe, while Columbus reached the western hemisphere because of a navigation error. The question remains whether Vespucci can be considered a true "hero" or something entirely different. Azerbaijani geographer Ramiz Deniz, for example, argues that "The name of the New World America [is] one of the greatest historical injustices and believes that all scientists of the world should take decisive measures to correct this mistake." In a new work presented to the public, Ramiz Deniz creates a unique opportunity to re-examine Vespucci's biography, his private and public activities, as well as the deep mark he left on the pages of history. I am convinced that in order to form a common opinion in this matter, new research, new literature will be required. And these stories were perhaps most vividly described by the Austrian famous writer Stefan Zweig (Fig. 16.9) in his book "*Amerigo. The Tale of a Historical Mistake.*" It is indeed surprising that the naming of America is considered one of the greatest historical injustices, so much so, that there is a collective call amongst scholars worldwide to rectify this error. In concluding the narrative of Amerigo Vespucci, it seems that his role, as suggested by Deniz, may be linked to other significant historical injustices. Indeed, it pales in comparison

Figure 16.8. Amerigo Vespucci awakens sleeping America (Engraving by Theodor Halle based on a drawing by Johannes Stradanus, 1630).

Figure 16.9. The famous Austrian writer Stefan Zweig
(1881–1942).

to one of the most catastrophic injustices of all—slavery, which surged to great heights after the era of "Great Discoveries" and the Great Sixth Intercontinental Migrations of Peoples.

16.3. *Us* and *Them* in the Context of Globalization and Colonization

During the modern era of globalization, European colonization proliferated at an exponential pace in the uncharted territories of America and Australia. Notably, the colonization of Africa, which had served as the cradle of the Homo family for thousands of years, was also incorporated into European colonizing ambitions. In the eyes of the ruthless colonizers, those they subjugated appeared primitive and backward, a misconception that overshadowed their historical significance as the progenitors of humanity.

Thus, a polarizing dichotomy emerged that would underscore the new globalized world—the binary between *us* and *them*. The colonizers who associated themselves with *us* maintained the belief that they were the harbingers of truth to the rest of the "uncivilized" world, armed both ideologically and militarily, while *they* were tasked with assimilating and internalizing this truth diligently. The psychological opposition underscored the enduring and pervasive nature of this duality within human society.

Guided by the tenets of the Abrahamic faith, particularly Christianity, the colonizers asserted their ideology as evident truth and themselves as its custodians. The missionary endeavors of figures like Bishop de las Casas, though perhaps numerically modest and unsuspecting at the time, reinforced the European narrative of superiority and divine mandate. It was not long before *they* were enslaved and used as commodities, ultimately becoming the economic backbone of a newly globalized world, with Africa emerging as the primary reservoir for human "wealth" in the form of enslaved labor. Furthermore, the comparison between indigenous Native Americans and enslaved Africans favored the latter, accentuating their economic benefit (Figs. 16.10–16.12). However, despite the apparent unanimity sought by the colonizers, divergent interpretations of religious and ideological tenets led to internal discord. Discrepancies in

Figure 16.10. Portuguese galliot with slaves. Illustration from the travel diary of Jan Huygen van Linschoten, 1596.

the true understanding of divine nature and its manifestation fostered a complex hierarchy in the *us* versus *them* paradigm, fueling a relentless dialectic between conformity and hegemony.

In the Russian context, Russian navigators played a marginal role in global discoveries, as their expeditions were largely confined to the northern reaches of Eurasia. While Russians made great headway along the Arctic coastlines, they did not venture into the western hemisphere, thereby paling in comparison to their European counterparts.

16.4. Russia amidst Globalization and Psychological Counter-Doubles

The history of slavery in Russia is intricately linked to the long-standing institution of serfdom, which continues to generate significant debate amongst scholars regarding its classification as a form of slavery.[1] In our analysis, serfdom unequivocally constitutes a form of slavery, comparable to the classic slave-owning societies, albeit distinguished by its own variations and nuances. The question remains, however, whether Russian slavery connects with the slavery that flourished in Europe. For example, how does one explain chronological coincidences such

---

[1] It's important to mention the recent books written by Boris Yuryevich Kerzhentsev (Tarasov), a major expert on this subject (for example, see *The History of Popular Slavery* [2020]; *Cursed Time: Russia in the XVII–XVIII centuries* [2013] and others). In his formulation, the history of serfdom in the Russian Empire is inextricably linked to Russia's fundamental identity—in his words, Russia is serfdom. There is little need to discuss how Russian literary classics or prominent artists addressed this grim aspect of history; the well-documented nature of these topics likely requires no further elaboration.

Figure 16.11. Transportation of African slaves to America. In comparison with the Portuguese slave ship (see previous figure), there is already a significantly larger number of "passengers."

Figure 16.12. Transportation of African slaves to America; in comparison with the Portuguese slave ship (cf. Figs. 16.10, 16.11), as well as with the previous one, a much larger number of unwitting "passengers" are visible.

as the 1497 decree by Tsar/Grand Duke Ivan III, often regarded as the true inception of Russian slavery, as documented in the *Sudebnik*? For example:

> It is known that until the XVI century, a peasant could leave the landowner's estate within a week before St. George's Day—November 26—and a week after it. However, everything was changed by Tsar Fyodor Ioannovich (1557–1598), who, after listening to his brother-in-law Boris Godunov, forbade peasants from passing from one landowner to another even on November 26 for the time of writing scribal books. In those days, it shocked people.

> By the way, Fyodor Ioannovich issued a decree in 1597, according to which the period of investigation of fugitive peasants was five years. If during this period the landowner did not find the fugitive, then the latter was assigned to the new owner.

> In 1649, Tsar Alexei Mikhailovich (nicknamed the Quietest, 1629–1676) issued a Council Code, according to which an unlimited period of investigation of fugitive peasants was announced. Plus, even the debt-free peasants could not now change their place of residence. At the same time, the tsar was quiet, and at about the same time, the famous church reform was carried out, which subsequently led to a split in the Russian Orthodox Church. (According to the text of Yevgenia Sakhno: https://scientificrussia.ru/articles/ kak-aleksandr-ii-otmenil-krepostnoe-pravo)

Thus, Russian slavery began in 1497, five years after Christopher Columbus reached America's shores and one year before Vasco de Gama's arrival. This brief period marked not only the beginning of the Age of Great Geographical Discoveries but also the beginning of conquests that spurred the rise of global slavery. Remarkably, a few centuries later, the Russian abolition of serfdom in 1861 nearly coincided with the end of slavery in the United States in 1865. How can we explain these synchronous dates—1497 as the beginning and 1861 as the end—within a global context? Despite numerous publications, there are few convincing explanations for this phenomenon. This intriguing question, although beyond the scope of this book, certainly merits further attention. We can only hope that future research will provide answers.

Serfdom and slavery remain a persistent thorn in Russia's side, particularly when it comes to its distinctive characteristics compared to slavery in the US and Europe. In the Euro-American context, the individuals were clearly defined by their ethnicity, skin color, and differing religious beliefs, which according to the colonizers, justified their subjugation. In Russia, however, the situation was markedly different: those subjugated to slavery were ethnically Russian, white-skinned, shared the same land, and were devout Orthodox Christians, like their enslavers. How then do we explain their oppression?

Another pressing question emerges for those contemplating this topic: how can one comprehend the stark contrast between global slavery and Russian serfdom? Will it be possible to arrive at a satisfactory explanation for this discrepancy, or should we await the emergence of a Russian cleric similar to de Las Casas who will provide deeper insight into this issue?

16.5. Ideological Divisions amidst Globalization

At the outset of what we have delineated as the post-Paleolithic era of the globalization of Homo cultures in the Modern Era, a conspicuous trend emerges of victors demanding complete acceptance of their ideology as the ultimate authority from those they subjugated. However, it is worth noting that this demand did not extend to other areas, such as novel techniques in industry (like metalworking or cultivation of vital crops like wheat). Instead, the oppressors used religion as a means of humiliation and a declaration of supremacy—"*You* have been defeated, *we* are superior to you. *Our deity* reigns supreme over yours, thereby confirming our righteousness!" Physical violence also supplemented spiritual violence, which, for many, inflicted deeper wounds than physical harm. Nonetheless, the underlying impetus was the pursuit of an all-encompassing ideological triumph, transcending mere physical conquest and aiming at unanimity.

The question, however, remains—is this validation significant or relevant? After all, the victors have emerged triumphant, while the oppressed lie in defeat. Yet the victors demand proof of their superiority by appealing to the ethereal realm, governed by unseen forces that hold sway over the vanquished from obscured heights. The answer, though seemingly straightforward, is compelling: victory had to be all-encompassing, requiring recognition from the higher powers of the defeated. Without this acknowledgment, the conflict would remain incomplete and unsolved, with numerous possibilities yet to unfold.

Whether this explanation is satisfactory remains uncertain. Nevertheless, it merits inclusion in the roster of inquiries for future researchers examining the themes explored in the chapters of the second half of this book. Let us hope for fortuitous outcomes in this endeavor.

In concluding this chapter and the second part of this book, I present a map that delineates the distribution of Abrahamic and Dharmic religions Hinduism, Jainism, Buddhism, and Sikhism (Fig. 16.13) in order to demonstrate the migratory patterns discussed in the last few chapters. We will also confine ourselves to a few pertinent observations: the depicted map reveals only the actions from the early twentieth century, thereby projecting forward beyond our current narrative. It offers a generalized overview and refrains from exhibiting the main trajectories of both Abrahamic and Dharmic religious spheres. Furthermore, we may supplement this map with another crucial cartographic representation related to our subject matter, detailing a more precise proportional distribution of the principal and global religious movements (Fig. 16.14).

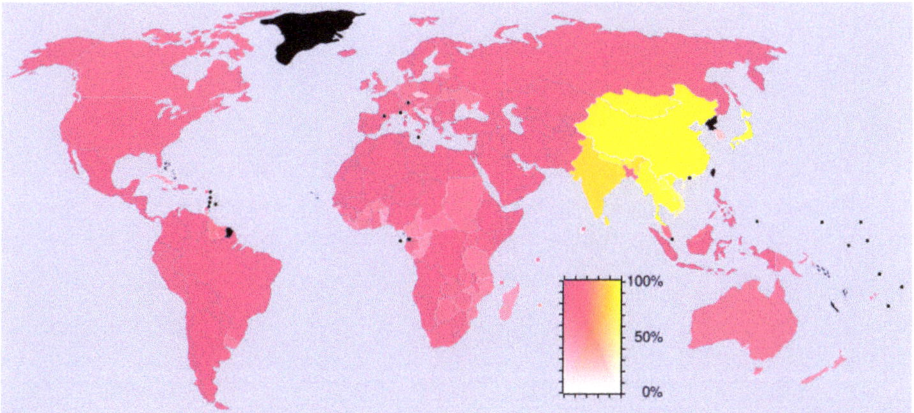

Figure 16.13. Distribution of Abrahamic and Dharmic religions across the continents of the Globe: purple and yellowish colors, respectively (schematic map).

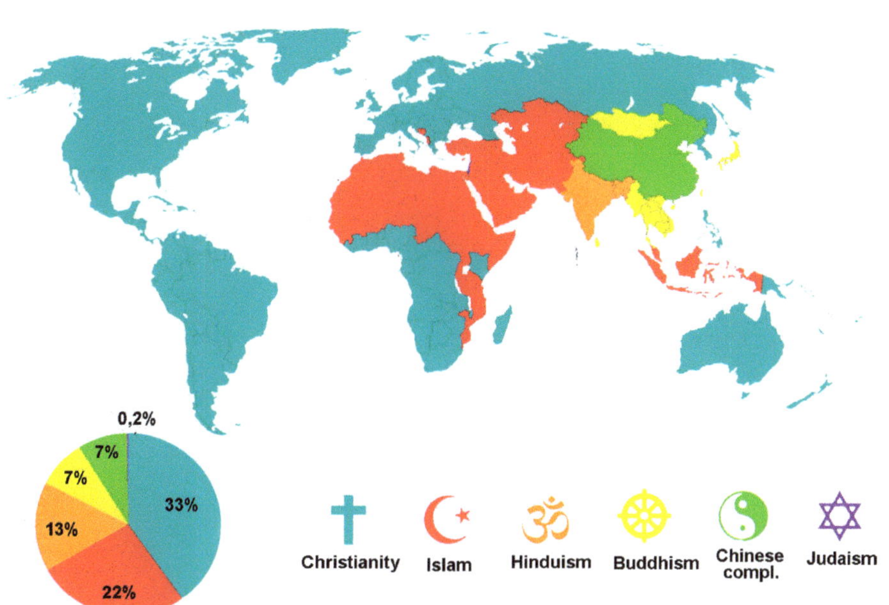

Figure 16.14. Proportional proportion of the six major religions of the Globe (schematic map).

Nonetheless, we shall explore such visual aids in greater depth at the conclusion of the third part of this book.

In the last chapter we included analysis of B. F. Porshnev's work, "Social Psychology and History." Here, the author, in alignment with Porshnev's findings, emphasized the profound significance of the timeless psychological contrast "us and them." This concept took center stage not only in the previous chapter but also in our current discussion. We might revisit certain of these themes briefly in the upcoming sections, acknowledging that we've already given ample focus to this pivotal aspect of social psychology.

Part Three

Reflections on the Essence and Architectonics of the Surrounding World

Chapter 17

Chaos as Progenitor

We now move the focus of our research to the ideological and cultural dimensions, which are typically considered the most difficult to interpret. As previously discussed, this area of interest is particularly saturated with religious beliefs and various speculative constructions. Thus, our discussion transitions from the tangible and evident geological aspects of our planet to the more abstract and interpretative realm of fantasies and myths. This shift immerses us in a unique, intangible aspect of individual existence that is crucial for understanding the broad spectrum of human life.

Many phenomena that are difficult to explain are often dismissed or interpreted as miracles, or evidence of a higher power. Such miracles demand resolution, a privilege granted only to those deemed virtuous or worthy, based on their proximity to divine entities. Indeed, narratives of miracles frequently populate accounts in ancient history.

In the subsequent chapters that comprise the second part of this book, we will encounter a fair share of such tales. We commence our discussion of a so-called Greek miracle, written not by a narrator overgrown with layers of ancient moss, but by a well-known French scientist and writer Ernest Renan.

17.1. The Greek "Miracle" by Ernest Renan against the Background of Christianity and Judaism

Ernest Renan (1823–1892) is a notable figure in Eurasian history, particularly for our discussion of chaos as a force of progeny. With an impressive career that saw the publication of over forty volumes, Renan's contributions are significantly noteworthy and merit a deep examination. Our review of his works, achievements, and persona will be supplemented by the ideas presented in S. F. Godlevsky's book *E. Renan: His Life and Scientific and Literary Activity* (1895), a monograph that remains significant over a century after its publication. Furthermore, we will consider various interpretations from encyclopedias, notably the 1994 article in the Brief Jewish Encyclopedia (KE) that is particularly relevant to the objectives of our study.

Figure 17.1. Ernest Renan (A. S. Adam-Salomon, 1776–1784) and a caricature of him (Andre Gill, 1767).

In 1865, Ernst Renan (Fig. 17.1) visited Athens for the first time at the age of forty-two, when he was already at the pinnacle of his literary career. Before setting foot in Greece, he had traveled and worked in Syria and Palestine as an archeologist collecting valuable materials on semitology. He was particularly interested in recognizing early Christianity as an exceptional creation of Jewish religious and national genius. Thus, Renan's special attention to Jewish ancient history and national character allowed him to highlight Christianity's sublime and ideal properties that contributed to its establishment as a world religion, with a specific focus on Jewish rejection of Christianity as its spiritual offspring (KE).

Renan's experience in Athens resembled that of a cardinal metamorphosis after he first laid eyes on the famous Acropolis (Fig. 17.2). He composed his famous "Prayer in the Acropolis" (*Prière sur l'Acropole*), in which he admitted:

> For a long time I no longer believed in a miracle in the literal sense; and the unique fate of the Jewish people, leading to Jesus and Christianity, seemed to me something very special. And so, next to the Jewish miracle, I had to put a Greek miracle [*voici qu'à côté du miracle juif venait se placer pour moi le miracle grec*], something that happened only once, something not seen before or after, but the influence of which will last forever . . . The impression Athens made on me was the strongest I have ever experienced. There is a place where perfection exists; no two: that's one thing. I've never imagined anything like this. It was a crystallized ideal of Pentelian marble that appeared before me. Until then, I believed that perfection was not of this world . . . For a long time I did not believe in a miracle, in the proper sense of the word; however, the single fate of the Jewish people, which led to Jesus and Christianity, seemed to me something completely separate. So, next to the Jewish miracle, a Greek miracle appeared for me, something that existed only once, has never been seen, will never be

seen, but whose action will last forever, I mean the type of eternal beauty, without any local or national spots. I knew well before my journey that Greece had created science, art, philosophy, and civilization; but I lacked scale. When I saw the Acropolis, I received a revelation of the divine, just like the first time I felt that I was living according to the gospel.

The Athenian Acropolis was constructed over many generations, yet the most iconic structures on this hill date back to the reign of the renowned Pericles, spanning from 443/444 BCE until his death in 429. More than five hundred years later, Plutarch (46–127 CE) narrated his impressions of the Acropolis and Pericles' role in its construction in his work *Comparative Biographies* about the Acropolis and the role of Pericles in these deeds:

Meanwhile, buildings grew, grandiose in size, inimitable in beauty. All the masters tried to distinguish themselves in front of each other by the elegance of their work; the speed of execution was especially amazing. The structures, each of which, as it was thought, would only be brought to an end with difficulty for many generations and human lives, were all completed during the flourishing period of the activity of one statesman. . . . The more surprising, therefore, are the creations of Pericles, because they were created in a short time, but for a long-term existence. In their beauty, they were ancient from the very beginning, and in their brilliant preservation they are still fresh, as if recently finished: to such an extent they always shine with some color of novelty and retain their appearance untouched by the hand of time, as if these works are imbued with the breath of eternal youth, have a soul that does not age! (Plutarch, 164–165)

Figure 17.2. The Acropolis of Athens.

Plutarch's account appears to affirm the Acropolis' enduring beauty that continues to inspire its viewers in our current age. However, his narrative prompts intriguing questions—the first concerns the identity of the observer who might experience such a profound reaction to the Greek marvels, almost akin to genuine prayer. Is this observer an artist, a poet, or a scientist?

The author's answer is predictable: anyone—an artist, writer, poet—but likely not a scientist. Such a sublime response does not align with the typical perception of a professional scientist, particularly in the latter half of the nineteenth century. The European sensibility at that time had already been well acquainted with the splendors of the Renaissance. The interest in antiquity, which had been sustained for over four centuries, had become somewhat mundane and commonplace amongst the educated public.

Renan, however, as a professional historian, linguist, and lecturer at the Collège de France, was more similar to his predecessors at the Platonic academies in Florence, Rome, and Naples that gathered enthusiasts of antiquity. The canons of ancient architecture heavily influenced the grand structure of that era. Moreover, the great Michelangelo had unveiled the sculpture of David almost three hundred years prior, which had become a bright symbol of antiquity—an era long since passed but still enduring in its greatness and influence. The year 1862 became quite a significant year in Renan's biography, as he was honored with presenting his views on the world to the wide scientific public in Paris. Godlevsky describes his efforts:

> At his first lecture on the role of Semitic peoples in the history of civilization, a huge crowd gathered, overflowing not only the auditorium, but also the corridors of the university building. Renan limited himself to general considerations and did not utter a single word that could offend the faithful. In his opinion, the Semitic peoples did not contribute anything original to science, politics, art, or philosophy.... [But here is the crown merit of the Semites]: they gave humanity a religion, which in its further development among the Indo-European peoples became the most perfect ideal of goodness, beauty, truth and justice. Renan called Christ the divine founder of the universal "religion of the spirit"—"a religion accessible to all races, standing above all castes, an eternal and unconditional religion."
>
> Despite the restraint and tact of the lecturer, an unimaginable hubbub arose in the clerical camp; an outrageous demonstration was made. Renan had to leave the department . . . At first, Renan's readings were only temporarily suspended, but after the publication of the "Life of Jesus" in 1863, his enemies already had formal arguments for accusing the scientist of unbelief and impiety: he rejected the divine nature of Christ and in general any manifestation of supernatural power in his life, exploits and miracles, trying to reduce all this to the influence of general historical reasons. . . (Godlevsky, 91–93)

Such was the assessment of Renan by adherents of the official Catholic doctrine. Additionally, in those years there were also a considerable number of supporters of the Jewish miracle, which is why violent attacks followed Renan from another side.

The desire for maximum objectivity, which Renan diligently demonstrates in his concept of the historical and spiritual vocation of Jewry, is actually embodied in his image of a "strange people" who differ from all other peoples and therefore cannot claim to be treated equally with others . . . [So he] on the basis of linguistic analysis gives a comparative psychoanthropological characterization of Semites (primarily Jews) and Aryans, and without hesitation puts the latter above the former in racial terms. . . . [In the above-mentioned lecture at the Collège de France], Renan argued that the Semitic peoples (that is, mainly Jews). They did not contribute anything original to science, politics, art, or philosophy. . . . [But most importantly], the Jews gave humanity a religion that in further development, that is, in the form of Christianity, brought to the Indo-European peoples perfect ideals of goodness, beauty, truth and justice . . .

Renan accompanies the recognition of these merits of Jewry to humanity with two reservations: firstly, he believes that this completely exhausts the contribution of Jews to world civilization; secondly, that these same great achievements of the Jewish people predetermined its national catastrophe. According to Renan, peoples should choose one of two destinies—the selfish pursuit of national and state order and well-being or selfless service to the universal ideals of goodness and justice. On this basis, Renan concludes that the final liquidation of the Jewish national statehood in the 1st century was inevitable and even beneficial, because thanks to this, Israel was freed from the false idea of religious law applied to state life and returned to its true vocation—to the religious and moral correction of mankind. According to Renan, Christianity exhausts all the valuable things that humanity has received from the Jewish people, and its entire purpose was originally to create Christianity.

Therefore, after the emergence of the teachings of Christ, the existence of the Jewish people does not make sense. He is nothing more than a dry branch, a creature that has exhausted its strength. "There is no more strange sight in history than the sight of this people turned into a ghost—a people who have lost their sense of business for a thousand years, who have not written a single page worthy of reading, who have not given us the right information about themselves . . ." However, in Renan's view of Jewry as a living corpse, which is only from stubbornness clings to life without having any historical, national, or creative grounds ("Israel will never create a state or philosophy. He will never have a developed secular literature . . ."), anti-Semites often found a reason to enroll him in their ranks.

Thus, Renan's works were assessed from two divergent perspectives—the first from admiration, the second from rejection, ridicule, and hatred, coupled with expressively mocking caricatures (Fig. 17.1b).

Nevertheless, we recall Renan's words: "I knew well before my journey that Greece had created science, art, philosophy, [and] civilization; but I lacked scale." Unfortunately, however, it becomes apparent that Renan's work possesses a significant shortcoming, namely, his departure from reality as opposed to his knowledge of antiquity. While he clearly demonstrates his familiarity with the works of great Greek thinkers—philosophers, geographers, mathematicians, historians, and others—these references only seldom appear in his writings and do not receive the attention they deserve. It is hardly reasonable to draw a direct comparison between the unparalleled beauty of the Acropolis and the greatness of Greek scientific achievements. As such, this straightforward comparison is not convincing.

Therefore, the question remains: might the essence of the true Greek miracle lie elsewhere?

17.2. The Millennial Areopagus of the Greek Sages

The peak of the true Greek miracle, of course, was considered differently. These were genuine creations of the thousand-year-old Areopagus of Greek thinkers, whose contributions spanned from the earliest centuries of the first millennium BCE to the first centuries of the Common Era. In the fourth century CE, the Roman imperial authorities—having long since incorporated Greece into the Roman Empire—forbade all religions except Christianity, thereby ushering in a new era of imperial uniformity. Alongside the prohibition of religion, the foundations of ancient science hostile to Christian doctrines also suffered a tremendous blow. Thus, the negative impact of these policies was only overcome nearly a millennia later during the Renaissance.

The names and accomplishments of these various scientists, while well-known to many, deserve to be discussed and their contributions remembered, as these esteemed thinkers laid the groundwork for modern Eurasian or, more precisely, European science. Without their foundational work, contemporary science would not have achieved such great success. However, the well-established nature of these figures is largely due to a more forgotten figure named Diogenes Diogenes Laertius or Laertius (Fig. 17.3). Details about his life are scattered and difficult to confirm—even his birth and death dates remain unknown. Generally, he is considered to have lived during the third century BCE and published one book called *Lives and Opinions of Eminent Philosophers* (Losev 1981, 10). This work, however, did not enjoy widespread success due to its unsystematic organization and sometimes frivolous characterizations of the philosophers which he profiled. In any case, this book's value should not be doubted—it remains a source that is referenced by many contemporary researchers.

In this chapter, we will initially recall the most prominent figures of Greek philosophy, who are considered the fathers of various fields such as history, geometry, and medicine. While we avoid quoting excerpts from their works, we will highlight only those figures whose thoughts warrant direct quotation, thereby aligning with the stated goals of this book. We will refer to these individuals chronologically, starting with the oldest associated with Greek city-polises to those living during the Roman Empire. Attempting to group them by their professional interests proved unfeasible, as most of these sages exhibited a broad range of interests across diverse disciplines. For example, the "father of geography" Strabo, whose work intersected with multiple scientific fields, opens his seventeen-volume *Geography* with the following words:

Figure 17.3. Diogenes Laertius and his book "The Life of Philosophers,"
published in Venice in 1701.

I believe that the science of geography, which I have now decided to study, as well as any other science, is part of the philosopher's circle of studies. That this view of ours is correct is clear on many grounds. After all, those who first took the liberty to do it were, according to Eratosthenes, in a sense philosophers: Homer, Anaximander of Miletus and Hecateus, his compatriot; then Democritus, Eudoxus, Dicaearchus, Ephorus and some other of their contemporaries. Philosophers were also their successors: Eratosthenes, Polybius and Posidonius. On the other hand, great scholarship alone makes it possible to study geography: it is peculiar exclusively to a person who is equally capable of considering things, both divine and human, the knowledge of which, as they claim, is philosophy. The benefits of geography are diverse: it is applicable not only for the activities of statesmen or rulers, but also for the science of celestial phenomena, phenomena on earth and at sea, animals, plants, fruits, and everything else that can be found in different countries. The usefulness of geography presupposes that the geographer is also a philosopher—a person who devoted himself to the study of the art of living, i.e. happiness. (Strabo, 7)

In presenting these thinkers chronologically, we will retain the integrity of the map showing the distribution of Greek monuments (Fig. 17.4). This map reflects the vast expanses of Greek colonization of the entire giant gulf of the Atlantic Ocean, specifically the entire Mediterranean with the more distant basin of the Euxine Pontus (Black Sea) and Meotida (Azov Sea).

We begin our presentation of these ancient Greek crown figures with three early natural philosophers (Fig. 17.5) from Miletus, west of Asia Minor. The earliest of this group is Thales

(640/624–548/545), who was considered one of the "seven wise men" responsible for laying the foundations of Greek culture and statehood. Thales was succeeded by his disciple Anaximander (611–546), the author of the ecumene map, and by his other discipline Anaximenes whose dates are similarly obscured but considered to be 585/560–525/502. Pythagoras of Samos (570–490; Fig. 17.6) known as the "Pythian broadcaster" lived on the island of Samos near the western coast of Asia Minor. He was a mathematician, mystic philosopher, and founder of the Pythagorean religious and philosophical school. Herodotus of Halicarnassus (c. 484–c. 425; Fig. 17.7) is a figure that does not require introduction for historians and rchaeologists, who was once referred to as "the father of history" by Cicero. Hippocrates (c. 460–c. 370; Fig. 17.8) was a famous physician and philosopher and author of the so-called "Hippocratic Oath."

Figure 17.4. Map of Greek (red) and Phoenician (yellow) colonies in the basin of the Mediterranean, Black, and Azov Seas (Wikimedia: Commons Atlas of the World—Atlas of Greece. Historical maps).

Figure 17.5. Images of Thales, Anaximander and Anaximenes.

Figure 17.6. Pythagoras of Samos.

Figure 17.7. Herodotus of Halicarnassus.

Figure 17.8. Hippocrates.

He was considered to be the "father of medicine." His name was mentioned in his works by both Plato and Aristotle. In the current age, doctors are required to take the Hippocratic oath after completing their education and entering medical practice.

Aeschines Socraticus (or Xenophon?), Socrates, Plato, and Aristotle are depicted in the fresco "The School of Athens" by Raphael and his students (1508–1511) in the Stanza della

Segnatura of the Vatican Palace (Fig. 17.9). More than 20 figures have been identified on this remarkable panel, and in the accompanying illustrations, these characters are arranged from left to right. This fresco is widely regarded as one of the finest works not only of Raphael but also of general Renaissance art.

Socrates (470/469–399) is a classic figure of Western philosophy famous for his dialogues; while he did not leave behind any of his own works, his thoughts and ideas are preserved in the writings of his followers. Accused of dishonor and corrupting the minds of the youth of Athens, he was sentenced to death by the court executions by poisoning.

Plato (429/427–347), a disciple of Socrates and a teacher of Aristotle was the first ancient Greek philosopher whose writings have been preserved in full.

Aristotle (384–322) is a disciple of Plato, and is considered the most influential philosopher of antiquity. His research concerned physics, biology, economics, psychology, linguistics, and other disciplines. Western science owes a great deal to his works, notably in its intellectual vocabulary and formulation of problems, as well as its research methods. Aristotle's philosophy also had a noticeable influence on almost all forms of knowledge in the West; it continues to be the subject of philosophical discussion to this day.

Euclid (c. 325–265; Fig. 17.10) is an ancient Greek mathematician, the author of the first extant theoretical treatises on mathematics; his main book "Beginnings" or in the Latin version "Elements," published in 1703 (Fig. 17.10, left). Many call him the "Father of Geometry." His research took place in Alexandria, Egypt (at the Alexandria School).

Archimedes (287–212; Fig. 17.11) was a scientist, mathematician, and engineer. He was born and lived most of his life in the city of Syracuse in Sicily. He was a notable inventor and made several important discoveries in the field of geometry, mathematical analysis, and laid the foundations of mechanics and hydrostatics. Archimedes' works were referenced by

Figure 17.9. Philosophers from the fresco "Athenian School" by Raphael; from left to right: Aeschines Socraticus (or Xenophon?) and Socrates, Plato and Aristotle.

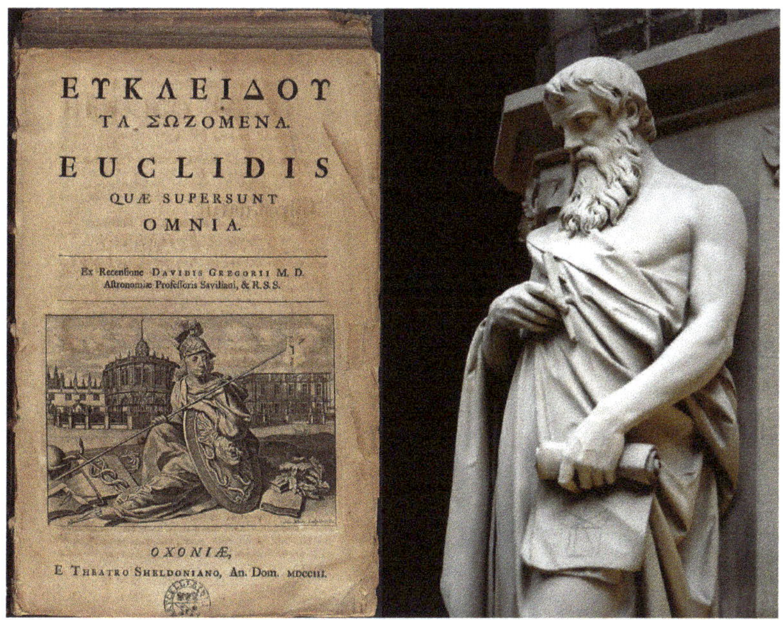

Figure 17.10. Euclid and his book "Beginnings," 1703 edition.

mathematicians and physicists of the sixteenth and seventeenth centuries such as Johannes Kepler, Galileo Galilei, Rene Descartes, and Pierre Fermat.

Eratosthenes of Cyrene (276–194; Fig. 17.12) was a mathematician, astronomer, geographer, philologist, and poet who served as the head of the Library Alexandria. He is the first famous scientist who calculated the size of the Earth. Strabo (c. 64/63 BCE–c. 23/24 CE; Fig. 17.13) was a historian, geographer of Roman Greece, and the author of the almost completely preserved seventeen-volume *Geography*. The monument to Strabo was erected in the Turkish villages Amasya in the Cappadocia Province, Asia Minor.

Claudius Ptolemy (c. 100–170 CE; Fig. 17.14) was a scientist of the Late Hellenistic period, cartographer-geographer, astronomer, optician, mathematician, and

Figure 17.11. Archimedes.

mechanic. His main work was carried out in Alexandria, Egypt. He is the author of the famous and classic monograph *Almagest*, which was a peculiar and relative result of the development of the science of celestial mechanics.

While only the top thinkers are included in the list above, there are noticeably more of them. When we discuss the challenges related to the progenitor of chaos, we will mention the names of those philosophers who should be included in this list of notable truth-seekers.

Figure 17.12. Eratosthenes of Cyrene and an illustration to his theory of calculating the scale of the Earth.

Figure 17.13. The monument to Strabo in Asia Minor (Amasia, Cappadocia) and the title page of his "Geography" in 17 books (edition of Casaubon, 1620).

Figure 17.14. Claudius Ptolemy and his scheme of geocentrism.

17.3. From Chaos to the Creation of the Earth and Other Worlds

The lines of the mythological poem "Theogony" likely resonate most vividly within this context. Its author Hesiod (Fig. 17.15) was born in western Asia Minor but grew up in Balkan Greece, where he penned the following poem around 700 BCE. It is as if Hesiod himself is in conversation with the Muses—the daughters of Zeus:

> Tell me everything: How the gods, how our earth was born,
> As the boundless sea appeared noisy, rivers,
> Stars carrying light and a wide sky above us;
> Which of the immortal givers of benefits was born from what,
> How did they divide the riches and honors among themselves,
> How they mastered for the first time the plentiful Olympus.
> From the very beginning, you tell me everything, Muses,
> And at the same time let us know what first of all originated
>
> . . .
>
> First of all, Chaos was born in the universe, and then
> Broad-chested Gaia, universal safe shelter,
> Gloomy Tartarus, lying deep in the bowels of the earth,
> And, among all the eternal gods, the most beautiful is Eros.
> (Theogony, 108–120)

According to Hesiod, in the narrative relayed by the Muses of Zeus, Chaos emerged as the primary entity, succeeded by Gaia (representing the earth), along with the gloomy Tartarus, ruler of the underworld, and only after them, the most beautiful Eros came into being.

However, Democritus (460–370; Fig. 17.16), one of the most revered and great Greek thinkers, presented his idea of the universe and the world as formed from indivisible atoms. In light of our current scientific understanding, his ideas appear much more rational, albeit devoid of Hesiod's colorful language. However, here too, Chaos reigns supreme:

> Bodies captured by the vortex continuously and all the time flowed to the center. In the same way, the earth was formed when the bodies that were carried away to the center came together. As for the shell-like cover surrounding the earth, it received an increment due to the addition of bodies from the outside: since he was spinning in a vortex, everything that came into contact with him joined him. Some of these bodies intertwined with each other, formed a strong connection, at first wet and mud-like, but after (such connections) dried up and (for some time) rotated together with the whole vortex, they ignited and finally received the nature of the luminaries . . .

Figure 17.15. Hesiod.

Figure 17.16. Democritus.

The worlds are infinite in number and differ from each other in magnitude. In some of them there is neither sun nor moon, in others—the sun and moon are larger than ours, in others—there are not one, but several of them. The distance between the worlds is not the same; in addition, there are more worlds in one place, less in another. Some worlds are increasing, others have reached full bloom, and others are already decreasing. In one place, worlds arise, in another—they are on the wane. They are destroyed by colliding with each other. Some of the worlds are devoid of animals, plants, and any moisture. (Vic 1979, 189–190, rev. 382; Lurie 1970, 288, rev. 349)

If we were to compare Democritus' words with the contemporary thinker R. Hazen's, we would indeed see many similarities (see Section 1.1). Around the same time that Democritus was writing the Xenophon (430–356) presented his "Economia" its lines reflecting prosaic, everyday life. It is worth mentioning that ancient Greek thinkers were not the only ones who soared to cosmic heights. Here, for example, is how Socrates' thoughts are presented in his conversation with Christobulus:

For in the first place it is to those who cultivate it that the earth yields means of sustenance, and of enjoyment as well. Next also it provides them with decorations for altars and statues, as well as for their own persons, and these very sweet to smell and to see. And there is much food too, some of which it produces, some of which it rears, inasmuch as the art of tending cattle comes into the province of agriculture. And thus have men sufficient, both for giving Heaven the sacrifice that is its due honour and for their own uses. But whilst the earth provides this abundance of good things ungrudgingly, it does not suffer those who are effeminate to reap it,f but accustoms them to endure with patience winter cold and the heat of summer. (Xenofont, Chapter 5: 2, 3)

However, now let us move to Imperial Rome, where the great Publius Ovid Naso (43 BCE–17/18 CE; Fig. 17.17) indulged in reflection in his Metamorphoses:

Before there was earth or sea or the sky that covers everything, Nature appeared the same throughout the whole world: what we call chaos: a raw confused mass, nothing but inert matter, badly combined discordant atoms of things, confused in the one place. There was no Titan yet, shining his light on the world, or waxing Phoebe renewing her white horns, or the earth hovering in surrounding air balanced by her own weight, or watery Amphitrite stretching out her arms along the vast shores of the world. Though there was land and sea and air, it was unstable land, unswimmable water, air needing light. Nothing retained its shape, one thing obstructed another, because in the one body, cold fought with heat, moist with dry, soft with hard, and weight with weightless things.

This conflict was ended by a god and a greater order of nature, since he split off the earth from the sky, and the sea from the land, and divided the transparent heavens from the dense air.Whenhe had disentangled the elements, and freed them from the obscure mass, he fixedtheminseparate spaces in harmonious peace. The weightless fire, that forms the heavens, dartedupwards to make its home in the furthest heights. Next came air in lightness and place. Earth,heavier than either of these, drew down the largest elements, and was compressed by itsownweight. The surrounding water took up the last space and enclosed the solid world. (Ovid: The Metamorphoses, Book I, 5–37)

Seven centuries later, Ovid underscored Hesiod's idea that chaos stands at the inception of everything. The brief excerpts from the writings of ancient Greek and Roman thinkers reveal a fascinating observation: the remarkable similarity in their accounts of the creation of the Earth and other worlds, as described by Democritus and Ovid, and echoed two millennia later by modern thinkers such as Robert Hazen (see Section 1.1). This thematic continuity is noteworthy, particularly as it resonates with contemporary interpretations of phenomenon like the Big Bang, wherein energy and matter emerge from an incomprehensible void, essentially

Figure 17.17. Publius Ovid Nazon (monument in Constanta, Romania).

evoking the notion of something emerging from nothing—a concept reminiscent of the diving presence referenced, albeit ambiguously, by Ovid.

However, within the framework of the ancient cosmological worldview, the narrative of human origins consistently occupied a central position.

17.4. Prometheus as the Creator of the Homo-Family

As mentioned above, the traditional Greek understanding of Chaos posits that it was born by a broad-chested Gaia—Mother Earth which symbolizes a refuge for all. The Greeks also saw Prometheus as arguably the most vibrant and expressive of the Titans, the creator and savior of the *Homo* family, who paid tragically for his deeds.

> Prometheus, mixing earth with water, molded people (Fig. 17.18) and gave them fire secretly from Zeus, hiding it in a hollow reed stalk. When Zeus found out about this, he ordered Hephaestus to nail the body of Prometheus to the Caucasian Ridge. This ridge is located in Scythia. There, chained to a rock, Prometheus stood bound for many years, and every day the eagle, arriving, pecked out his liver blades, which grew back overnight (Fig. 17. 19). Prometheus bore such a punishment for stealing fire, until Hercules later freed him. (Apollodorus, 1:7, 1)

Prometheus created humanity, which propagated towards the most diverse, unknown corners of the earth. It was on these vast spaces that the Homo family fended for itself, engaging in both great feats and debasing sins.

Figure 17.18. Titan Prometheus sculpts a man out of clay (in the center); goddess Athena instills a soul into the newborn *Homo* (right). Roman sarcophagus, second century BCE; Prado Museum, Madrid.

Figure 17.19. An eagle pecks out the liver of Prometheus; on the left, an Atlas holding a ball of Earth on its back (a black-figure painted kilic; Arkesilas, ca. 560/550 BCE; Vatican Museum).

G. N. Volkov (1933–1993) was correct in asserting that ancient philosophy formed the foundation for science. In his book *At the Cradle of Science*, he argued that

> the unique peculiarity of early ancient philosophy in the world history of spiritual culture is primarily that it was the initial stage of the development of theoretical, systemically organized science as a whole. But that is why modern science recognizes its infant prototype in early ancient Greek philosophy. That is why it is of much greater interest to us in terms of content than, say, after the Aristotelian systems. (Volkov 1971, 12)

Nevertheless, one of the opening segments of his book is entitled "The Greek Miracle," evoking echoes of Renan's sentiments. Moreover, Volkov aligns this miracle with Pericles' assertion that "We will be a subject for surprise both for contemporaries and for posterity." Evidently, Pericles' foresight is accurate, as there is indeed much within the Greek civilization to evoke wonder and astonishment.

Now, we travel from the west back to the east, to the valleys of the Yellow and Yangtze Rivers, to the Han enclave—already mentioned in this book and fantastic in its remoteness—where we encounter a world completely different from the one we left west of Eurasia.

17.5. Taoism of the Celestial Empire

We commence this discussion by referencing the ideas of Chinese thinkers about the universe and the place of the Middle State within it:

> Ancient Chinese thinkers imagined the universe something like this. The boundless round sky dominates everything, the square-shaped earth stretches below, and in the center of it is the Middle state, i.e. China, which was also called the Celestial Country. According to ancient Chinese views, the world

at first was a chaos consisting of the smallest particles, qi. These particles existed in a disordered, rarefied state, in the form of a formless fog. In the future, there was a process of "separation" of qi: light, light particles rose up, and heavy, dark ones fell down. From the light, light particles, called yang, the sky was formed; from the heavy, dark parts, called yin, the earth. The whole nature was created out of chaos gradually: first the formation of the sky was completed, then the earth, only after that man appeared. All objects and phenomena originated from the relationship of the light principle of yang and dark yin. This relationship generates movement and peace in nature and in life, heat and cold, light and darkness, good and evil. One should not argue about what is better, what is worse—movement or peace, heat or cold, light or darkness, good or evil: both are necessary, one is unthinkable without the other, one passes into the other, therefore it can be both—depending on the time and situation. (Sidikhmenov 2000, 3)

In Ancient China at the turn of the first millennia, there coexisted three dissimilar religions—Confucianism, Taoism, and Buddhism. If we were to proceed further into the future, this triad would be reduced to two, with Buddhism effectively disappearing from the stage. Indeed, against the background of the other two religions, the Indian roots of Buddhist canons manifested themselves in China about five centuries later. The early doctrines of Confucianism and Taoism took place essentially simultaneously since both the founders of the teachings— Confucius (551–479) and Lao Tzu (c. 571–c. 471)—also established their doctrines in China at virtually the same time.

We will certainly note, among other things, a number of important points. Firstly, if the figure of Confucius (Fig. 17.20) does not cause much controversy, then Lao Tzu (Fig. 17.21), at best, is revered as a semi-legendary person, with whom many strange and, in fact, difficult to verify tales are associated.

Secondly, many consider the inclusion of not only Confucianism, but also Taoism in the category of religions to be equally controversial. It is believed that these are most likely religious and philosophical teachings, very different in many basic details of their structures from, say, the Western canons of the Abrahamic religions.

Finally, Werner points out the following concerning Confucianism:

Confucius was not concerned with the problem of the origin of the world. He did not reflect on the origin of existence, nor on the problem of the end of the world. Confucius was not interested in the origin of man, he did not care about his future. He did not study physics or metaphysics, perhaps believing that they were on the other side of life. (And also, he believed, there was no need to worry about them, in any case, no more than about other supernatural things, physical forces or monsters. How can we serve beings on the other side of existence if we don't know how to serve people? We feel that something invisible and mysterious feels, but their essence and meaning are so significant that a person cannot comprehend them). (Werner 2007, 50)

Figure 17.20. Confucius (monument in Yushima
Cathedral, Japan).

Figure 17.21. Lao Tzu (monument in the Qingyuangshan Mountains; Wikipedia).

Thus, it seems reasonable to focus our analysis solely on the figure of Lao Tzu (Fig. 17.21), as he is credited with the creation of the Tao te Ching or the Book of the Way and Virtue.

Before we explore the tenets of the Tao te Ching, it is important to mention the famous Russian sinologist L. S. Vasiliev (1930–2016), who asserted that this treatise

> outlines the foundations of Taoism, the philosophy of Lao Tzu. At the center of the doctrine is the doctrine of the great Tao, the universal Law and the Absolute. The Tao dominates everywhere and in everything, always and infinitely. Nobody created him, but everything comes from him. Invisible and inaudible, inaccessible to the senses, permanent and inexhaustible, nameless and formless, it gives rise, name and form to everything in the world. Even the great Heaven follows the Tao. To know the Tao, to follow it, to merge with it—this is the meaning, purpose, and happiness of life. Tao manifests itself through its emanation—through te, and if Tao generates everything, then te nourishes everything. (Vasiliev 2000, 334)

In the words of the Tao te Ching itself:

> Here is a thing that arises in chaos, born before heaven and earth! O soundless one! O formless one! She stands alone and does not change. It works everywhere and has no barriers. She can be considered the mother of the Celestial Empire. I do not know her name. Denoting it with a hieroglyph, I will call it tao; arbitrarily giving it a name, I will call it great. The great is in endless motion. What is in endless motion does not reach the limit. Without reaching the limit, it returns [to its source]. That is why the tao is great, the sky is great, the earth is great, and the sovereign is also great. There are four great ones in the universe, and among them is the sovereign.
>
> Man follows the [laws] of the earth. The earth follows the [laws of] heaven. Heaven follows the [laws] of the tao, and the tao follows itself. . . . The Tao gives birth to one, one gives birth to two, two give birth to three, and three give birth to all beings. All beings carry yin and yang, are filled with qi and form harmony. People don't like [names] "lonely," "brave," "unhappy." Meanwhile, the Gunas and vanas call themselves by these [names]. Therefore, things rise when they are belittled, and they are belittled when they are exalted. What people teach, I also teach: the strong and cruel do not die by their own death. This is what I am guided by in my training. (Blocks 25 and 42; translated by Yang Hin-shun)

Significantly, the lines of the treatise itself underscore that the Tao arises again in chaos, before the creation of heaven and earth. The geographical distance between China and Greece signifies the importance of this similar understanding of chaos as a primordial and all-consuming phenomenon. Similarly, the lines of the Tao te Ching again demonstrate a certain consonance with the philosophical reflections of R. Hazen, which we discussed at the beginning of the book. Ultimately, in summarizing the assessments of ancient thinkers, it is impossible not to notice obvious differences between them: against the background of the vividly irrational coloring of the lines of the Tao te Ching, the reflections of Western philosophers seem to be more rational in nature.

Chapter 18

The Ancient Ecumene and Its Riddles

In the following chapter, we will continue our discussion about antiquity. Firstly, however, it is important to mention that the previous chapters' analysis did not intend to persuade readers that nearly all ancient Greek philosophers viewed chaos as the primary progenitor of everything in the earthly realm. This is decidedly not the case, as there was no consensus among the sages of the ancient Athenian Areopagus. For example, Anaximenes posited that everything depended on the boundless air, stating that air was the beginning of existence from which everything arises. However, he argued that air is perceived beyond sensory experience, through an inferential understanding that is difficult to comprehend and accept. Aristotle, on the other hand, saw the origin of all things as matter and motion, and God as the first mover.

The reasons for the author's focus on the Chaos theory as a significant aspect of Greek philosophy are threefold. First, there was indeed a relatively widespread presence of diverse variations of the "chaotic" theory among the thinkers of the Areopagus. Second, to point out the similarity between this theory and the modern hypothesis of the universe, as highlighted in the works of R. Hazen (see, for example, Sections 1.1 and 9.3) is of special academic interest for the remaining chapters of this book. Finally, the fundamental resemblance between this hypothesis and the book of Taoism of the Han enclave, despite the vast geographical and cultural distances separating Greece and China, is remarkable. The significant isolation of European civilizations during that period precluded any mutual exchange of information, making the development of fundamentally similar concepts independently noteworthy.

Now, we will discuss the Greek ideas on the concept of the ecumene as the habitat of the Homo species on our planet.

18.1. Pangaea and Hellas in Its Center

In Greek, the word *Pangea* means "the whole earth." The phenomenon of Pangea has already been discussed in the first chapter (Figs. 1.5 and 1.6) in the context of this supercontinent during the Mesozoic geological era (ca. 228–200 million years ago; cf. Wegener 1929; Hain 2010).

It is evident that the Hellenic thinkers of the first millennium BCE considered Pangaea from a different perspective. The world at that time was the ecumene of three continents—Europe, Asia, and North Africa (then referred to as Libya)—interconnected both geomorphologically and ethno-culturally. These three continents now comprise the lands of the planet's eastern hemisphere. For the thinkers of that era, the Hellenic Ecumene did not proceed past the southern edge of the hemisphere into South Africa. Hellenic scholars expressed their understanding of Pangaea mainly in written descriptions as opposed to in maps. The maps, which already appeared as reconstructions of ancient narratives, were only created ages later.

In this section, we will briefly discuss the geographical hypotheses that appear most important in demonstrating the dynamics of the Hellens' worldview in chronological order. Perhaps the most indicative element of the early stages of this process is the map of Anaximander of Miletus (611–546 BCE; Fig. 18.1), which, despite its lack of detail, represents the entire ecumene with three continents (Hain, 1960). Unfortunately, the map's reconstructor is unknown to us. Anaximander was succeeded by Hecateus, also of Miletus (550–490; Fig. 18.2), who is believed to have refined several details in Anaximander's three-part landscape, an improvement reflected in the map's depiction. However, this map's reconstructor also remains known. (Adkins 2008, 303–304; Ancient Geography, 139, 251, 344; see also, for example: "Pangaea," Online Etymology Dictionary; Vergilius Maro, Publius, Georgicon, 4:462; Lucan, Pharsalia, 1:679).

Let us recall the "father of history" Herodotus and his map, discussed earlier (see Chapter 4, Fig. 4.1). We noted that Herodotus map's outline was noticeably different in many essential details from the previous maps and that its design was also likewise connected to an unknown reconstructor. Notably, on the maps of Anaximander, Hecataeus, and Herodotus depicting Pangaea, Greece occupies a central or near-central position. This placement likely reflects the author's intent to emphasize Hellas as the defining center of the ecumene, ultimately reflecting the prevailing idea of Greece's paramount position in

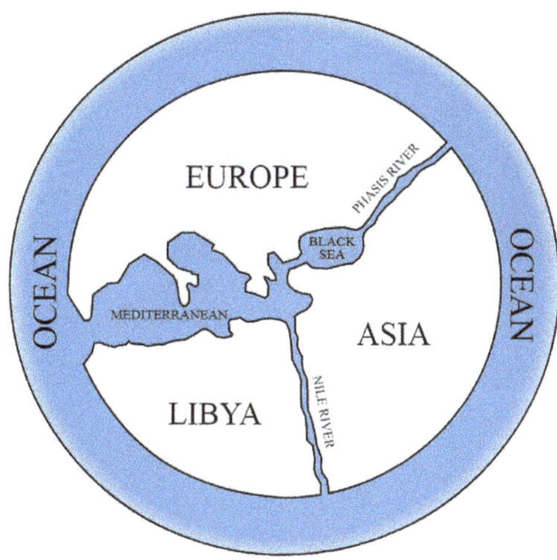

Figure 18.1. Pangaea Anaximandra.

Figure 18.2. Pangaea Hecatea.

the world. Furthermore, Mount Olympus—the home of the Greek deities, led by Zeus—might have been considered even more important (see Fig. 18.3). The ancient thinkers' views of Pangaea do not cause much controversy in modern history. It is, however, vital that we represent the creation of these maps accurately, by highlighting the work of both the creators and the reconstructors. By doing so, we can better comprehend the creations that have transcended centuries.

Figure 18.3. Olympian Areopagus of Greek gods at the head with Zeus over sacred Olympus; depicts a heated argument between Athena and Poseidon for the title of a new city in Attica: Athens, the most famous city in Greece, won and received her name (M.-J. Blondel, 1848, Louvre, Paris).

18.2. An Overview of the Originators of Pangaea Concepts and Their Interpreters

For our discussion, we begin with an examination of the creator-reconstructor pair Claudius Ptolemy (Ptolemaeus) and Donnus Nicolaus Germanus. Claudius Ptolemy (c. 100–170 CE) is a notable figure within geographical science. In his book *Claudius Ptolemy* (1985), V. A. Bronshten argues:

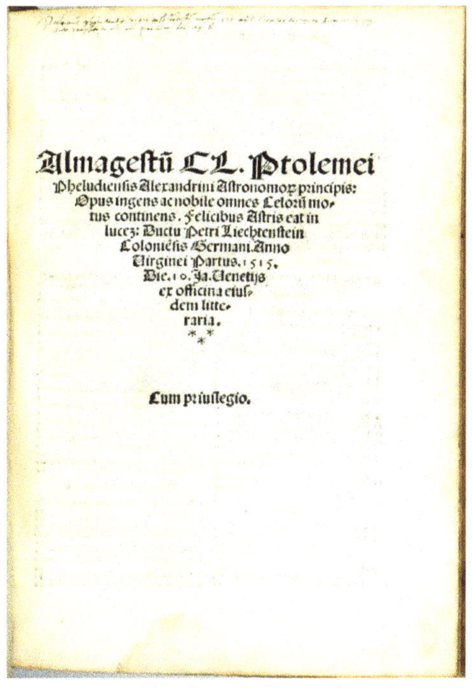

Figure 18.4. Claudius Ptolemy, the title cover of the book "Almagest," 1515.

> The famous Alexandrian astronomer, mathematician and geographer of the II century P. E. Claudius Ptolemy is one of the largest figures in the history of science of the late Hellenistic era. In the history of astronomy, Ptolemy had no equal for a whole millennium—from Hipparchus (second century BCE) to Biruni (tenth to ninth century CE).

History has treated Ptolemy's personality and writings in a rather strange way. There is no mention of his life and work among historians of the era when he lived. Even the approximate dates of Ptolemy's birth and death are unknown, as are any facts of his biography. Ptolemy was lucky in another way. Almost all of his major works have survived and have been appreciated by posterity, from his younger contemporaries to the astronomers of our days. Ptolemy's main work, widely known as the Almagest (Fig. 18.4), was translated from Greek into Syriac, Middle Persian, Arabic, Sanskrit, Latin, and later into French, German, English and Russian. Until the beginning of the seventeenth century, it was the main text book of astronomy.

Many readers usually associate the name of Ptolemy with the so-called "Ptolemy world system," where the Earth is located in the center, and the Moon, Mercury, Venus, the Sun, Mars, Jupiter and Saturn orbit around it in circular orbits. . . . Ptolemy's geocentric system is contrasted with the heliocentric system of Copernicus, who made a truly revolutionary revolution by placing the Sun at the center of our planetary system and reducing the Earth to the position of an ordinary planet, and allegedly eliminated epicycles, showing that they were needed only to represent the movement of the Earth around the Sun.

Ptolemy's Geography (1482), also called the Geographical Guide, consists of eight books. In the first book, Ptolemy sets out his ideas about the globe and

235

gives the concept of geographical coordinates: latitude and longitude. Ptolemy's Geography is a huge volume, not much inferior to the Almagest. This book provides the longitudes and latitudes of approximately 8000 localities. The book is illustrated with 27 maps: one general and 26 by region. Unfortunately, the original Ptolemaic maps have not reached us, so we can only judge them by copies and reconstructions. His views on geographical science are based on a mathematical approach to mapping and determining the positions of cities and other points. Ptolemy sharply criticizes the country-specific approach based only on the use of qualitative descriptions of travelers, without a mathematical basis.

Despite numerous errors in the depiction of certain geographical regions, Ptolemy's work retained its significance almost until the seventeenth century. With the beginning of printing, in the period between 1475 and 1600, forty-two editions of this book were published. The first edition of Geography (without maps) was published in 1475 in Vicenza (Italy) in a Latin translation by Jacopo d'Angelo, made back in 1406. Two years later, in 1477, the first edition with maps was published in Bologna. In 1482 and 1486. Nikolai Herman published two editions of Geography in Ulm (Germany). These editions were prepared very carefully and exceed many subsequent editions in terms of the accuracy of reproducing the names and coordinates of settlements. In all editions of the Geography, up to the Venetian edition of 1511, a translation by Jacopo d'Angelo is used. Only in the Roman edition of 1478 the text was checked with the Greek original by its publisher Domizio Calderino, who made some amendments to the text.

An important stage was the Strasbourg edition of 1513, prepared by Martin Waldseemuller and Matthias Ringman on a completely new basis. First of all, they found Greek manuscripts of the Geography in Basel and in Italy, according to which they made a number of corrections to its text. When preparing the publication, they set out to show the differences between ancient and new geography and for this purpose included 20 new maps, which the famous geographer, traveler and polar explorer Adolf Eric Nordenskiold (1832–1901) called "the first new atlas of the world." (Bronshten, Introduction and Chapter 12)

Following Ptolemy, we turn to the reconstructor of his works, the Benedictine monk Donus Nicholas Herman (c. 1420–1490). It may seem surprising to place both individuals in a single, relatively brief, section of the chapter. However, without the diligent efforts of this notable figure (Fig. 18.5), it is unlikely that we would have been able to fully appreciate Ptolemy's creations. Brown Llyod asserts:

Ptolemy's maps remained inaccessible to us, and we cannot say with certainty that they simply disappeared in the course of time. It is possible that everything—in the very outcome—could be limited only to data on the geographical coordinates of those eight thousand points, which were taken into account by the cosmographer monk (Figs. 18.6 and 18.7).

Figure 18.5. Claudius Ptolemy, the title cover of the book "Geography," published in Ulm, 1482; the figures of Pope Paul II and kneeling in front of him are depicted monk Donnus Nicholas Herman.

Figure 18.6. Claudius Ptolemy, the title cover of the book "Geography," stored in the Vatican Library.

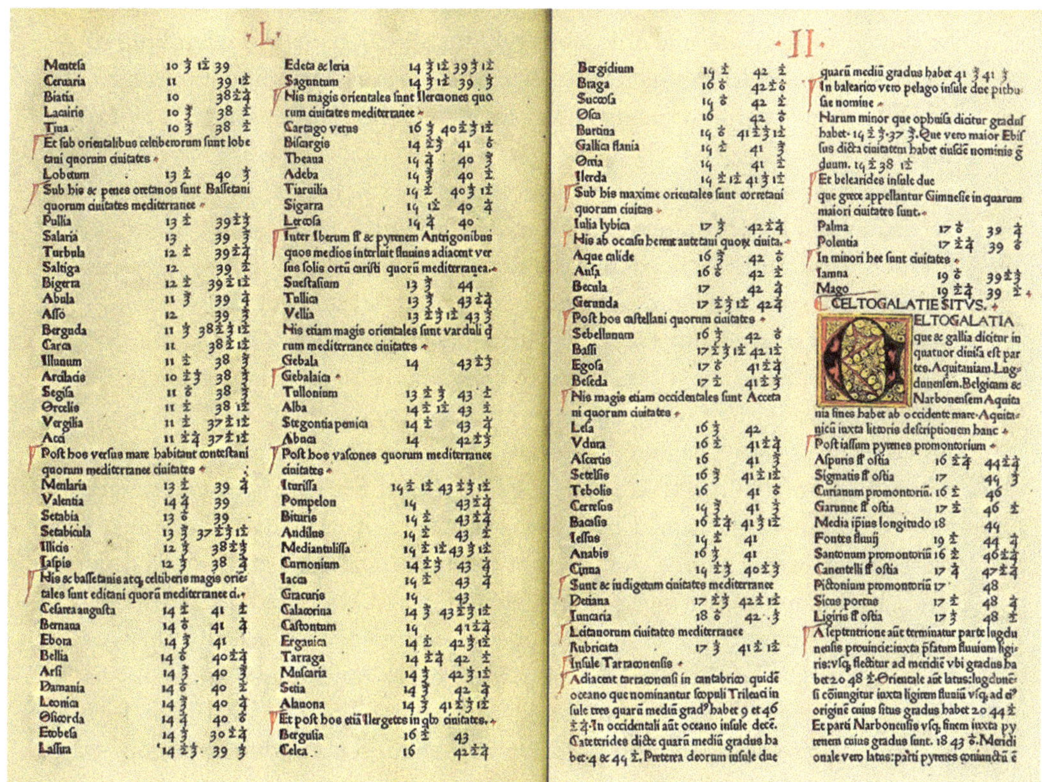

Figure 18.7. Claudius Ptolemy, "Geography," published in Ulm, 1482, publication of data on geographical coordinates.

However, here again we will give the floor to someone who, for example, has thoroughly studied all the intricate plots of the creation of maps—Brown Lloyd (Lloyd 1980; Lloyd 2013).

One of the handwritten editions of "Geography," based on the Latin translation of [Jacob] Angel, edited by another scientist of the fifteenth century—Donnus Nikolaus Germanus, or Nicholas the German, a Benedictine monk from Reichenbach. He did for Ptolemy's maps what the Angel did for his text. He himself was an ardent admirer of the Alexandrian geographer and approached this task with great modesty and verbose apologies to the author and his patron Pope Paul II. He ventured to finalize the maps and redrawn them anew in that complex spherical projection that Ptolemy himself first recommended and then left . . . Donnus read Ptolemy's text very carefully. When he got to chapter 24 of the first book, where Ptolemy invited all cartographers who were not too lazy and had sufficient knowledge to draw maps in his spherical projection, he accepted the challenge and began to think about "how we ourselves could get some fame." He came to the conclusion that the best way to do this is to redraw Ptolemy's maps without changing anything in them except the projection. However, in no version of the text—neither in Greek nor in Latin—did he find good descriptions of various regions, "how many and what kind of people live there, what villages, cities, rivers, harbors, lakes and mountains, or under what region of heaven they lie, or in which direction to look for them."

The copyists seem to have overlooked many small details that Donnus, according to his own statements, found in Ptolemy's writings, for example, the description of the borders between provinces and countries. Donnus reduced the size of the picture—that is, the map of the world, "which was previously too large and exceeded the usual size of books"—to a size that was supposed to make its study more convenient. In all other respects, he left the cards as they were. Donnus produced many maps, and the maps he redrawn in the Ptolemy—Donnus projection served as a model for printed editions of the atlas . . . The borders of the maps themselves, with the exception of the world map, were usually not curved, and the maps were isosceles trapezoids. Apart from the modification of the projection, Ptolemy's maps practically did not change when transferred from hand-painted parchment to a wooden cliche or a copper mold. All the changes were due solely to the need to reflect the color shades and designations of the original map on the black-and-white engraving so that the picture remained understandable. The work on the maps has taught the engravers a lot! (Lloyd, 284–286)

Nicholas Herman and his followers recreated a considerable number of maps, many of which have been published in various editions. Here, however, we will focus on only three examples (Figs. 18.8–18.10) that prominently feature the shores of the Baltic Sea, referred to as the Sarmatian Ocean, and the Sea of Azov (called Palus Maeotis or the Meotis swamp). According to Herman, this region marked the border between Europe and Asia. The maps are notable for their differences, both in overall outline and in various important details. Of particular interest is the position of the Riphean Mountains, which played a significant role in delineating Europe and Asia, which will be discussed in more detail in the following sections.

Ptolemy's descriptions were important for Europeans because they provided the basis for world maps that differed significantly from the earlier reconstructions of Anaximander and Hecateus. The earliest series of world maps or ecumenes, created according to the ideas of Hellenic geographers and Ptolemy's descriptions, likely included the examples provided in this section (Figs. 18.11 and 18.12).

> For researchers later than Nicholas Herman, we also know the names of two reconstructors of the general map of the ancient geographer Posidonius (139/135–51/50 BCE). The Flemish mathematician, cartographer, courtier of King Louis XIII Pierre Berthier or Petrus Bertius (Pierre Bertius; 1565–1629) together with the French engraver and publisher Melchior Tavernier (Melchior Tavernier; 1594–1665) proposed in 1628 their own reconstruction of the map of Posidonius, in the form that the Ecumene was presented to the scientists of Hellas in the century before Ptolemy. It is curious that this map appeared in 1905 even in the Library of Congress of the USA (Fig. 18.13).

Thus, the map of Pangaea had been created and expressed in drawings, with three main continents outlined—Europe, Asia, and Libya (Africa). However, many questions arise from these formulations, which are laid out below.

18.3. The Question of Europe

One of the most intriguing mysteries of world history concerns the origins and significance of the title "Europe." What accounts for her acquisition and elated title? When compared to Gaia, the goddess from the dynasty of the most eminent Greek deities, the goddess Europe perhaps may not seem as illustrious. Nevertheless, she can be portrayed in a highly majestic manner as the queen of the Earth (Fig. 18.14). To delve deeper into this enigma, we turn to the grammarian and synthesizers of Greek myths Apollodorus of Athens (c. 180–120), who provides us with an account of the noble lineage of Agenor, from which it appears the enigmatic figure of Europe emerged:

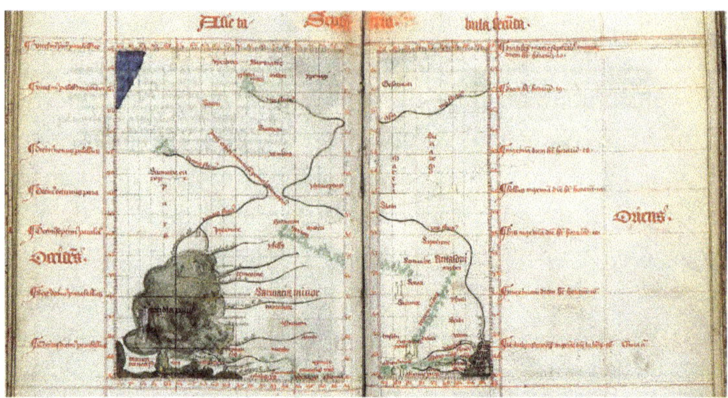

Figure 18.8. Claudius Ptolemy, "Geography," map of "Little Sarmatia" with the Sea of Azov (Meotian swamp).

Figure 18.9. Claudius Ptolemy, "Geography," map of Europe eighth (Octava Evro).

Figure 18.10. Claudius Ptolemy, "Geography," map of Europe eighth; edition of Sebastian Munster.

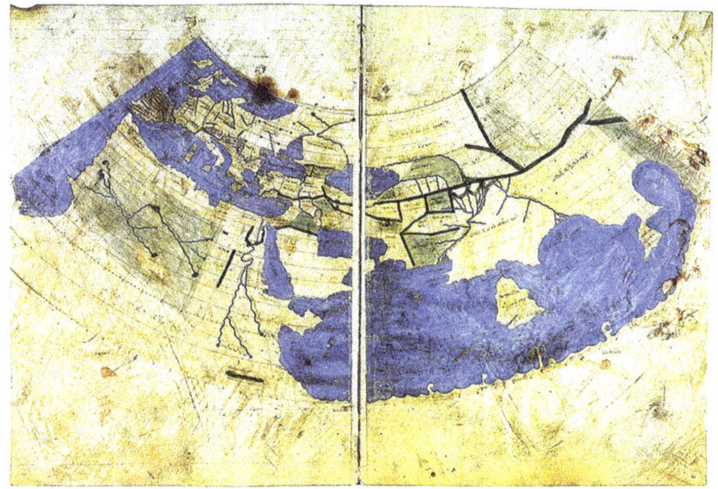

Figure 18.11. Byzantine Greek world map reflecting the first (conic) projection of Ptolemy: Code Vaticanus Urbinas Graecus 82; preface. The reconstructor was a monk, Maxim Planud, ca. 1300 (https://commons.wikimedia.org/wiki/File:Ptolemy-World_Vat_Urb_82.jpg).

Figure 18.12. Printed map of 1482, recreated on the basis of Ptolemy's descriptions by Nikolai Herman, by engraver Johann Schnitzer (1482 Cosmographia Germanus.JPG).

Figure 18.13. Ancient Pangaea, Posidonius, reconstruction by Pierre Berthier, as well as Mehior Tavernier.

Livia gave birth to two sons from Poseidon, Bela and Agenor...Agenor, arriving in Phoenicia, married Telephassa, and he had a daughter Europa and sons Cadmus, Foinicus and Cilic. Some, however, argue that Europa was not the daughter of Agenor, but of Foynik. Zeus, who fell in love with her, turned into a meek bull, and letting Europe sit on his back, sailed with her to the island of Crete. After Zeus shared a bed with her there, she gave birth to Minos, Sarpedon and Rhadamanthus. However, Homer claims that Sarpedon was the son of Zeus and Laodamia, the daughter of Bellerophon. After the disappearance of Europa, her father Agenor sent his sons on a search, ordering them not to return until they found Europa. Together with them, the mother of Europa, Telefassa, went in search of them, as well as Phasos, the son of Poseidon, according to Ferekid, the son of Cilic. After a long search, they became convinced that it was impossible to find Europe and, refusing to return home, settled in different places... (Apollodorus, 3:1, 1)

However, even after consulting Apollodorus, we remain intrigued by the enduring question: what is the underlying reason for such an exalted, iconic title? Apollodorus, for some reason, seems unconcerned with this question. After all, in comparing the many Greek mythological heroines, Europe—so abruptly faded into obscurity—does not appear particularly remarkable, aside from her captivating presence that enticed Zeus.

I would like to explain, however, that Europe has not gone anywhere and has not dissolved at all. It turned out to be quite easy to find her: after all, she managed to fit her more than two thousand-year-old image into myths, legends, monuments and even money. Moreover, it is possible that this seemingly strange memory of her will be eternal. It is possible, for example, to illustrate the memory line about this unique lady with a picture (Fig. 18.15). It presents the plot of the main adventure of her life—from the red-figure crater Asteas (350–320), to the painting by Valentin Serov in 1910, and up to the bright modernist abstract symbol "The Abduction of Europe" by the Belgian avant-garde sculptor Olivier Strebel (Olivier Strebelle) on the square named after her—that is, Europe—one from the central and quite expressive squares of Moscow. And finally, of course, her indispensable image with Zeus the bull reminds us of her on coins, though not very weighty—only two euros (Fig. 18.16).

Figure 18.14. Europe as a Queen; "Cosmography," ed. Sebastian Munster, 1570 (1488–1552).

Figure 18.15. Europe and the bull-Zeus: from the red-figured crater of Asteya to the square named after her in Moscow.

243

Figure 18.16. Two euro coin with the image of Europe on the bull-Zeus.

Still, the lingering question of why this woman became the symbol for the whole continent persists. Could this role not have been claimed by her own son, Minus, the son of Zeus himself and the future ruler of Crete? Indeed, the son of the supreme deity of the Hellenes was notorious for his cruel disposition and ultimately met his demise in a bath of boiling water prepared by his enemy Daedalus. The figure of the Cretan lord Minos often appears unattractive in the works of various artists, as, for example, is depicted by Michelangelo in the fresco "The Last Judgment" (Fig. 18.17).

For our purposes, however, the exploration of the theme of Europe and its continent leads us to the works of Herodotus. While we have extensively discussed this topic in connection with Europe and the enigmatic figure of Europe herself (see Chapter 4.1), it is as equally compelling and important to consider the portrayal of another noble lady: Asia.

18.4. The Question of Asia

According to Herodotus, Asia is Prometheus' wife. In other versions, however, she is not his wife, but his mother.

In following Herodotus' approach, who himself found it impossible to ascertain the origins of the name, we will speculate about the name "Asia" and its historical roots. In fact, "Asia" did not truly correspond to the continental (as opposed to maritime) bulk of the world, despite Herodotus' belief that Europe, with its immense scale, equaled another pair of Earth's continents, effectively constituting half of Pangaea envisioned by the Hellenes (Fig. 4.1). Incidentally, Herodotus himself sarcastically remarked, "It is ridiculous to see how many people have already drawn maps of the earth, although none of them can even correctly explain the outlines of the earth. . . . And they consider Asia to be equal in size to Europe" (Herodotus 4:36).

Figure 18.17. Minos in Michelangelo's fresco "The Last Judgment," Sistine Chapel, 1537–1542.

In discussing the question of how the Greeks separated Europe from Asia and how they drew the border between these continents, it becomes clear that they did so by dividing the world of the Hellenes from the world of the barbarians. In other words, they separated the world of the Scythians and their related steppe peoples from the world of western Europe. The domain of nomadic pastoralists, so alien to Hellas, spanned across the extensive expanses of the Steppe Belt of Eurasia—a giant double enclave that we have previously explored in several chapters of this book (see, for example, Sections 3.3, 6.4, and 7.7). However, it is evident that during Herodotus' lifetime, the Hellenes were entirely unaware of the existence of such a vast steppe double enclave:

> That was the world, as Aeschylus (525–456) expressed it in his tragedy "Prometheus Chained," when the servants of the ruler of all the gods Zeus—Power and Hephaestus—led the chained Prometheus to the Scythian Caucasian rock to which Hephaestus was supposed to chain this glorious titan—and according to one version even the cousin of Zeus—for many years:
>> Here we have come to the distant borders of the earth
>> Into the spaces of the Scythians,
>> into the wild empty wilderness . . .

However, Homer, who was the originator of many of the views held by ancient thinkers around the world, felt the oppressive threat of the Scythian world encroaching on the enlightened Hellenes:

> We have finally crossed the deep-flowing Ocean.
> There is a country and a city of Cimmerian men. Usual
> Twilight and fog there. Never the radiant sun
> Does not illuminate the rays of the people who inhabit the edge of that,
> Does it leave the earth, entering the starry sky,
> Or it descends from the sky, heading back to earth.
> A sinister night surrounds a tribe of unhappy people.
> (Odyssey 11:13–19)

In the Hellenic view, the "Cimmerians," are the people of Asian spaces, the predecessors of the Scythians west of the Steppe Belt near the Euxine Pontus.

Let us, however, return to Aeschylus, wherein the chained martyr Prometheus, while sympathizing with the Elder Oceanida, says:

> From here you are confused by sunrise
> You will direct the step through the virgin land of the untilled
> And you will come to the nomadic Scythians. They live
> Under the free sun on carts, in boxes
> Braided. Behind his shoulders is a well—aimed bow.
> Don't go near them! . . .
> (Aeschylus, Prologue: 1, 2; 3, 707–712)

The Hellenes certainly viewed the barbarian Scythians as completely alien to their native Hellas and as hailing from a wild, northern wilderness, where people lived and moved in wicker boxes (Fig. 18.18).

> Scythia begins beyond Thrace in the place where the sea forms a bay and where the Istrian flows into the sea. . . . This is the primordial Scythia, it begins from the mouth of the Istra. . . . The northern parts of Scythia, extending into the mainland, up the Istrian, border first with the Agafirses, then with the Neurs, then with the Androphages and, finally, with the Melanchlens. . . . If we take Scythia for a quadrangle, two sides of which are stretched to the sea, then the line going into the country, in length and width will be exactly the same with the seaside line. For from the mouth of the Istra to Borysthenes [Dnieper] is 10 days' journey, and from Borysthenes to Lake Meotida [Sea of Azov] is another 10 days, and then from the sea inland to the Melanchlens living above the Scythians, 20 days' journey. I take the daily transition in 200 stages. Thus, the transverse sides of the [quadrilateral] The Scythians make up 40,000 stages, and the longitudinal ones going inside the mother are still the same. Such is the magnitude of this area. (Herodotus 4, 99–101, 118)

According to the Greeks, it was the mouth of the Istra (Danube) that delineated the coveted border between two worlds—Europe, the cradle of Hellas, and Asia, the homeland of the Scythians. It was on this border that the Persians, at the behest of their ruler Darius, met their demise in their pursuit of the elusive Scythian horsemen, and subsequently built their famous bridge over the Istrian. Ultimately, Herodotus' narrative reveals that this structure was half-destroyed by Darius's allies. Interestingly, the preserved half of the bridge serves as

Figure 18.18. Greek colonies on the Black and Azov Seas.

a salvation for the unsuccessful Persians, who managed—albeit ingloriously—to escape the vast, desolate, and wild Scythian steppes. However, it was at this point that the beginning of Asia was established, of which the Scythians and their kindred tribes became the masters (see Figs. 4.1 and 4.2).

The Istrian-Danube Delta as a boundary between the European and Asian continents was hardly convincing, a point that is insignificant to our discussion. Indeed, the border required a more impressive, fundamental outline, which is why the idea of replacing it with the Ripeian/Riphean Mountains gained traction. The originator of this idea, however, remains unknown, similar to the case of Herodotus's formulation of the three continents of Pangea.

18.5. The Chimera of Riphei/Ripei

For us, the identity of the true initiator of the Riphean idea holds little significance. What truly matters is that the Riphean mountains did not exist in Eurasia; they were a product of fiction, fantasy, and myth. Yet, this legend has endured through the ages, captivating the minds of many for at least two and a half millennia and thereby holds a special interest for us. It has not faded into obscurity but continues to intrigue people to this day, and perhaps will do so in the future. However, the current situation regarding the Riphean/Riphei concept is remarkably curious:

> Anyone who deals with the ancient and medieval geography of Northern Eurasia, one way or another faces the problem of localization of the mythical (or semi-mythical) Ripaean (in Latin spelling, often Riphean) mountains . . . supposedly located in the north of Eurasia. Known to ancient Ionian science, Herodotus, the Ripeian Mountains firmly entered into the ideas of ancient man about the northern outskirts of the Ecumene. . . . Moreover, throughout the Middle Ages and in subsequent centuries up to the VI century. the existence of the Ripeian Mountains somewhere in Northeastern Europe or Northern Asia did not cause much doubt. Most of the medieval maps contained an indication of the Ripeian Mountains. In Modern European science, the Ripeian Mountains were identified differently: they were understood as the Tien Shan, the Altai, the Urals, the Caucasus, the Carpathians, even the Alps. . . . It is known that Swedish scientists of the VII century . . . believed that the Ripeis were mountains in Sweden; they were also placed in Poland, Lithuania, on the Middle Russian Upland.

This text was published in 2017 by a specialist closely connected with this issue (Podosinov, 35). The author of this text is correct in asserting that many individuals face the problem of localizing the mythical Ripei/Riphei. Therefore, let us turn to another point of view expressed by the prominent Russian historian L. A. Yelnitsky:

We already know that the real basis of the idea of the legendary Ripeian Mountains for the ancient Greeks could most likely be only the Caucasus with its highest, wild and covered with eternal snow peaks. There were no other similar mountains in the field of view of the Greeks in the 4th century BCE. We also know that the very name of Ripeyev is connected with the Caucasus through the biblical names Rafa and Rifat, localized according to the instructions of the "Book of Jubilees" in the Caucasus. This ancient Eastern localization of the Ripei passed to the Ionians, as evidenced by Pomponius Mela and Pliny, who depicted the Ripeian Mountains as a continuation of the Caucasian Ridge. The opinion that Aristotle expressed about the Ripeian Mountains also goes back to the same ancient Eastern cosmological ideas. . . . He depicts the Ripeian Mountains as a hill in the north of the earth's surface, behind which the Sun hides at night. . . . On the famous ancient Roman map of the world by Vipsania Agrippa, the Ripeian Mountains were depicted stretching along the northern tip of the inhabited earth, from the Atlantic to the Eastern (Aeoi) Ocean. . . . Herodotus also does not mention the name of the Ripeian Mountains in his history for some reason, but, talking about the northern neighbors of the Scythians, mentions the inaccessible mountains in the northeast, beyond the lands of the Argippeans and Issedones, at the foot of which lie unknown countries and fantastic people live in whose existence he does not believe. . . .

As a result of our examination, it was possible to make sure that Aeschylus in his Prometheus did not draw an arbitrarily distorted picture of the northern countries, but exactly the one that determined the state of science of his time: in the north of Europe, in the imagination of the Greeks of that time, the Ripeian Mountains towered—they are Caucasian, first investigated under their ancient Eastern name . . . In these mountains the river Hypanis-Facis [Don] originates, separating the European continent from the Asian one. (Yelnitsky, 60, 67)

At the same time, it is interesting that the "father of geography" Strabo, exactly two millennia ago, spoke quite unambiguously about the Ripei and the boundaries of countries connected to these imaginary mountains:

However, I cannot specify the exact boundaries. Familiarity with these countries makes it necessary to attach importance to those who composed the mythical "Ripean Mountains" and "Hyperboreans," as well as to all these inventions of Pytheas from Massalia about the country along the ocean coast, which he covered with his information from astronomy and mathematics. Of course, we ignored these people. (Strabo, Geography, Book 7: 3.1)

The discussions continued to grow and extend, scarcely straying from the confines of ancient Pangaea, where the elusive mountain ranges of the Ripheans were rumored to have existed

Figure 18.19. L. A. Yelnitsky's reconstruction of ancient Pangea (Yelnitsky: 66). Usl. designations: 1—Mediterranean Sea, 2—Black and Azov Seas, 3—Caspian Sea, 4—The Red Sea, 5—the Ripeian Mountains, 6—the Caucasus, 7—the Cassiterid Islands, 8—Cape Soloent, 9—the Istrian River (Danube), 10—the Nile River, 11—the Tanais-Fasis River (Don).

(Figs. 18.9, 18.10, 18.13). Nonetheless, there is no discernible rationale for us to delve into the unfathomable whirl of this mythical fantasy. Therefore, it is prudent to yield the floor once more to those specialists who have delved into this topic.

It is interesting to note that L. A. Yelnitsky's study of the mysterious mountain country resulted in his own drawing of the Ecumene-Pangea map, which prominently features the Ripei, significantly stretched in a latitudinal position (Fig. 18.19). It appears that the map's author perceived the Ripeian mountain range as particularly suitable for locating the source of mighty rivers, which flowed either southward to the Euxine Pontus or northward toward the Northern Ocean (Septentrionali oceano). However, there is one notable exception: the Istrian-Danube, whose sources are believed to lie far to the west within the Alps and the Black Forest, according to some speculations. It is there, according to the descriptions in Appolonius of Rhodes' poems (295–215 BCE), that the Ripeian mountains are said to exist:

> There is a river among those rivers, the boundless horn of the Ocean,
> Equally wide and deep for cargo ships is passable,—
> The "Istrom" who named the river traced it far:
> She dissects the fields without counting, first remaining
> Whole; its sources are far beyond the places where
> the Cold Boreas breathes, in the Ripeian Mountains they make a noise, falling down,
> But only it will reach the border of the Thracians and Scythians,
> How it splits in two; the first sleeve directs
> Our sea has its own current, and the other—in a deep bay,
> Where far into the mainland does the Trinacrian Sea extend,
> It is close to the earth adjacent to yours,—if they truly say,
> That the river Aheloy runs out of your country.
> (Apollonius of Rhodes, Canto 4, 282–293)

Concerning the Rige—the far western source of the Istra-Danube—about six hundred years after Apollonius of Rhodes, a similar notion, though less detailed, was reiterated by one of the founding fathers of the Christian church, St. Basil the Great (c. 330–379 CE).

> And from the summer west, Tartis and Istria come out from under the Pirinean Mountain, of which the first flows into the sea beyond the Pillars, and the Istrian, flowing through Europe, flows into the Pontus. And why list other rivers generated by the Riphean Mountains lying beyond the inner neck of Scythia? (Basil the Great, conversation 3)

As we proceed to the next logical conclusion, we take both a giant spatial leap from the Alps to Hindustan and a temporary leap from the days of Apollo of Rhodes to our current age, to trace the origins of the Ripeian legend:

> Rivers flow in golden channels there, lakes with a golden bottom shine like the moon. Sacred streams originate from these reservoirs, the purest and most transparent, because their waters rush over the purest gold. Here the best of the rivers is the "heavenly Ganga," the "daughter of Mandara," the source of all the waters of the earth. From here and into the human world, the gods sent great streams. Indian legends about the sacred mountains of Meru are incomparably richer than the information about the Ripeian Mountains, which we know only from fragmentary mentions of ancient writers.
>
> But even such heterogeneous materials give us the basis for new comparisons of these two traditions. Both the Ripeian Mountains and Meru were located in the far north, they stretched along the northern edge of the world. Both reach the heavens; the luminaries revolve around their peaks. Meru is the abode of the gods, as well as Rips are the mountains of the gods, the Greeks identified them with their gods—Apollo and Zeus. The golden Ripeian Mountains shone like the peaks of Meru. Rivers flowed from the Ripeian Mountains, as well as from Meru, in golden channels—let us recall at least the "Plutonic gold-flowing stream" in Aeschylus. The great rivers of the earth also originate from these torahs. Some of the coincidences can, of course, be considered common to the mythological representations of various peoples. But in the comparison we have made, very significant details are also similar. However, the analogies do not end there . . .

These words belong to two major Russian scientists, the Indologist academic G. M. Bongard-Levin, and the Iranist scholar E. A. Grantovsky in their coauthored book *From Scythia to India. Ancient Arias: Myths and History* (2001). They argue that,

> We also have more characteristic, specific correspondences. And at the same time, it is not individual plots that coincide, but the entire cycle of interrelated representations in direct connection with each other. High mountains, scary monsters, and a happy paradise abode occupy a certain

place here. But these are not just high, but great mountains in the north with descriptions characteristic of both traditions . . . The area on the outskirts of the "land of the blessed" is impassable and shrouded in darkness . . . It ends—both in the Indian and in the ancient tradition—with the Northern Ocean and such a specific feature as "polar" phenomena. Scientists have drawn attention to some of these essential correspondences, for example, to the idea of the northern mountains: the Meru—among the Indians and the Ripeian Mountains—among the Greeks. Coincidences were often explained as follows: the idea of the great northern mountains, over which the luminaries move, arose among the Greeks themselves in the 7th century BCE under the influence of the geographical and astronomical views of the Egyptians; initially, the Hellenes linked the Ripeian Mountains with the mountain ranges of neighboring Thrace, and then, when their geographical horizons expanded significantly, with the distant Zascythian mountains in northern Europe; only much later, during the era of significant influence of Greek and Roman astronomy and geography in the countries of the East, ideas about the northern mountains penetrated into Iran and India. Otherwise, the famous German scientist Kiessling wrote, it will have to be considered a miracle that the Indians gave the Measure the same features and the same astronomical significance as the Greeks to the Ripeian Mountains. (Bongard-Levin, Grantovsky, 35–45)

In the context of this discussion, we are particularly intrigued by the power of myths and their ability to captivate generations, not only those who lived during the era of myth production or those simply interested in the myth's contents, but also prominent specialists in ancient history. It seems that, in this case, the latter may not be particularly concerned with verifying whether the Ripeian myth has any truth to it. The phantasm begins to take on its own image. The demise of a myth typically occurs when interest wanes completely—as fascination fades, so does the myth's power.

18.6. The Chimera of Hyperborea

The chimera of Ripei/Riphei evident in first millennium BCE documents does not stand as the only mystery of the era. Another, perhaps more intricate and remarkable chimera, Hyperborea, is closely associated with intertwined in it. This mystery, derived from ancient Greek thinkers and set against the backdrop of the Ripei, may hold greater interest for those fascinated by this subject. Translated from Greek, hyper refers to something of the north, beyond the north wind, or situated in the north. In essence, it represents the land of the Hyperborean people who resided north of Boreas in the far reaches of the ancient Ecumene.

On maps that either directly or indirectly identify Hyperboreans' enclaves, their locations are depicted side by side with the Ripeians (Figs. 18.9, 18.10, 18.20, 18.21). However, all these maps are reconstructions based on ancient Greek geographic-cosmographic descriptions, created by Renaissance scholars who attempted to interpret the works of ancient sages. Of course, we are compelled to trust these interpretations, despite the fact that their basic source descriptions have been separated from these maps for over a millennium. It is often impossible to ignore the compilation of eye-catching, ludicrous details in these fantastical representations. For instance, the Hyperborean ridge appears to press against Muscovy from the north, yet these mountains cannot be located. Similarly, the Hyperborean ridge, seamlessly transitioning into the multi-thousand-kilometer Ripei/Riphei range, divides the entire Eastern European plain known as Sarmatia into two parts—European and Asian—and raises the same question (Fig. 18.9). There are even depictions of certain islands in the Arctic Ocean attributed to the Hyperborean people (Fig. 18.21)—but what are these islands? However, perhaps most intriguingly, all of this sparks our interest when reconstructing the main stages of development in geographical science as applied to specific regions and continents of our planet.

The Hypeborean chimera is distinguished by another intriguing and mysterious facet: who, after all, were these Hyperboreans allegedly living in northern Europe, and more specifically, what kind of people were they? Herodotus explained them as such (4, 32–36):

> Nothing is known about the Hyperboreans to the Scythians or other peoples of this part of the world, except the Issedonians. However, as I think, the Issedons also know nothing about them; otherwise, perhaps, the Scythians would have told about them, as they tell about one-eyed people. Nevertheless, Hesiod has news of the Hyperboreans; Homer also mentions them in the Epigones (if only this poem really belongs to Homer).

Figure 18.20. The position of the Hyperborean mountains on the Asia VII map; reconstruction: Tommaso Porcacchi, Padua, 1620 (https://upload. wikimedia.org/wikipedia/commons/4/4a/Thomas_Porcacchi._Tavola_ Settima_Dell%27Asia_Tabula_Asiae_VII._Padua_1620.jpg.

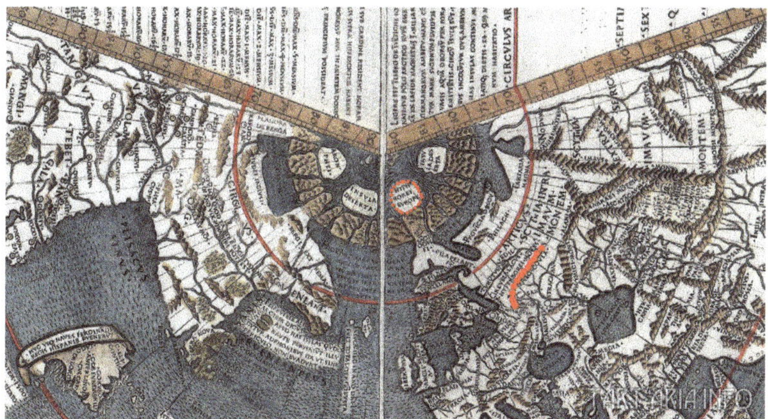

Figure 18.21. Clipping from the world map of 1507–1508 by the Flemish cartographer Johannes Ruysch (c. 1460–1533). The red stripe marks the Hyperborean Mountains that overlap Muscovy from the north; the red dots in a circle, in all probability, indicate some islands in the Arctic Ocean, referred to as the Hyperborea of Europe.

The Delosians tell much more about Hyperboreans. According to them, the Hyperboreans send sacrificial gifts wrapped in wheat straw to the Scythians . . . For the first time, the Delosians say, the Hyperboreans sent two girls named Hyperocha and Laodice with gifts. Together with them, five Hyperborean citizens were sent as counselors for the safety of the girls . . . However, when the messengers did not return to their homeland, the Hyperboreans were afraid that misfortune could befall the messengers every time and they would not return home. Therefore, they began to bring sacred gifts wrapped in wheat straw to the border of their possessions and pass them to their neighbors with a request to send them to other nations. And I know xactly what they do. In honor of these Hyperborean girls who died on Delos, girls and boys there cut their hair . . .

So, enough has been said about Hyperborea. I do not want to mention the legend of Abaris, who, as they say, was also a Hyperborean: he wandered all over the earth with an arrow in his hand and at the same time did not eat anything (*I do not believe in the existence of Hyperboreans at all*) [italics added—E.Ch.].

It appears particularly curious and significant that the father of history, when discussing the Hyperboreans, admits his disbelief in their eistence. Similarly, another prominent figure in geography, Strabo, who lived four centuries after Herodotus, shared the same doubts. In fact, he regarded both the Hyperborean Mountains and the Hyperborean people themselves as fabrications.

For the ancient Greeks, however, the idea of a remote, fantastical, and beautiful country was compelling and attractive. One of the most prominent and pioneering creators of this fantasy was Hecataeus of Abdera. However, little remains of his work "On Hyperborea," and what *does* survive consists mostly of fragments found in various later authors. The fantasy introduced by

Hecataeus of Abdera likely contributed to the formulation of intriguing perspectives regarding the remarkable land of Hyperborea and its people among numerous thinkers in subsequent centuries. However, even before Hecataeus, the famous poet Pindar (522/518–448/438 BCE) wrote in his "Olympic Odes" that

> none of the mortals, either by sea or by land, can find a wonderful road to the dwellings of the Hyperboreans. One hero Perseus shared their feast with them, coming to their homes; he found them bringing to God a glorious hecatomb of donkeys. There are endless festivals, hymns are heard that delight the heart of Apollo, and he laughs, seeing the lustful impudence of these animals. The cult of the muses is no stranger to Hyperboreans; from everywhere choirs of young girls gather to the sound of peace and the sweet sounds of flutes, and, crowned with a golden laurel, they indulge in the joy of holidays. This holy tribe knows neither diseases nor weakness of age; it lives far away from (hard) labors and battles, not fearing Nemesis' vengeance. Danae's son got in here . . . led by Athena, and killed the Gorgon here. . . . Nothing seems improbable to me. . . . The Istrian River [Danube] originates in the country of the Hyperboreans. . . . Hyperboreans everywhere are considered to be dedicated to Apollo. They are called Hyperboreans from a certain Athenian Hyperborea, as Phanodemus says. Philostephanus says that Hyperboreus was a Thessalian, and others produce them from Pelasgus Hyperboreus, the son of Phoroneus and Perimela, the daughter of Aeolus. Ferenik says that the Hyperboreans descended from a titanic family, and Hecateus tells a different story. (Latyshev, 51, 52)

Gellanik of Mytilene (c. 485–400 BCE) also discussed Hyperboreans living beyond the Ripeian Mountains:

> The concept of justice is familiar; they do not eat meat and eat only vegetation and fruits. Those who have reached the age of sty, they take them out of the gate and thus get rid of them.

> The Christian theologian and philosopher Titus Flavius Clemens (c. 150– c. 215), known to us as Clement of Alexandria, reported about Hyperborea and the work of Hellanik in his famous book "Stromata." Clement himself was quite familiar with classical Greek philosophy and literature and taught at the Catechetical School of Alexandria. (Clement of Alexandria, 117)

The Roman encyclopedist Pliny the Elder (22/21–79 CE), who lived almost half a millennium after Pindar, Hecateus, and Hellanica, wrote even more fantastic lines about the Hyperboreans:

> Beyond the Ripeian Mountains and Aquilon lives, if one can believe it, the blessed people of the so-called Hyperboreans. They reach a very advanced age and are famous for legendary miracles. Here, it is believed, are the loops on

which the world turns, and the farthest edges of stellar orbits. [These places are] illuminated by the Sun for six months, but not as ignorant people say, from the vernal equinox to autumn. For [Hyperboreans] the sun rises once every year, on the day of the summer solstice, and sets once, on the day of the winter solstice. Their area is warm, [the climate] is happily proportioned, without any harmful miasma. Forests and groves serve as homes for Hyperboreans. Both separately and collectively they worship the gods, never knowing strife and resentment. They die only when they are sated with life: having given a feast and anointed their old age with precious ointments, they throw themselves from a certain rock into the sea. Oh, the most blessed kind of celebration! (Pliny the Elder 4, 89)

However, in the Roman Empire at the end of the fourth century, the supreme powers declared that citizens were only permitted to entertain thoughts within the strict boundaries of Christian dogma. Any other speculation was considered hostile to the truth, and those who promoted them faced punishment. A millennium later, when Europeans ventured into Hyperborean land (whether through thick forests or along the coastal tundra of the Arctic Northern Ocean), the reality they encountered was vastly different to the idyllic picture painted by ancient thinkers. Yet, the mythical land of the Hyperboreans, where strife and disease were conquered and all was harmonious, had long been forgotten by the European pioneers.

Still, these fantasies refused to fade completely. Sometimes, they resurfaced unexpectedly, taking on strange forms and occupying unforeseen positions, as evidenced by the following instances.

18.7. From the Bowels of the Earth to the Heights of Space

The era of a new time has dawned, marked by a resurgence in scientific knowledge and understanding of the planet. It quickly became evident that this new science needed to build its conclusions upon the observations of ancient times. Consequently, the words, titles, and definitions widely employed by ancient thinkers began to resurface in the minds of scientists and scholars.

Geologists, for example, revisited the ancient notions laid out by Greek thinkers—such as Tartarus lying deep in the bowels of the earth and the Riphean mountains attributed to both him and Gaia. According to ancient tradition, Tartarus emerged as the third entity in the Universe after Chaos and Gaia, among the highest and most significant of the Gods. Geomorphologists chose to designate one of the early stages of Earth's formation as Riphean, a topic explored in some detail earlier in this book. Tatarus was depicted as a stern, ruthless, and ferocious deity. Perhaps it was his formidable nature that dissuaded scientists from incorporating him into the nomenclature of terrestrial periods.[1]

---

[1] The Riphean as a geological period (erythema) was isolated, by the way, by the Soviet geologist N. S. Shatsky in 1945 in the Southern Urals.

Figure 18.22. The heavy stone Sisyphus turning in the underworld and observing behind him is the lady of the underworld Persephone. Black-figure amphora, ca. 530 BCE.

Figure 18.23. Sisyphus; Abraham Jans van Diepenbeck, 1655; Warburg Institute Library, London. In the lower part there is a text from Homer's Odyssey corresponding to this image.

Figure 18.25. Riphean Mountains on the Moon; satellite image.

Figure 18.24. Selena, Capitolini Museum, Rome, Italy.

The enduring martyrdom of Sisyphus, condemned to eternal toil by pushing a boulder from the depths, is also linked to Tatarus. The Sisyphisian labor—endless and futile in its struggle—is still a term people use today (Figs. 18.22 and 18.23). Perhaps only the arduous labors of miners can be compared in some ways to his efforts.

However, the Riphean presence extended beyond the depths of the Earth. Unexpectedly, it ascended and found itself on the surface of the moon—the celestial Selene (Fig. 18.24). There, terrestrial astronomers discovered a certain mountain range (Fig. 18.25) near the lunar Ocean of Storms and named it Ripheus. This discovery proved immensely intriguing—for millenna, the Riphean mountains were sought on Earth, depicted on maps whenever feasible, yet they only became tangible in our solar system. Such occurrences, it seems, are not entirely uncommon.

Judaism: Tracing the Origins of the Abrahamic Trio of Faiths

One of the main objectives of our research is to discern the distinctive features and peculiarities of ideological-cultural frameworks of the Modern Era. However, it is imperative to remember that the ideological constructions of societies, as a rule, typically delve into the intricate depths not only of previous centuries but of the current millennia as well. This reality is evident in the trajectory of the fundamental triad of the Abrahamic faith religions. Therefore, we shall metaphorically journey back through the past two millennia to explore the origins of this triad.

19.1. Judaism as the Origin of Monotheism

The monotheism of the three religious doctrines bound together by the title "Abrahamic" had a fundamental impact on the ideological foundations of numerous ethnic and cultural groups across the globe. The original roots of Abrahamic monotheism can be traced back to a remarkably proximal and spatially compressed region in the Iranian-Anatolian enclave, specifically in the smallest of all the sub-enclaves discussed—the Palestinian enclave (see Section 6.5). It is in this geographical area that the cradles of two Abrahamic religions were born: first Judaism and later Christianity. The triad was completed by Islam, which was established in relative proximity to the origins of the first two religions, within the Arabian enclave.

The original history of the Jews and Israel has been espoused in numerous sections of the Bible, with the so-called "Shestodnev" occupying the initial position. The prophet Moses is often considered to be the original monotheistic prophet and the author of the entire Torah (also called the Pentateuch), even though the Bible rarely refers to him as a prophet. However, in describing Moses' life in the book of Deuteronomy, it is written that,

> Israel has never had a prophet like Moses, for the Lord knew Moses by sight. The Lord sent Moses to perform great miracles in the land of Egypt, and these miracles were seen by Pharaoh, and all his entourage, and all the Egyptian

people. No prophet has ever performed such great and amazing miracles as Moses performed before the eyes of the entire Israelite people. (Deuteronomy 34:10–12)

At the same time, the absence of any reliable historical information about Moses' life in ancient sources of the pre-Hellenistic period (excluding the Bible) raised many doubts among biblical scholars about whether Moses actually existed. Some researchers have concluded that Moses is a fictional, legendary figure and that his narrative is purely a result of mythological creativity. However, numerous scholars acknowledge that biblical legends contain elements rooted in historical events, and in the case of Moses, there is a basis for such events associated with his persona, although pinpointing the exact nature of his deeds remains challenging. Nevertheless, according to the traditional narrative, Moses was a divine creation, a product of the Lord's work over six days. Therefore, we will commence this chapter by exploring this pivotal aspect of Judaism.

19.2. Hexameron: at the Origins of All Things

The sixth day, which marks the beginning of the Bible in the book of Genesis, is arguably the original and most significant foundation of the Abrahamic triad of religions. It is commonly believed that many texts of the Old Testament were canonized following the return of the Jews from Babylonian captivity at the end of the sixth to fourth centuries BCE. In any case, it is likely that the six days first appeared in the Second Book of Moses, Exodus:

> In the beginning God created the heavens and the earth. The earth was without form, and void; and darkness was on the face of the deep. And the Spirit of God was hovering over the face of the waters. Then God said, "Let there be light"; and there was light.
>
> And God saw the light, that it was good; and God divided the light from the darkness.
>
> God called the light Day, and the darkness He called Night. So the evening and the morning were the first day. Then God said, "Let there be a firmament in the midst of the waters, and let it divide the waters from the waters."
>
> Thus,God made the firmament, and divided the waters which were under the firmament from the waters which were above the firmament; and it was so. And God called the firmament Heaven. So the evening and the morning were the second day. Then God said, "Let the waters under the heavens be gathered together into one place, and let the dry land appear"; and it was so. And God called the dry land Earth, and the gathering together of the waters He called Seas. And God saw that it was good.

Then God said, "Let the earth bring forth grass, the herb that yields seed, and the fruit tree that yields fruit according to its kind, whose seed is in itself, on the earth"; and it was so. And the earth brought forth grass, the herb that yields seed according to its kind, and the tree that yields fruit, whose seed is in itself according to its kind. And God saw that it was good. So the evening and the morning were the third day.

Then God said, "Let there be lights in the firmament of the heavens to divide the day from the night; and let them be for signs and seasons, and for days and years; and let them be for lights in the firmament of the heavens to give light on the earth"; and it was so.

Then God made two great lights: the greater light to rule the day, and the lesser light to rule the night. He made the stars also. God set them in the firmament of the heavens to give light on the earth, and to rule over the day and over the night, and to divide the light from the darkness. And God saw that it was good. So the evening and the morning were the fourth day.

Then God said, "Let the waters abound with an abundance of living creatures, and let birds fly above the earth across the face of the firmament of the heavens." So God created great sea creatures and every living thing that moves, with which the waters abounded, according to their kind, and every winged bird according to its kind. And God saw that it was good. And God blessed them, saying, "Be fruitful and multiply, and fill the waters in the seas, and let birds multiply on the earth."

So the evening and the morning were the fifth day.

Then God said, "Let the earth bring forth the living creature according to its kind: cattle and creeping thing and beast of the earth, each according to its kind"; and it was so. And God made the beast of the earth according to its kind, cattle according to its kind, and everything that creeps on the earth according to its kind. And God saw that it was good. Then God said, "Let Us make man in Our image, according to Our likeness; let them have dominion over the fish of the sea, over the birds of the air, and over the cattle, over all the earth and over every creeping thing that creeps on the earth."

So God created man in His own image; in the image of God He created him; male and female He created them. Then God blessed them, and God said to them, "Be fruitful and multiply; fill the earth and subdue it; have dominion over the fish of the sea, over the birds of the air, and over every living thing that moves on the earth."

And God said, "See, I have given you every herb that yields seed which is on the face of all the earth, and every tree whose fruit yields seed; to you it shall be for food. Also, to every beast of the earth, to every bird of the air, and to everything that creeps on the earth, in which there is life, I have given every green herb for food"; and it was so.

Then God saw everything that He had made, and indeed it was very good. So the evening and the morning were the sixth day. (Genesis 1)

In general, throughout the course of the six days, God evaluated His own creative works with the word "good," but the culmination of His achievements led Him to consider them "very good." Now, in a notably simplified rendition of Josephus from his famous book *Jewish Antiquities*, let us delve deeper into the significance of the six days:

In the beginning, God created heaven and earth. And since the latter was not visible, but hidden in deep darkness, and the spirit [of God] hovered over her, the Lord commanded the light to be created. Having surveyed, after the appearance of the latter, all matter in its totality, He separated light from darkness and gave the latter the name of night, and called the first day, and called the beginning of the emergence of light and its cessation morning and evening. Thus arose the first day; but Moses says: one day. Although I would be able to explain the cause of this phenomenon even now, however, since I promised to provide an explanation of the causes of all phenomena in a special essay, I postpone explaining this until then. After that, on the second day, He spread the sky over everything, because He found it necessary to separate the sky from everything else as something independent, and surrounded it with a crystal, in which, for the benefit of irrigation of the earth, He included water and dampness very conveniently. On the third day, He created the earth by spreading the sea around it. On the same day, He immediately summoned plants and seeds from the ground. On the fourth day, He decorates the sky with the sun, moon and other luminaries, determining their movements and paths [in the sky], so that the changes of time are determined thereby. On the fifth day, He created floating animals and birds, assigned the depth of the sea to the first, and the air to the second, and brought them together, respectively, sexually, for the sake of reproduction of offspring, location and multiplication of their kind. On the sixth day, He created four-legged animals, making them males and females. On the same day He created man. And so, says Moses, in all these six days the world arose with all its contents, and on the seventh [the Lord] rested and rested from his labors. Hence, we abstain from our labors on this day, calling it sabbaton: this name in Hebrew means rest. (Flavius I, 1, 1)[1]

---

[1] There are two opposite assessments by Jews of the personality of Josephus (Fig. 19.2). Some believe that Josephus was a traitor who left his people in trouble, joined the enemy camp, and became an apologist for Rome.

The text of the sixth day also appeared in the same order for Christians. For Muslims, the canon of six days is much shorter, and therefore departs from the original:

> There is no doubt that your Lord is Allah, Who created the heavens and the earth in six days. Then He rushed to the heavens of heaven. The day is covered by the night that follows it. And the sun, and the moon, and the stars are subject to His will. And isn't it given to Him to create and command? (Surah Al-Araf, verse 54)

19.3. Moses: The Prophet of Judaism and Deeds

The origins of Judaism were allegedly rooted in the second millennium BCE, although there exists no reliable grounds for such statements. However, it is certainly held that the first (or early) temple of Yahweh in Jerusalem was built in the tenth century BCE. Moses, as the true prophet of Jerusalem, is believed to have lived even earlier.

According to the Torah, Moses (Fig. 19.1) is the founder of Judaism, which united the Israeli tribes into a single people during the thirteenth century BCE. He is revered, not only as a prophet but also as a liberator-leader and legislator. According to the Electronic Jewish Encyclopedia, the image of Moses is endowed with unique features that are absent in epics of other cultures. He was born in Egypt, and it was there that the Almighty recognized his special talents, bestowing him with extraordinary abilities and even revealing His hitherto unknown name—Yahweh. The Lord also commanded him to rescue all his tribesmen from Egyptian slavery. It was an extremely difficult task, which Moses managed to accomplish by going to the same almost dry Red Sea:

> Having offered up these prayers, Moses struck the sea with his staff, which parted from this blow and, retreating before the Jews, gave them the opportunity to leave on a dry path. . . . The Egyptians at first thought that the Jews had lost their minds, going to an obvious death. When they saw that the Jews had walked a considerable distance without harm, without encountering any obstacles or difficulties on their way, they decided to rush after them in pursuit, hoping that the sea would part in front of them. . . . And so, when the whole Egyptian army was in the very middle of the sea, the latter closed again and the waves, swollen by the winds, collapsed back on the Egyptians with all their strength and surged over them. . . . The latter all perished, so that there was not even a person left who could announce to the rest [the inhabitants of Egypt] about the disaster that had befallen the army. . . . The Jews were unable to

Figure 19.1. Moses. Sculptor Michelangelo Buonarroti;
in the sculptural complex of the tomb of Pope Julius I,
work of 1513–1515. It is noteworthy that the famous
artist depicted horns on the head of Moses at the famous
Micheangelo sculpture, becoming an unwitting victim of a
poor translation of the Vulgate in the fourth century from
ancient Greek into Latin, where the sun rays shining on the
prophet's head could also be translated as horns.

Figure 19.2. Marble bust with the alleged image of Josephus
(Ny Carlsberg Glyptotek, Copenhagen).

restrain themselves from joy at the sight of their miraculous salvation and the
death of their enemies and were now firmly convinced that they would hence-
forth be free, since the people who oppressed and held them in slavery had
been destroyed, and the Lord God had so obviously patronized them. . . . They
spent the whole night in fun and songs. Moses composed a hymn of praise in
honor of the Eternal, in which he glorified and thanked the Lord God for his
mercy. (Flavius 2:16, 2–4)

Although the Jews were saved in such a miraculous way, they soon lost heart during their
journey to Mount Sinai, because the country through which they had to move was completely
deserted, and there was nothing necessary to maintain life in it.

The extreme shortage of water oppressed everyone here, so that not only could
there be no question of any amenities for people, but there was no way to deliver
the necessary pasture to the cattle. This country is a continuous sandy desert
without the slightest sign of oases. (Flavius 3:1, 1)

Therefore, not all Jews following Moses were satisfied, which caused discontent and
even riots.

Figure 19.3. Moses receives from the Almighty the stone tablets of the covenant: And when God stopped talking to Moses on Mount Sinai, he gave him two tablets of revelation, stone tablets on which it was written with the finger of God (Ex 31:18); Raphael Santi, painting of loggias in the Vatican Palace, 1508–1517.

And so the people began to accuse their leader again and reproach him, putting on him all the grief they experienced and all the hardships . . . [After all] during their thirty-day journey, the people ate all the supplies they had taken with them . . . and were close to complete despair. Thinking only of their present plight and completely forgetting about the support that the Lord God had already repeatedly given them, as well as about the valor and wisdom of Moses, the Jews became enraged against their leader and were already preparing to stone him as the main culprit of their present calamity. (Flavius 3:1, 3)

Indeed, this marked the apex of the renowned exodus in the desert, which spanned forty years and the inception of the monotheistic doctrine of Judaism. Nonetheless, the journey was not always without challenges:

Moses turned 120 years old when he completed his earthly journey. And he died there . . . in the land of Moab . . . and was buried in the valley . . . opposite Beth-Pegor, and no one knows the place of his burial even to this day. (Deut 34:5–6; see also Figs. 19.3–19.6)

19.4. Foreskin Circumcision as Evidence of God's Choice

For the heirs of Abraham in the Jewish faith, circumcision of the infant male's foreskin became a necessary practice in designating themselves as Jews and identifying as God's chosen people:

Figure 19.4. Moses shows the tablets of the covenant to the people; the tablets were the work of God, and the writings inscribed on the tablets were the writings of God (Ex 32:16); Raphael Santi, painting the loggias in the Vatican Palace, 1508–1517.

Figure 19.5. Moses in anger when he saw that his people were worshipping the golden calf; "he approached the camp and saw the calf and the dancing. . . . He was inflamed with anger and threw the tablets out of his hands and broke them under the mountain; and took the calf that they had made and burned it in the fire, and he wiped it to dust, and scattered it on the water, and gave it to the children of Israel to drink" (Exodus 32, 19, 20; biblical engravings by Yu. Sh. von Carolsfeld, 1852–1860).

Figure 19.6. Torah scroll, Pentateuch.

And God said to Abraham, "But you keep my covenant, you and your descendants after you in their generations. This is my covenant, which you must keep between me and between you and between your descendants after you . . . let all the male sex be circumcised among you: circumcise your foreskin: and this will be a sign of the covenant between me and you. Eight days from birth, let every male child born in your house and bought with silver from some foreigner who is not of your seed be circumcised in your childbirth. Be sure that he who is born in your house and bought with your money will be circumcised, and my covenant on your body will be an everlasting covenant. But the uncircumcised male who does not circumcise his foreskin [on the eighth day], that soul will be cut off from his people because he has broken my covenant." (Gen 17:9–14)

19.5. Solomon Constructs the First Temple

Next in the historical narrative of the Jewish people emerges an equally legendary figure (after Saul and David)—the king of Israel, Solomon (Fig. 19.7). As many Jewish sources assert, Solomon possessed more wisdom than anyone else on Earth:

And the Lord gave Solomon wisdom and a very great mind, and a vast mind, like the sand on the seashore. And the wisdom of Solomon was above the understanding of all the sons of the East and all the wise Egyptians. He was wiser than all people. . . . And he spoke three thousand proverbs, and his song was one thousand and five. (2 Samuel 4:29–32)

The idea of building the Temple of the Lord first came to King David, Solomon's father. However, the Lord said to David:

Your son, whom I will put in your place on your throne, he will build a temple to my name. . . . In the four hundred and eightieth year after the departure of the children of Israel from the land of Egypt, in the fourth year of Solomon's reign over Israel . . . [Solomon] began to build a temple to the Lord. (I Kings 5:5, 6:1)

The construction of the temple proved to be difficult for Solomon and his followers, and he was forced to seek repeated help from the Phoenician king Hiram, who resided in the city

Figure 19.7. On the left: King Solomon of Judea, tenth century BCE; on the right, Jerusalem during its destruction in 586 BCE by the Babylonian king Nebuchadnezzar II (Nuremberg Chronicle, fifteenth century).

of Tyre (now Sur in southern Lebanon). The monarchs enjoyed a close friendship, which was reflected in their correspondence (2 Kings 5:1–18). In Solomon's words:

> You know that David, my father, could not build a house for the name of the Lord his God, because of wars with the surrounding nations, until the Lord subdued them under the soles of his feet; now the Lord my God has given me rest from everywhere: there is no enemy, and there are no more obstacles; and behold I intend to build a house in the name of the Lord my God, as the Lord said to my father David. . . . Therefore, order the cedars from Lebanon to be cut for me; and, behold, my servants will be with your servants, and I will give you the payment for your servants, which you will appoint; for you know that we have no people who could cut down trees like the Sidonians.

It might not seem very flattering to admit to the king, who serves as the epitome of wisdom and whose intellect rivaled the vastness of sand on the seashore, that among his subjects, there were none skilled enough even to fell trees properly. Nonetheless, Hiram's response carries a favorable tone:

> I have listened to what you sent to me for, and I will fulfill all your desire for cedar and cypress trees; my servants will bring them from Lebanon to the sea, and I will bring them by rafts by sea to the place that you will appoint for me, and there I will lay them down, and you will take them; but you also fulfill my the desire to deliver bread for my house.

> And King Solomon imposed a duty on all Israel; the duty consisted of thirty thousand people. And he sent them to Lebanon, ten thousand a month, alternately; they were in Lebanon for a month, and two months in their house. . . . Solomon also had seventy thousand heavy-bearers and eighty thousand stone-cutters in the mountains, besides three thousand three hundred superiors appointed by Solomon over the work to supervise the people who

did the work. And the king commanded to bring large stones, expensive stones, for the foundation of the house, stones made. And Solomon's workers and Hiram's workers hewed them. (1 Kings 5:8–9, 13–17)

Having conceived a large-scale construction program, Solomon was completely dependent on the Tyrian king Hiram, who surpassed him both with his wealth and with the necessary craftsmen in his skill. From Hiram, Solomon received both construction materials, and experienced workers, and artisans, and, finally, gold, in turn paying him for it exclusively in kind or by ceding part of the territory. All this does not speak either in favor of the imaginary illustrious riches of Solomon, nor in favor of the high economic development and economic prosperity of the country in his time (Tyumenev, 26).

Still, all must be concluded:

Solomon built his house for thirteen years and finished his whole house. And he built a house of Lebanese wood, a hundred cubits long, fifty cubits wide, and thirty cubits high, on four rows of cedar pillars. . . . And a platform of cedar was laid over the logs on forty-five pillars, fifteen in a row. There were three rows of window jambs; and three rows of windows, window against window. (2 Samuel 7:1–5)

Upon completion of this grandiose construction, Solomon turned to the Almighty with a heartfelt prayer:

I have built a temple to house You, a place to dwell with You forever. . . . And may these words, with which I have prayed before the Lord, be close to the Lord our God day and night, that He may do what is necessary for his servant, and what is necessary for his people Israel, from day to day, so that all nations may know that the Lord is God and there is none but Him. . . . And the king and all the Israelites with him offered a sacrifice to the Lord. And Solomon offered twenty-two thousand cattle and one hundred and twenty thousand sheep as a peace offering that he had offered to the Lord. So the king and all the children of Israel consecrated the temple to the Lord. (2 Samuel 6:1–3; 8:20–63)

The first temple is believed to have existed for 364 years. In 597 BCE, the Babylonian ruler Nebuchadnezzar II is credited with consolidating power and sending his soldiers to capture Jerusalem in 586 with the intent to destory its main attraction—the legendary First Temple of Yahweh. However, there is a considerable lack of information that has been preserved about the appearance of the early temple of Judaism:

The conquered Jews were taken into captivity, which lasted for about seven decades. However, the time of the Lord's command has come, as the First Book of Ezra narrates:

In the first year of Cyrus king of Persia, in fulfillment of the word of the Lord from the mouth of Jeremiah, the Lord stirred up the spirit of Cyrus king of

Persia, and he commanded to declare throughout his kingdom, verbally and in writing . . . the Lord God of heaven gave me all the kingdoms of the earth, and He commanded me to build Him a house in Jerusalem, that in Judea. Whoever is among you, of all his people, may his God be with him, and let him go to Jerusalem, which is in Judea, and build the house of the Lord God of Israel, the God who is in Jerusalem. And all those who remain in all places, wherever he lives, let the inhabitants of that place help him with silver and gold, and other property, and cattle, with a willing gift for the house of God that is in Jerusalem. And the heads of the generations of Judah and Benjamin, and the priests and Levites, rose up, every one in whom God had stirred up his spirit, to go and build the house of the Lord, which is in Jerusalem. The whole company together consisted of forty-two thousand three hundred and sixty persons, besides their servants and their female servants, of whom there were seven thousand three hundred and thirty-seven; and with them two hundred singers and singers. (1 Ezra 1:1–4; 2:64–65).

19.6. The Construction of the Second Temple

The construction of the Second Temple began on the ruins of the first, with the date presumed to be around 539 or 538 BCE (Figs. 19.8, 19.9).

In the first year of Cyrus' reign (i.e., the seventieth since the eviction of our people from their homeland to Babylon) The Lord God took pity on the captivity and suffering of those unfortunate people, according to how he predicted to them through Jeremiah the prophet on the eve of the destruction of the city [Jerusalem], namely, that after seventy years of slavery to Nebuchadnezzar and his descendants, he would return them back to their homeland, and they would rebuild the temple and enjoy their former well-being. And now the Lord has granted them all this, having inspired Cyrus to write and send out the following proclamation throughout Asia: "Thus says King Cyrus: after almighty God granted me the kingdom over the whole earth, I became convinced that he is the same Deity worshipped by the people of Israel. He foretold my name through the mouth of the prophets and announced that I would rebuild His temple in Jerusalem in the land of Judea." (Flavius 11:1, 1)

While the First Temple existed for 364 years, its second iteration existed for almost twice as long—585 years until it was destroyed by fire in 70 CE by Romans led by Titus. Both temples were built on a hill, which is still referred to as the Temple Mount.

Around the same time, many considered the Babylonian captivity to be an undeniable turning point in the development of Jewish religious and national consciousness. In this context, scholars refer to the Ktuvim, or the third and final part of the Tanakh (the Jewish Holy Scripture or the Old Testament). Included in this section is the tenth book, the so-called First Book of

Ezra (Figs. 19.10 and 19.11). Ezra's actions are commonly linked to the canonization of several fundamental tenets of Judaism found in the Tanakh.

Figure 19.8. Construction of the Second Temple in Jerusalem.

Figure 19.9. The Temple Mount in Jerusalem, modern appearance; it is believed that the First and Second temples of Yahweh were built here.

Figure 19.10. Ezra the Scribe. "When the sacred books were destroyed in the fire of war, Ezra repaired the damage."

Figure 19.11. Ezra with the sacred texts in front of the audience (G. Dore).

19.7. The Phenomenon of *Eretz Israel*

Eretz Israel (Hebrew) translates to "the land of Israel." Adherents of Judaism will assert that Israel is the Promised Land, destined by the Almighty to the people of Israel according to his Covenant, thereby making the land inseparable from its people. It was in this land that the cradle of monotheism was formed, and the people of Israel carefully cherished and upheld this monotheistic tradition. In his book *The Heretic*, Mikhail Veller argues that

> the Jews were the first to come to Monotheism. If we remain in the positions of an atheist and a materialist, an honest analyst, we can talk about the increased creativity of the collective conscious and unconscious people of Israel. Roughly speaking, the Jews invented and invented the one God; and the fruitfulness of this brilliant idea cannot be overestimated. On the core of Monotheism, our entire Judeo-Christian civilization was created and exists, as well as the Muslim one. (Veller, 7–8)

Another significant facet of the phenomenon is the preservation of monotheistic ideals and the longstanding and unwavering adherence to it that has spanned nearly three millennia. Maintaining fidelity to a primordial worldview and one's special place within it is difficult considering the many changing norms and social expectations. This is evidenced by the countless examples of victors imposing new rules and doctrines to suppress the oppressed's cherished doctrines. However, the victors could not always succeed in enforcing their demands—consider, for example, the history of the Great Mongol Empires of Eurasia. At times, the oppressed would renounce their past willingly, believing it would be easier to join the ranks of their oppressors and assuming that their divine protectors were stronger. Nevertheless, every oppressed ethnic group must inevitably face the often humiliating test of loyalty to their heritage and culture. The resilience of the Israelites in maintaining their monotheistic beliefs through millennia of trials and tribulations is a testament to the profound strength of their cultural and religious identity.

For the Jewish people, this test was repeated across many generations. We will only briefly list these trials, highlighting the periods of domination by various conquerors. The beginning of this process is attributed to the return of the Jews from Persian captivity and the construction of the Second Temple. Consequently, the total chronological length of such comparisons approximately equaled two and a half millennia, when the Israelis succeeded in canonizing their doctrine into the form of the Torah.

Although the Persian ruler Cyrus II allowed the Jews to restore their destroyed First Temple, this did not imply the end of Persian rule, which would endure for over 200 years from 538 to 322 BCE. However, after Alexander the Great defeated the Persians, he imprisoned the Jews under Hellenic rule in Macedonia, which lasted until 63 BCE. Following this, the supreme power shifted to Rome, first as a republic until 27 CE, then as the Roman Empire until 320 CE. The split of Christianity and the Roman Empire transferred power to Byzantium in the fourth century. From 639 to 1099, Palestine was under Arab control, until they were defeated by Crusader armies which would subsequently establish their dominance for the next 200 years until they were defeated by the Malmuks in 1291.

Next came the rise of the Ottoman Empire, under whose rule Palestine and its inhabitants remained for about four centuries. The significant year of 1917 brought glimpses of relief from oppression for the Jewish people. Finally, on May 14, 1948, the independent Jewish state was established in Tel Aviv, one day before the end of the British mandate of Palestine.

Thus, two and a half millennia have passed in the unique history of Eretz Israel. It is essential to mention the history of the Hasmonean dynasty here as perhaps the only period when Israel achieved independence, even if only for a short time.

19.8. Hasmoneans in the History of Eretz Israel

In 167 BCE, an armed uprising of the Israelites broke out against the rulers based on the canons of Hellenism prevailing in Palestine:

> In 167 BCE, the king of Seleucid Syria, Antiochus IV Epiphanes, forbade, under threat of death, the execution of the laws of the Torah, in particular, circumcision and Sabbath observance. Along with the ban on the practice of Jewish worship, he ordered the introduction of pagan cults in Judea; the Jerusalem temple was desecrated and turned into a sanctuary of Olympian Zeus. These decrees and the religious persecutions that followed them, unprecedented in the ancient world, not only did not lead to the desired results, but also provoked armed resistance from the Jewish population . . .

> As a result of this consolidation, the religious and ethnic unity of the non-Hellenized population of Eretz Israel became a factor that the Roman conquest failed to change. The name "Judea," which in the past designated a part of Eretz Israel, has become the name of the whole country. The success of the Hasmonean uprising guaranteed the preservation of Judaism as the religion of the Jewish people, which, in turn, had a decisive influence on the further history of the Jewish people and the formation of monotheism in Western culture (Electronic Jewish Encyclopedia).

The period of the Hasmonean reign lasted a little more than a hundred years from 140 to 103 BCE. The Jews, however, failed to retain power.

The Hasmoneans' activity halted the Hellenization of the Jews and contributed to the religious assimilation of the Semitic peoples who inhabited Judea. The preservation of Judaism as the national religion of the Jews became an important factor that predetermined the entire future history of the Jewish people. It seems that the authors of the Electronic Jewish Encyclopedia are partly correct in discussing the decisive impact of this dynasty on the history of Israel. Of course, the Hasmoneans served as an important step that led to the preservation of the foundations of Jewish doctrine.

The Israelites held the belief that, particularly since the Babylonian captivity, dispersion into foreign lands and maintaining isolation through diasporas—where several to many families could come together in an unfamiliar environment—was essential for preserving the true faith.

Figure 19.12. God is the Creator of the Universe (Jan Brueghel the Younger, 1670 (?). (Brueghel Jan II God creating.jpg)).

> Libraries have been written about Scattering and diaspora. Sometimes Jews were allowed to live on a par with everyone. But the general vector is a defeat in civil and property rights. Restriction in professions, positions, place of residence. Humiliation and injustice. (Veller, 10)

This issue remains one of the most painful for Jews, and, as Mikhail Veller rightly noted, entire libraries are devoted to it. However, at the end of this section and chapter, for those interested, we will reference just one of the books published relatively recently, thereby limiting our participation in this vast and boundless topic.[2]

It is remarkable, however, that Christianity emerged from the depths of Judaism, linking many dramatic issues throughout the extended history of Judaism. Nonetheless, the Israelites maintained an unwavering faith in the supreme and unshakeable strength of Yahweh (Fig. 19.12), as expressed by prophet Isaiah:

> The word that was in the vision to Isaiah, the son of Amos, about Judea and Jerusalem. And it will come to pass in the last days, the mountain of the house of the Lord will be set at the head of the mountains and will rise above the hills,

---

[2] *Israel and the Diaspora: New Perspectives*, ed. Natan Aridan and Gabriel Sheffer, Israel Studies. Vol. 10. No. 1 (2005).

and all nations will flow to it. And many nations will go and say, Come, and let us go up to the mountain of the Lord, to the house of the God of Jacob, and He will teach us his ways, and we will walk in his paths; for the law will come out of Zion, and the word of the Lord from Jerusalem. And He will judge the nations, and will rebuke many tribes; and they will turn their swords into plowshares, and their spears into sickles: the people will not lift up the sword against the people, and they will no longer learn to fight. (Isaiah 2:1–4)

Chapter 20

Christianity: The Abrahamic Triad in Development

20.1. The Early Schism in Abrahamic Monotheism: The Birth of Christianity

Let us turn briefly to the dramatic and explosive splits that often signify the emergence of ideological innovations and subsequent changes in the social fabric of culture. Indeed, the first of the most sensitive disruptions occurred in the same land of Palestine in the first century of the new era.

The birth of Christianity and the formation of Christian doctrine is inextricably linked with the figure of Jesus Christ. The four canonical Gospels—Matthew, Mark, Luke, and John—are the most significant texts of this era, and arguably for the whole of human history, and were canonized in the fourth century (although the Gospel of John was most likely incorporated later). The Gospel accounts of Jesus's life and deeds are integral to the arguments in this book in that they present a framework for understanding the composition and development of some integral aspects of the Abrahamic religions over more than a millennium and a half. Within the scope of the following arguments, it is important to outline the main milestones of the formation of this triad itself. Thus, we will avoid extensive discussion on these subjects, as debates on these topics are limitless and endless in world literature.

Therefore, to achieve clarity in this task, we will focus on excerpts from the Gospel of Matthew, which seem to the author to be most appropriate for the objectives of the book.

The Birth of Jesus

The Nativity of Jesus Christ occurred in this matter: after the betrothal of his mother Mary to Joseph, but before they were married, it was discovered that she was pregnant by the Holy Spirit. Joseph, who was righteous and wished to protect her honor, planned to divorce her quietly. When he was struck by this thought, however, an angel of the Lord appeared to him in a dream and said: "Joseph, son of David! Do not be afraid to take Mary your wife, for what is born in her is of the Holy Spirit; she will bear a Son, and you will call his name Jesus, for he will save his people from their sins." (1:18–21)

When Jesus was born in Bethlehem of Judea in the days of King Herod, the magi from the east came to Jerusalem (Fig. 20.1) and said: where is the born King of Judea? for we have seen His star in the east and have come to worship Him. When Herod the king heard this, he was alarmed, and all Jerusalem was with him . . . Then Herod, secretly calling the magi, found out from them the time of the appearance of the star and, sending them to Bethlehem, said: go, carefully investigate about the Baby and, when you find it, inform me so that I also go to worship Him . . . (Matt 2:1–8)

However, Herod was dishonest about his intentions to worship the newborn King of Judea. Understanding this, an angel of the Lord appeared in a dream to Joseph, warning him of Herod's true intentions. When Herod realized he had been outwitted by the wise men, he was furious and ordered the massacre of all male infants in Bethlehem and its vicinity who were two years old and younger. This fulfilled what had been spoken through the prophet Jeremiah, who said: "A voice is heard in Ramah, weeping and sobbing and a great cry (Fig. 20.2); Rachel weeps for her children and does not want to be comforted, because they are not there." (Matt 2:16–18)

Many years had passed since the birth of Jesus before a Baptism was required.

Figure 20.1. The Virgin Mary with Jesus and the adoration of the Magi (Raphael Santi, 1508–1517, Vatican Palace, Stanze di Raffaello, wall paintings).

Below, all Biblical references are from the Gospel of Matthew.

Baptism—The Figure of John the Baptist

Then Jerusalem and all Judea and all the surrounding Jordan went out to him and were baptized by him in the Jordan, confessing their sins. And when John saw many Pharisees and Sadducees coming to him to be baptized, he said to them, "Offspring of vipers! who inspired you to run away from future anger? create a worthy fruit of repentance . . . I baptize you with water for repentance, but He who follows me is stronger than me; I am not worthy to carry His shoes; He will baptize you with the Holy Spirit and fire; His shovel is in His hand, and He will cleanse His threshing floor and gather His wheat into the granary, and He will burn the straw with an unquenchable fire. Then Jesus comes from Galilee to the Jordan to John to be baptized by him (Fig. 20.3). But John restrained Him and said, "I need to be baptized by You, and do you come to me?" But Jesus answered and said to him, Leave it now, for thus it behooves us to fulfill all righteousness. (3:5–9)

Figure 20.2. The beating of infants by servants of Herod II, the figure of the king rises above all; Siena, Church of San Agostino (Matteo di Giovanni, 1482).

Figure 20.3. The Appearance of Christ to the people (painting by Alexander Ivanov, 1837–1857; Tretyakov Gallery).

The Sermon on the Mount

And Jesus went about all over Galilee, teaching in their synagogues and preaching the gospel of the kingdom, and healing every disease and every infirmity in people. And a rumor spread about Him throughout Syria; and they brought to Him all the infirm, possessed with various diseases and seizures, and possessed, and sleepwalkers, and the paralyzed, and He healed them. And a multitude of people followed Him from Galilee and Decapolis, and Jerusalem, and Judea, and from beyond the Jordan. (4:23–25)

When He saw the people, He went up on the mountain; and when He sat down, his disciples came to Him. He opened His mouth and taught them, saying: Blessed are the poor in spirit, for theirs is the kingdom of heaven. Blessed are those who weep, for they will be comforted. Blessed are the meek, for they will inherit the earth. Blessed are those who hunger and thirst for truth, for they will be satisfied. Blessed are the merciful, for they will be pardoned. Blessed are the pure in heart, for they will see God. Blessed are the peacemakers, for they will be called the sons of God. Blessed are those who are cast out for the truth, for theirs is the Kingdom of heaven. Blessed are you when they will revile you and persecute you and slander you in every way unrighteously for Me. Rejoice and be merry, for your reward is great in heaven: so they also persecuted the prophets who were before you (5:1–12) . . . When He came down from the mountain, a lot of people followed Him. (8:1)

Awaiting Punishment from the Rulers and Coming Resurrection

And from some point . . . Jesus forbade His disciples not to tell anyone that He is Jesus Christ. From that time on, Jesus began to reveal to His disciples that He must go to Jerusalem and suffer a lot from the elders and chief priests and scribes, and be killed, and on the third day rise again. And, having called Him away, Peter began to contradict Him: Be merciful to Yourself, Lord! may this not happen to You! But he turned and said to Peter: Get away from Me, Satan! You're a temptation to Me! because you don't think about what is Divine, but what is human. Then Jesus said to His disciples, "If anyone wants to follow me, deny yourself, and take up your cross, and follow me, for whoever wants to save his soul will lose it, and whoever loses his soul for My sake will find it. . . . Verily I say to you: there are some standing here who They will not taste death, as they will already see the Son of Man coming in His kingdom. (16:20–28)

Jesus Anticipating his Betrayal and then Told His Disciples

You know that in two days it will be Easter, and the Son of Man will be delivered to the crucifixion. Then the chief priests and the scribes and the elders of the people gathered in the courtyard of the high priest, named Caiaphas, and decided in council to take Jesus by cunning and kill him; but they said: only not on the feast day, so that there would be no disturbance among the people. (26: 2–5)

Then He comes to His disciples and says to them: Are you still sleeping and resting? Behold, the hour is at hand, and the Son of Man is delivered into the hands of sinners; arise, let us go: behold, he who betrays me is at hand. And while He was still speaking, behold, Judas, one of the twelve, came, and with him a multitude of people with swords and stakes, from the chief priests and elders of the people. The one who betrayed Him gave them a sign, saying: Whoever I kiss, He is, take Him. And immediately approaching Jesus, he said: Rejoice, Rabbi! And kissed Him. But Jesus said to him, "Friend, why have you come?" (26:45–50)

Crucifixion

When morning came, all the chief priests and elders of the people had a conference about Jesus to put Him to death; and having bound Him, they took him away and delivered Him to Pontius Pilate, the governor. . . . But Jesus stood before the ruler. And the ruler asked Him: Are you the King of the Jews? Jesus said to him: You speak. And when the chief priests and elders accused Him, He did not answer anything. Then Pilate said to Him: Do you not hear how many are testifying against You? And he did not answer him a single word, so that the ruler was very surprised. . . . On the feast of Easter, the ruler had the custom of releasing one prisoner to the people, whom they wanted. . . . So when they were gathered together, Pilate said to them: Whom do you want me to release to you: Barabbas, or Jesus, who is called Christ? for he knew that they betrayed Him out of envy. . . . But the chief priests and elders stirred up the people to ask Barabbas, and to destroy Jesus. . . . Pilate, seeing that nothing helps, but confusion increases, took water and washed his hands before the people, and said: I am innocent in the blood of this Righteous Man; see you. And all the people answered and said, "His blood is on us and on our children." Then he released Barabbas to them, and, having beaten Jesus, he gave him up for crucifixion. . . . And when they came to a place called Golgotha, which means: the place of the forehead, they gave Him vinegar mixed with bile to drink; and after tasting it, he did not want to drink. Those who crucified him divided his clothes, casting lots; and, sitting, guarded Him there; and put an inscription over His head, signifying His guilt: This is Jesus, the King of the Jews (Fig. 20.4). Then two robbers were crucified with Him, one on the right side and the other on the left. (27:1–38)

Figure 20.4. Raising the cross with the crucified Jesus
(Jorg Bray the Elder, 1524; Museum of Fine Arts, Budapest).

Ascension

After the Sabbath, at dawn on the first day of the week, Mary Magdalene and the other Mary came to see the coffin. And behold, there was a great earthquake, for the angel of the Lord, who came down from heaven, came and rolled away a stone from the door of the tomb and sat on it; his appearance was like lightning, and his clothes were white as snow; being afraid of him, the guards trembled and became like the dead; the angel, turning his speech to the women, he said: do not be afraid, for I know that you are looking for Jesus crucified; He is not here—He has risen, as he said (Fig. 20.5). Come, see the place where the Lord lay, and go quickly, tell His disciples that He has risen from the dead and is going before you in Galilee; there you will see Him. Here, I told you. And coming hastily out of the tomb, they ran with great fear and joy to tell His disciples. And when they went to tell His disciples, behold, Jesus met them and said, Rejoice! And they came and took hold of His feet and worshiped Him (28:1–9). But the eleven disciples went to Galilee, to the mountain where Jesus had commanded them, and when they saw Him, they worshiped Him, while others doubted. And Jesus drew near and said to them, "All authority in heaven and on earth has been given to me. Go, therefore, and teach all nations, baptizing them in the name of the Father, and of the Son, and of the Holy Spirit, teaching them to observe all that I have commanded you; and lo, I am with you all the days to the end of the world. Amen." (28:16–20)

Figure 20.5. Jesus Christ after being removed from the cross on Mary's lap: "Pieta," sculpture by Michelangelo Buonarroti, 1498–1499; Vatican, St. Peter's Basilica. (Note that there are no descriptions in the Gospels of a situation in which Mary holds Jesus removed from the cross on her knees.)

20.2. Peter and Paul: The First Apostles of Christianity

The first century of the new era is widely recognized as beginning with the year of Jesus Christ's birth, marking a significant milestone at the turn of the century from BCE and CE. However, this century proved to be the most challenging, even tragic, for the birth of the Christian faith. The prophet's life lasted for precisely thirty-three years and ended by crucifixion. Another thirty-three years later, the same cruel fate awaited Jesus's disciples, Peter and Paul. In Orthodoxy and Catholicism, these two figures are perhaps revered for their devout service to the Lord and diligence in the spread of the Christian faith.

In discussing the chronological framework of these blood-stained events in human history, we will identify the discussions about the birth of Jesus Christ and his crucifixion and acknowledge the endless debates surrounding these notable events. These dates hold immense significance for devout Christians and fuel ongoing discussions. However, for our research, these specific dates are not important, so we will not delve deeply into these issues. Instead, we will consider the most widely accepted dates—33 CE as the year of Jesus's crucifixion and the executions of Peter and Paul in 66 and 67. The author does not insist on these exact years but notices an interesting pattern, namely that the first century CE is divided into roughly three equal segments of thirty-three years each, highlighting the profound impact of these events on the history of Christianity.

> But what is behind this strange triad? It's not clear. It was only in the year of the hundredth that the famous Jewish historian Josephus Flavius died, and the great Claudius Ptolemy, a geographer, astronomer, astrologer, mathematician, was born. But do they have anything to do with the subjects that occupy us? It seems that this question may already drag us into mystical depths. Therefore, we will stop here.

It should be noted that the first apostles, despite both being Jewish, came from distinct backgrounds. Peter, originally Simon (Shimon), hailed from the region near Lake Genisaret (Tiberias) and grew up around fishermen. It was in this setting that he first met Jesus Christ, whose personality captivated Peter's community. Peter quickly turned from a fervent admirer of Jesus to his faithful disciple.

> And having called His twelve disciples, He gave them power over unclean spirits to cast them out and heal every disease and every infirmity. The names of the twelve Apostles are these: the first Simon, called Peter, and Andrew, his brother, James Zebedee and John, his brother . . . (Matthew 10:1–2)

> And of those twelve, Peter became the most beloved disciple of Jesus, and he changed his name from Simon to Peter: "I say to you, you are Peter, and on this rock I will build my Church, and the gates of hell will not overcome it; and I will give you the keys of the kingdom of heaven: and whatever you bind on earth will

be bound in heaven, and whatever you loose on earth will be loosed in heaven."
(Matthew 16:18–19)

However, when Christ shared with his disciples the terrible torments that were unexpectedly revealed to him, Peter took him aside and began to rebuke him, saying, "Far be it from you, Lord! This shall never happen to you!" But he turned and said to Peter "Get away from Me, Satan! You're a temptation to Me! Because you don't think about what is Divine, but what is human." (Matthew 16:22–23)

Usually on these phrases of Christ ("get away from Me, Satan!") emphasis is avoided, but it is extremely difficult to get past these quite ungracious words. Probably, the first steps of the new creed, even among its adherents, did not always proceed smoothly.

The life of the second original apostle Paul seems to be outlined with completely different contours. He was born and raised in the Jewish diaspora, although geographically close to indigenous Israel. He appeared from Cilicia, the southeastern region of Asia Minor. His original name was Saul, and he had never seen or met Christ personally. Unconditionally and initially rigid adherence to the tenets of Judaism made Saul an active persecutor of the new trend, which he saw as a vile heresy against the seemingly indestructible canons of Jewish truth. One of the most notable episodes in his early life was the participation in the stoning of the heretic Stephen (Fig. 20.6). From the standpoint of Judaism, this act was a rightful punishment of a heretic; however, from the perspective of Christianity, Stephen is revered as the first great martyr who courageously held up his faith:

> Being filled with the Holy Spirit, looking up to heaven, he saw the glory of God and Jesus standing at the right hand of God. . . . But [the people in the crowd], shouting with a loud voice, covered their ears, and with one accord rushed at him, and, leading him out of the city, began to stone him. The witnesses laid their clothes at the feet of a young man named Saul, and stoned Stephen . . . Saul approved of killing him. In those days there was a great persecution of the [Christian] church in Jerusalem; and all, except the Apostles, were scattered to various places in Judea and Samaria. . . . And Saul tormented the church, entering houses and dragging men and women into prison (Acts 7, 8).

However, a divine intervention occurred quite quickly after this event. On his way to Damascus, where the trial of a group of heretical Christians was also to take place, Saul suddenly heard a voice coming from nowhere: "Saul! Saul! Why are you persecuting me?" This voice not only startled and blinded him, but he immediately recognized whose voice it was. Brought blind to Damascus, he was healed on the third day by the Christian Ananias and was baptized by him simultaneously. From then on, Saul became one of the most influential Christian preachers. When he managed to convert the Proconsul Sergius to Christ in Cyprus, he changed his name and became known to all as Paul.

The most significant traces of his deeds in the development of Christianity, of course, include fourteen epistles to converts (Fig. 20.7), as well as four journeys in which Paul, with varying degrees of luck, managed to spread the canons of Christianity among pagans alien to this faith.

Figure 20.6. Young Saul during the beating stones of the Great Martyr Stephen, filled with the Holy Spirit; Saul is sitting upstairs, and Stephen's clothes are on his knees (Rembrandt van Rijn, 1625).

Paul begins his epistles, or rather his instructions, with the usual greeting and blessing, and then calls everyone to unity. "I beseech you, brethren," he writes, "in the name of our Lord Jesus Christ, that you all speak the same thing, and that there be no divisions among you, but that you be united in one spirit and in one thought.... For it has become known to me about you, my brethren, that there are disputes between you. I mean what you say: "I am of Paul"; "I am of Apollos"; "I am of Cephas"; "and I am of Christ." Is Christ divided? Did Paul crucify himself for you? Or were you baptized in the name of Paul?" (1 Corinthians 1:10–13)

Religious scholars and linguists generally agree that Paul was the author of the epistles, although this consensus does not extend to all fourteen usually attributed to him. In fact, most scholars reject his authorship of the Epistle to the Hebrews. Among the other thirteen, there is no dispute about Paul's authorship regarding, for example, "Romans," "Corinthians," "Galatians," and so on. However, opinions vary regarding the remaining epistles, largely delineated by linguistic analysis. S. I. Sobolevsky, for example, argues that "a Jew, originally from Jews, a great connoisseur of his native language, he could not express deep feelings in a foreign language, and did not care especially about words when relatively there was no sense of danger."

In practice, it was Peter who spread the dogmas of Christianity among the Jews, particularly those adherents of Judaism who saw ambiguities in their original doctrines and were open to developing its canon. Paul, on the other hand, focused his efforts on an ethnic group of pagans, far removed from Judaism. It is quite possible that this particular division contributed partially to Paul's success in his voluntary mission. Peter, however, faced much more hardship in converting the Jews, as the Jewish ideological bonds turned out to be much stronger.

Figure 20.7. Saint Paul composing his addresses/ epistles to the newly converted to Christianity (Valentin de Boulogne, 1620s).

The two leading apostles of Christianity thus pursued different paths in their efforts to spread the teachings of Christ. Artistic depictions of iconography, murals, and frescos, often portray these two figures as close friends and allies in their struggle for Christian truths. Written sources, however, present a diverging picture. Although there is limited written evidence, one vivid example is found in Paul's "Epistle to the Galatians":

> When Peter came to Antioch, I openly opposed him, saying that he had brought condemnation upon himself. Because until the people from Jacob came, he ate with the Christians from the Gentiles. And when these people came, Peter began to shun the Gentiles and eat separately from them, fearing the Jews. Other Jews were hypocritical with him, so that even Barnabas was involved in this. And when I saw that they were wrong, they were not acting according to the gospel truth, I said to Peter in front of everyone: "If you, a Jew by birth, live in a pagan way, and not in a Jewish way, then how can you force Gentiles to Judaize?"

> Although we are Jews from birth, and not "sinful Gentiles," [however] having learned that a person is justified not by observing the Law, but only by faith in Jesus Christ, we also believed in Christ Jesus in order to be justified by faith in Christ, and not by observing the Law, because no one is justified by observing the Law … But if, despite our striving to be justified in Christ, we Jews ourselves turn out to be the same sinners as the Gentiles, does this mean that Christ is the patron saint of sin? No way! (Gal 2:11–17; translated by Kulakov).

It is noteworthy that a certain response to this rather harsh conversation was a painting by Guido Reni, in which Paul, as it were, makes a reprimand to Peter (Fig. 20.8, left).

Thus, Paul openly denounced Peter's actions in Antioch. In this case, we can observe that there was more to be seen in their relations than previously assumed, which contrasted with Paul's demands for unanimity, as voiced in his epistle to the Corinthians.

Figure 20.8. Paul reproaches the repentant Peter (Guido Reni, 1603–1604); right: images of the apostles in the catacombs of ancient Rome, fourth century.

20.3. Execution of the First Apostles of "Shameful Destruction"

However, as centuries progressed, Christianity became hostile not only to Judaism but also to the prevailing ideology of the Roman Empire, causing further unfriendly responses from its imperial rulers. Thus, many believed that this "shameful calamity," as it was often described, needed to be eradicated by the root. This struggle typically manifested in either the expulsion of Christians or the physical elimination of their leaders.

Executions of prominent Christians occurred during the reign of the infamous Emperor Nero. Paul, for example, was ruthlessly beheaded—not burned or crucified on the cross—because of his Roman citizenship, which he considered a relief (Fig. 20.9). Peter, on the other hand, was crucified, as had become the customary practice for executing high-profile ideological dissidents. At the same time, Peter requested that he be nailed to the cross upside down to separate his execution from that of his great Teacher, as he felt unworthy of a similar death (Fig. 20.10).

The cruel executions of Peter and Paul, however, did not preclude the spread of the pernicious Christianity in any way but may have furthered its cause. Their martyrdom contributed to the perception of the two apostles as united figures who fought for the high truths of Christianity and who sacrificed their lives as the chosen ones of God. They are often depicted together in Roman catacombs, where many converted Christians sought refuge. Over the subsequent centuries, the belief that they were executed together and on the same day (June 29 according to the Julien calendar or July 12 according to the Gregorian one) became established as fact, thereby reinforcing the belief in their great friendship and indestructible bond of Christian faith.

20.4. The Generation of Hatred in the Ideological Facets of Cultures

It was from those times that the chronology of events within the framework of Christian doctrine essentially began. It cannot be said that Judaism had not experienced similar

Figure 20.9. Execution of Paul; a fresco, from the Lateran Palace in Rome, 1278–1279. On the right is the church "Three Fountains" on the site of the alleged execution of the apostle (Abbazia delle Tre Fontane, Roma).

Figure 20.10. Peter's execution: at the request of the apostle, he was crucified upside down on the cross (Michelangelo da Caravaggio, 1601); the place of the apostle's supposed burial (the Papal altar in the so-called Vatican grottoes under St. Peter's Basilica); St. Peter was proclaimed the first Pope by the Catholic Church.

developments prior to this significant split. For example, different sects within Judaism, such as Sadducees, Pharisees, Zealots emerged, but in terms of the significance of their differences, these initial discrepancies cannot be directly compared with the formation of Christianity—the second branch of the Abrahamic triad. From the standpoint of Judaism, this second branch has always been regarded as a heresy.

However, this ideological innovation was perceived much more severely in the Roman Empire. Christianity was born within the territorial and ideological confined of the empire, thereby intertwining its early fate with Judaism, and was often marked by dramatic and tragic events. For example, the Great Fire of Rome during in 64 CE during Nero's reign which lasted six days left an indelible mark on the early history of Christianity, as the Roman leadership placed principal blame on the Christians. The reaction of the renowned historian Publius Cornelius Tacitus (c. 50–c. 120) to these events is particularly interesting, as he may have been an eyewitness to these events as a child, and provided a critical account, although the details of his birth and early life remain unknown:

> But neither by human means, nor by the generosity of the princeps, nor by appealing to the deities for assistance, it was impossible to stop the rumor that dishonored him [Nero] that the fire was set on his orders. And so Nero, in order to overcome the rumors, sought out the guilty and committed the most sophisticated executions of those who, by their abominations, had incurred universal hatred and whom the crowd called Christians. Christ, from whose name this name comes, was executed under Tiberius by the procurator Pontius Pilate; suppressed for a time, this pernicious superstition began to break out again, and not only in Judea, where this pernicious thing came from, but also in

Rome, where everything most vile and shameful flocks from everywhere and where it finds adherents. So, first those who openly recognized themselves as belonging to this sect were captured, and then, according to their instructions, a great many others were exposed not so much in the villainous arson as in hatred of the human race. Their killing was accompanied by mockery, for they were clothed in the skins of wild animals so that they would be torn to death by dogs, crucified on crosses (Fig. 20.11), or those doomed to death in the fire were set on fire after dark for the sake of night lighting. (Tacitus, Annals, XV:44)

Tacitus' opinion regarding Christianity as a shameful and malicious superstition reflected the typical ideas of Romans at the time, many of whom adhered to their own traditional religion and rejected the postulates of Christianity that were so alien to them. For example, the prominent thinker-philosopher Porphyry of Tyre (232/233–304/306) presented an essay of 15 books *Against Christians*, in which he reported, for example, his horror at the fact that Christians eat—even symbolically—the blood and flesh of Christ as part of the most fundamental Christian rites, the Eucharist.

20.5. Christianity in the Roman Empire: Fundamental Changes

As time passed, by 313, the rhetorician Lactantius reported that he and Constantine Augustus, who became the emperor Constantine the Great, decided that

first of all, we should make arrangements for those who have preserved the worship of God that we grant both Christians and everyone else the opportunity to freely follow whatever religion they wish, so that Divinity, whatever it may be on

Figure 20.11. "The Lights of Christianity or the Torches of Nero"
(G. I. Semiradsky, 1876; Krakow. National museum).

the heavenly throne, could be in favor and mercy to to us and to all those who are under our authority (Wikipedia).

In 325, the First Council of Nicaea was convened, and only after this general consultation (Fig. 20.12) that the persecution of Christians noticeably subsided. By 380, Emperor Theodosius I Augustus, Augustus Gratian, and Valentinian issued an "Edict to the people of the city of Constantinople," which stated:

> We wish that all the nations, which Our Grace wisely governs, live in the religion that the divine Peter the Apostle transmitted to the Romans, as it, being established by himself, testifies to this day . . . we must confess, in accordance with the apostolic instruction and the teaching of the Gospel, the one Deity of the Father and the Son and the Holy The Spirit is in equal majesty and in the Holy Trinity. We order those who obey this law to accept the name of catholic Christians, and We determine that the rest, the insane and insane, must suffer dishonor associated with heretical teaching, and their gatherings should not accept the name of churches, and that they must first suffer Divine punishment, and then punishment from Our actions, which We have undertaken at the behest of heaven (Wikipedia; http://gmir.ru/calendar/1307.html).

Figure 20.12. The First Ecumenical Council of Nicaea in 325 (the city of Nicaea near Constantinople, now Iznik, Turkey). At the bottom in a special chair is the Byzantine Emperor Constantine the Great, and at the top, Pope Sylvester I.

In 391, a new edict was issued:

> No one is given the right to make sacrifices, no one should make a detour around [pagan] temples, no one should honor temples. Everyone should know that Our law prohibits entry into pagan temples, and if anyone tries, despite Our prohibition, to commit certain cult acts in relation to the gods, then let him know that he will not be able to escape punishment, even using special signs of imperial favor. The judge . . . must compel the impious violator of the law who entered the desecrated place to pay a fine of fifteen pounds in gold to Our treasury.

The Roman Empire announced that Christianity was being incorporated into law as a state religion and that all subjects were to observe its canons. Ironically, these canons had previously been considered vile and forbidden. In 448, Emperor Theodosius II ordered the destruction of all the lists of the treatise "Against Christians" written by Porphyry of Tyre, who had recently under the banner of imperial authorities fiercely denounced the abominations of the malicious superstitions of Christianity.

In 395, a dramatic split of the Roman Empire occurred dividing it into the Western and Eastern empires (Fig. 20.13). Byzantium as the Eastern Empire, would endure, despite many challenges to its rule, for over a millennium until it was conquered by the followers of the last branch of the Abrahamic triad—the Muslims. The Western Empire eventually disintegrated, fragmenting into territories engaged in frequent battles with each other (Fig. 20.14). Despite

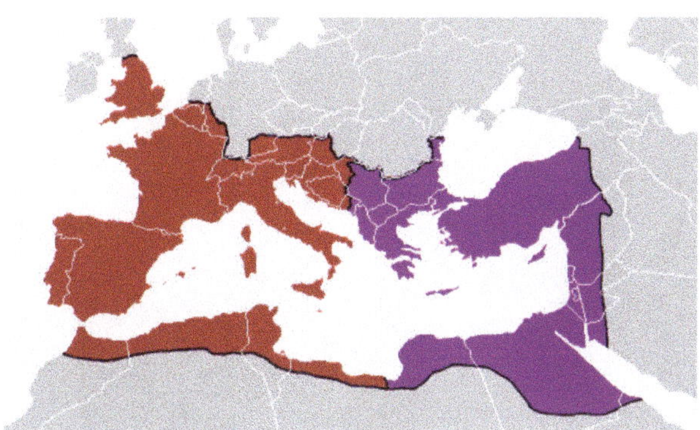

Figure 20.13. The territories of the Roman Empires: Western and Eastern (Byzantine) according to the split of 395, known as the Great Schism.

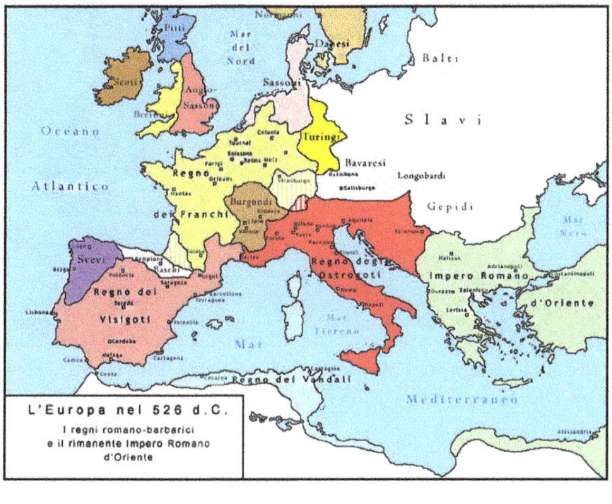

Figure 20.14. The collapse of the Western Roman Empire; various formations on its former territory.

this fragmentation, however, one significant exception stood out: Rome managed to preserve its original Catholic shrines. This legacy is now embodied in the geographically small but symbolically powerful Vatican City, an enclave that remains pivotal in the role of global Christianity, which is very noticeable, however, only with the overlapping layouts of many channels of world politics.

20.6. The Great Schism and the "Ship of Salvation" in Orthodoxy

From the moment of the Roman Empire's split, an evident schism began to appear in early Christianity. By the middle of the sixth century, the disagreements had become so cruel, even insulting, that the need for an official divorce became clear.

The Great Schism or the Great Schism of Christianity, it seems, without any hesitation, researchers define the year 1054. Then Pope Leo IX sent legates to Constantinople, led by Cardinal Humbert. The immediate reason was the closure in 1053 of the Latin churches in Constantinople by order of Patriarch Michael Kerulari, when these shrines were thrown out of the tabernacles of the Holy Gifts, prepared according to the Western custom from unleavened bread, and trampled underfoot. Papal legates were expelled from the capital of Byzantium at the same time. In the same year 1054, both the Byzantine and Roman churches, still revered as one, anathematized each other, that is, a curse.

Figure 20.15. The siege of Constantinople by the Crusaders in 1204 (Miniature from The Chronicles of the Emperors by David Aubert, 1470s).

The Catholic and Orthodox Churches continued to harshly criticize and abuse each other for a millennium after the schism of 1054. In 1203, during the Fourth Crusade against the Muslims, the Catholic crusading army broke apart from the Orthodox Christians. By 1204, they had celebrated the fall of the Byzantine capital Constantinople (Fig. 20.15 and 12.16) and proclaimed this city the capital of the new Latin Empire, which existed, however, for a brief time until 1261.

Adherents of Orthodoxy vigorously defend themselves, which is reflected in the texts of the modern Orthodox Newspaper:

> European countries are increasingly called post-Christian, they are reproached for the loss of their identity and faith. This is facilitated by the cult of consumption and so-called universal values, scandals within the Catholic Church, the ordination of women and the permission of same-sex marriages in some Protestant Churches. In this situation, many Europeans look with hope to Orthodoxy, which firmly stands for the preservation of its age-old traditions. It is Orthodoxy that becomes the ship of salvation for those who do not want to lose faith in Christ. The Russian Orthodox Church is not only a home for our compatriots who find themselves abroad, but increasingly also a home for indigenous Europeans returning to their spiritual roots. This situation is perhaps very vividly illustrated by the depicted "Ship of Salvation," surrounded on all sides by its evil enemies (see Fig. 20.17).

Figure 20.16. Victorious crusaders enter Constantinople (engraving by G. Dore, 1877).

Figure 20.17. Adherents of Orthodoxy on the Ship of Salvation, surrounded by innumerable and diverse enemies (https://www.mgarsky-monastery. org/author/svyashchennik-ioann-valentin-istrati).

Figure 12.18. Patriarch Nikon with the clergy at the New Jerusalem Monastery (D. Vuchters (?), 1660–1665).

However, Orthodoxy was not immune to the fervent zeal for eliminating dissenters, particularly those who opposed the church reforms initiated by Patriarch Nikon in the 1650s and 1660s (Fig. 20.18). Those who visibly identified as Old Believers by using the double-cross sign (Fig. 20.19) were persecuted and burned. The ideological leader of the Old Believers, Protopop Avvakum, also faced this severe punishment (Fig. 20.20), ultimately ending his earthly life at the stake (Fig. 20.21).

The Great Schism had an extremely strong impact on the methods of spreading and promoting tenets of both the Catholic and Orthodox branches. Orthodoxy moved with the explorers to the east and northeast, up to the distant Pacific coasts of the East Siberian Enclave. Catholics,

Figure 20.19. Burning of Old Believers in log cabins specially prepared for this purpose. Old Believers go to this terrible death, raising their hand with a double-finger—a sign confirming their unwavering loyalty to the canons of the original religion and distinguishing them from adherents of Orthodoxy reforms with the indispensable requirement of a triple-finger (Wikipedia).

Figure 20.20. Protopop Avvakum in Siberia (E. D. Maltsev, 1973).

Figure 20.21. Protopop Avvakum at the stake in Pustozersk, 1683 (P. Myasoedov, 1897).

Figure 20.22. Hugs: Meeting of Pope Francis with Patriarch Kirill
of the Orthodox and All Russia at Havana International Airport
on February 12, 2016 (photo by Thanassis Stavrakis).

on the other hand, faced a formidable goal of conquering the vast transoceanic spaces from relatively near Africa to the unknowingly remote transoceanic America and Australia at that time.

However, only 962 years have passed (!), and the primates of both Christian branches, Pope Francis and Patriarch Kirill of Moscow and All Russia, agreed to meet and met on February 12, 2016—in Cuba (!), at the Jose Marti Havana International Airport (Fig. 20.22). The meeting lasted more than two hours. The result was the signing by the Pope and the Patriarch of a joint statement of 30 points. The last paragraph of the document sounded like this: Filled with gratitude for the gift of mutual understanding shown at our meeting, we turn with hope to the Most Holy Mother of God, appealing to Her with the words of an ancient prayer: "We run to Your mercy, Virgin Mother of God."

Official representatives of both churches hastened to attribute the meeting of the pontiff and the patriarch to historical ones, although many do not agree with this in any way.

However, a whole millennium before the burning of schismatics in Russia, an extremely painful surprise for both Judaism and Christianity was born and began to spread rapidly, in the form of the third Abrahamic branch: Islam.

Chapter 21

Islam: The Final Block of the Abrahamic Triad

21.1. The Prophet Muhammad as the Messenger of Allah

Muhammad is the third iconic figure of the Abrahamic triad of monotheism, following the Jewish prophet Moses and the Christian prophet Jesus as the heralds of new divine truths. Unlike the former two prophets, whose earthly existences are continuously disputed, Mohammad was a historical figure credited as the originator of the Islamic doctrine. The Koran is revered as his main creation, comprised of Muhammad's sermons based on his revelations from Allah.

Muhammad was born in Mecca in 570 or 571 CE and died abruptly in 632. His 62-year lifespan was closely intertwined with the enclave of Arabia, classified as the focal center of the Islamic religion. Muhammad's life was marked by constant spiritual and physical struggles, information which is derived from a variety of narratives and concentrated in the book of Ibn Hisham with the relatively lengthy title *The Life of the Prophet Muhammad, narrated from the words of al-Bakkai, from the words of Ibn Ishaq al-Muttalib*. Ibn Hisham's book is quite remarkable for several reasons. First, Ibn Hisham was likely the editor of the book as opposed to its author. He collected information about the Prophet Muhammed (otherwise known as the Messenger of Allah) from Ibn Ishaq (704–767/768), whose life closely overlapped with the prophet's. Second, his accounts provided a more comprehensive account than Ziyad al-Bakkai's (d. 799) or, notably, Ibn Hisham's (d. 834). Al-Bakkai's role appears secondary in this trio of authors as he only managed to preserve information collected by Ibn Ishaq and transmit it to Ibn Hisham for further refinement. Therefore, this invaluable historical resource is often attributed to the work of Ibn Ishaq—Ibn Hisham, as noted by Kudelin (2017).

The book is distinguished from other works by another important circumstance, namely its references to individuals who relayed information directly or indirectly to the Prophet. Additionally, when considering the book of Ibn Ishaq-Ibn Hisham, it is important to note its similarities in the description of other central religious figures, Moses and Jesus Christ. The disparities that emerge between these accounts are striking and cannot be overlooked.

21.2. The Book of Ibn Ishaq—Ibn Hisham: The Early Days of Muhammad

We will get acquainted, albeit briefly, with the main sections of the Book of Ibn Ishaq-Ibn Hisham, in which he details the biography of the Prophet Muhammad from his birth to his death. Let us begin with a passage describing Muhammad's appearance at the prime of his being:

> Ali ibn Abu Talib said about the qualities of the Prophet: he was not tall, slender and was not short, was of medium height; his hair was not very curly and not smooth, neatly combed; was not full and was not cheekboned; his skin was white-pink, and his eyes were black; was he has long eyelashes, broad-boned, broad-shouldered, with small hair on his chest, devoid of vegetation, with thick arms and legs. (169)

This description deviates from those of Jesus Christ or Moses. Indeed, it is hard to imagine them being described in similar terms, as big-boned, with thick arms and legs. Muhammad's death is also described peculiarly:

> Aisha [the Prophet's wife] said: "And he brushed his teeth . . . with a toothpick as thoroughly as ever before, then put it down. I found that the Messenger of Allah was heavy on my lap, and I began to examine his face: his gaze became fixed. . . . And then the Messenger of Allah died. (632)

In these vivid descriptions, Muhammad appears much more human than the Judeo-Christian formulations of their prophets. Let us also examine the initial signs that Allah presented to the child still in his mother's bomb. Amina, Muhammad's mother, narrates:

> We have nothing left. We couldn't sleep all night because our baby was crying from hunger, and I had no milk in my breasts. . . . I took it and came back with it. When I put him on my lap, on my saddle, my breasts filled with milk, and he drank milk until he was full. His brother drank with him until he was full. Then they both fell asleep. (59–60)

We know very little about Muhammad's early childhood, although it is revealed that his clan was part of the Quraish tribe, with whom he would later clash in battle. For a short while, Muhammad was educated by the Bedouins of the Banu Sa'd tribe. He lost his mother at the age of seven and was adopted by his uncle, whom he accompanied on trade expeditions to Syria.

When Muhammad turned 25, he was noticed by Khadija, a wealthy forty-year-old widow. In 595, Muhammad became her third husband, raised six children with her (two boys and two girls of their own, in addition to her children from previous marriages), and enjoyed a happy marriage until she died in 619. After her death, he went on to have multiple marriages:

He left behind nine wives. The total number of women whom the Messenger of Allah married was thirteen: Khadija is the daughter of Huwaylid—she was the first woman whom the Prophet married . . . The Prophet married Aisha, the daughter of Abu Bakr al-Siddiq in Mecca when she was seven years old. And he began to live with her in Medina when she was nine or ten years old. Aisha was the only virgin the Prophet married. She was married to the Prophet by her father Abu Bakr. The Prophet gave four hundred dirhams for it. (634)

Again, it is worth noting the absence of similar references to virgin wives of nine and ten years old in the biographies of Solomon or Jesus Christ.

21.3. Muhammad and the Angel Jabrayil

Nevertheless, the main interest in the Prophet's life is how he approached his faith that would eventually inspire countless people across the globe.

Ibn Ishaq narrated: "Abd al-Malik ibn Ubaydullah told me, who kept in his memory the stories of knowledgeable people. The Messenger of Allah, when Allah wanted to honor him and his prophecy began, went out according to his need, went far from houses into mountain gorges and into the depths of valleys in the vicinity of Mecca. When the Messenger of Allah passed by a stone or a tree, they would certainly say to him: "Peace be upon you, O Messenger of Allah!" The Messenger of Allah looked around, looking to the right, to the left, back, and saw nothing but a tree or a stone. And so the Messenger of Allah continued to see and hear for as long as Allah wished. Then Jabrayil came to him and brought him the mercy of Allah when he was in the cave of Hira in the month of Ramadan . . .

[However,] the night came in which Allah honored him with the honor of bringing his mission and mercy to His servants. Jabrayil came to him with an order from Allah. The Messenger of Allah said: "Jabrayil came to me when I was sleeping, with a piece of silk and a book in it, and said: "Read!" I said: "I don't read." He started strangling me with this book so that I thought it was death. Then he let me go and said, "Read it!" I said: "I don't read." He started strangling me with this book so that I thought it was death . . . But in the end, the Messenger of Allah managed to read and pronounce the right words: In the name of your Lord, who created Man from a blood clot . . . your most Merciful Lord, who taught kalam, taught man what he did not know . . . Then Jibrail left me, and when I woke up, it was as if these words were imprinted in my heart . . . Then I heard a voice from the sky saying: "O Muhammad! You are the Messenger of Allah, and I am Jabrayil." I raised my head to the sky and looked. And so Jabrayil in the image of a man (Fig. 21.1), closing his legs, closed the entire horizon. (92–94)

This marks the commencement and progression of Muhammad, the Messenger of Allah, delving into the truths of Islam, which were previously unknown to any inhabitants of the Earth at that time.

21.4. Battles for Islamic Truths with Jewish and Christian Participation

Now we approach a crucial next step: an enlightened Messenger who endeavors to impart his revelations to his confidants. Regrettably, his initial successes were not remarkably fruitful; the first adherents to Islam—Muslims—proved to be few. Nevertheless, among them stood his faithful wife Khadija. Soon thereafter, opponents emerged, predominantly pagans.

> They began to take hostile actions towards those who converted to Islam and followed the Prophet. Each clan opposed its Muslims, began to block their way, beat them, expose them to hunger, thirst, the scorching heat of Mecca when the heat increased; especially those whom he considered weak in his faith, dissuaded them from this religion. Some of them succumbed to temptation under the weight of the trials that fell on them, others stood firm on their own. Allah held them back. (128)

Interestingly, Jews and Christians came to the aid of the first Muslims:

> Jewish priests and Christian monks talked about what they found in their books, which spoke about the qualities of the Prophet and the peculiarities of his time

Figure 21.1. Left: Muhammad receives his first revelation from the angel Jabrayil (Vellum miniature from the book of Jami al-Tawarih Rashid al-Din, published in Tabriz, Persia, 1307; collection of the Library of the University of Edinburgh, Scotland). Right: the cave of Hira near Mecca, where, according to beliefs, the meetings of the future Prophet with the angel Jabrayil took place (now it is one of the most visited places in the vicinity of the holy city for Islam).

and what was said about him in the time of their prophets. And as for the priests from the Arabs, demons from the devil came to them with such messages. The priest or priestess from time to time reported some of his deeds, but the Arabs did not pay any attention to it until Allah Almighty sent him and until the events they reported happened and they saw them with their own eyes. (76)

Ibn Ishaq: "Asim ibn Omar ibn Qatada told me from the words of the men of his people who said: "Truly, among the circumstances that converted us to Islam, with the grace of Allah and His right way, was what we heard from Jewish men. We were infidels, worshipping idols. They were admirers of the Book, they had knowledge that we didn't have. Between us and between them, the anger did not stop."

The Prophet endeavored tirelessly to introduce Islam to the local Arabian tribes but faced formidable obstacles and encountered little fortune in his efforts.

Then the Prophet came to Mecca. And his people became even more opposed to him and his religion, with the exception of a few low-born who believed in him. During the fairs, the Prophet began to speak openly to the Arab tribes, calling them to Allah, telling them that he was a Prophet sent by Allah. He asked them to believe him, to protect him, so that he could explain what Allah sent him with . . . (183)

When the Messenger of Allah saw what trials and torments his companions were subjected to and from which he was spared, thanks to his place with Allah and the protection of his uncle, Abu Talib, when he realized that he was unable to save them from these torments, he told them: "Maybe you will go to the lands of Ethiopia: there is a king there, where no one is persecuted. This is the land of truth. You will be there until Allah makes your lot easier for you." Then some of the Muslim companions of the Messenger of Allah went to the land of Ethiopia, fearing temptation, fleeing to Allah with their religion. This was the first hijra (migration) in Islam. (131)

21.5. The Great Hijra: Migration from Mecca to Medina

Subsequently, the idea emerged to depart Mecca for Medina, then known as Hijira. It is worth noting that in early literature, this event is often portrayed as an escape to Medina, the distance of which was considerable—about 350 kilometers in a straight line (Fig. 21.2). Nonetheless, Medina's proximity to the lands of the Ansar people, whose tribes declare their allegiance to Islam, made it a desirable destination:

Ibn Ishaq said: "When Allah allowed him to fight and when this part of the Ansars followed him, having converted to Islam, taking under the protection of the Prophet and the Muslims who followed him, who found shelter with the Ansars, the Prophet ordered his companions from among the muhajirs from his people and those Muslims who were with him in Mecca, to go to Medina, to move to it and join his brothers in the faith from among the Ansar. He said, "Allah, the All-powerful and All-powerful, has made for you brothers and a home where you will be safe." They went there in groups. The Prophet remained in Mecca, awaiting permission from his Lord to leave Mecca and move to Medina. (183)

Finally, the time has come for the resettlement of the Prophet himself,—his wife Aisha said: The Messenger of Allah always came to Abu Bakr's house either early in the morning or late at night. On the day when Allah allowed the Prophet to migrate and leave Mecca from his people, he came to us at noon at an hour at which he had never come. When Abu Bakr saw him . . . he got up to meet him, gave way to the bed on which he was sitting. The Prophet sat down . . . and said, "Allah has allowed me to leave and move." Abu Bakr said, "Together, O Messenger of Allah." The Prophet replied: "Together." (212)

Relocating to Medina was a significant day for the Islamic lunar calendar, similar to the birth of Jesus Christ. However, when counting time according to the Christian calendar, the lunar Islamic zero day falls on July 16, 622:

This is how, according to contemporaries, the famous flight of the prophet from Mecca to Hasrib (Medina) took place; Hijra is the event that separates the era of Islam, the true faith, from paganism—Jahiliya [stupidity]. In 637, Caliph Omar recognized this moment in all fairness as the starting point of the entire

Figure 21.2. Hijra, the flight from Mecca to Medina (https://www.flickr.com/photos/internetarchivebookimages/14580517757/).

epoch; and to this day, all Muhammadans have been counting time since this momentous day. Indeed, the nature of Islam and its spread, especially in Arabia, are closely connected with the migration of the prophet. In Mecca, Muhammad was only the spiritual head of a persecuted and persecuted minority. From now on, he is the main steward in the fate of not only the community, which soon filled the majority of the inhabitants of Iasriba, but also the state structure into which this community has transformed. (Muller 2004, 138, 139)

21.6. The Prophet in Medina and the Great Victory at Badr

Thus, the Messenger of Allah found himself in Medina, which marks a decade-long phase dedicated to deepening his understanding of Islamic sacred truths and achieving success in spreading these messages across the Arab lands. Perhaps the most important success was in his journey to Badr, bestowed upon Muhammad by Allah in March 624. The town of Badr is located about 150 kilometers from Medina. However, the preparation for the battle with the pagans, as well as the battle itself, proved to be distinctly different from the pursuit of lofty objectives.

> Hearing that Abu Sufyan was returning from Syria, the Prophet called on the Muslims to attack them, saying: "Here is a caravan of Quraish. It contains their riches. Attack them, and maybe with the help of Allah you will get them!" People were excited: some quickly gathered and came, while others did not come. This is because the latter thought that the Messenger of Allah was not capable of such a battle. (281)

But many were mistaken—the Prophet turned out to be very gifted with such deeds.

> The Prophet took a handful of pebbles, went to the Quraish and, saying: "May their faces become ugly!" threw them pebbles. He ordered his companions: "Advance!" And there was a complete rout. Allah Almighty sent death to the bravest Quraish, and the most noble of them were captured. The Prophet saw on the face of Saad ibn Mu'az dissatisfaction with what the companions of the Prophet were doing. Then the Prophet asked him: "By Allah, Saad, are you as if dissatisfied with what our people are doing?" Replied: "Yes, dissatisfied, by God, O Messenger of Allah. This is the first battle that took place at the behest of Allah against the pagans. It would be preferable for me to kill as many as possible rather than leave them alive." (296)

21.7. The Section of Extracted Wealth

The victory (Fig. 21.3) led to the division of all that was obtained from the crushed enemy:

The Prophet ordered all the loot collected in the army to be put together. There were disagreements between Muslims. Those who collected the loot said: "This is our prey." And those who fought with the enemy and got it, said: "If it were not for us, you would not have collected it; we distracted the enemy from you so that you could collect the loot." And those people who guarded the Prophet, fearing the enemy's attack on him, began to say: "We also have the right to prey; we wanted to kill enemies when Allah showed us their backs; We could also pack things when there was no one to protect them, but we were afraid for the Messenger of Allah, we were afraid that the enemy would attack him, and we defended him. So you have no more rights to loot than we do." (303)

This victory turned out to be so significant that its mention was included in the Koran:

Ibn Ishaq said: "When the events in Badr passed, Allah Almighty sent down the Surah al-Anfal from the Quran in full (8:1). It also talks about disagreements among Muslims about the spoils: "They ask you about the spoils, say: "The spoils are at the disposal of Allah and the Messenger. Fear Allah, make peace among yourselves, obey Allah and His Messenger, if you are believers." (317, 318)

Indeed, the focus of this surah predominantly rested on the necessity of resolving disputes concerning the distribution of spoils among the newly converted Muslim victors. Still, the Badr victory's significance remains undisputed. Even in Russian literature, the title of one of the most notable books on this subject, the translators seemed to intentionally underscore the Battle of Badr: *Ibn Ishaq—Ibn Hisham. The Life of the Prophet. The Great Battle of Badr* (trans. A. B. Kudelin and D. V. Frolov).

21.8. Endless Battles

The following eight years of the Prophet's life were spent in endless battle. The authors of the book *The Life of the Prophet* tried to clarify the chronicle of his fights:

The total number of campaigns and battles in which the Prophet took part reaches twenty-seven.... The Prophet fought in nine campaigns and battles; in Badr, at Uhud, al-Khandak, in

Figure 21.3. The Battle of Badr. Miniature from "Siyar-i-Nabi" (The Life of the Prophet), 1595.

battles against Banu Qurayza and Banu al-Mustalik, in Khaybar, during the conquest of Mecca, in Hunayn and at-Taif . . .The number of expeditionary detachments sent by the Prophet is thirty-seven. (597–598)

Not all battles ended in victory for Muhammad. Some caused only bitterness and tears, as in the case of the battle at Uhud (Fig. 21.4) next to Medina:

> The Muslims opened up, and the enemy struck them. It was a day of great misfortune and trial. Many Muslims were killed. The enemy came close to the Prophet. Stones were fired at him, and one stone hit him in the face: a front tooth was knocked out, a cheek was wounded, a lip was cut. . . . [The Prophet said:] And if Allah gives me victory over the Quraysh in some place, then I will definitely mock thirty men of them. . . . When the Muslims saw the grief and anger of the Prophet because of what they had done with the corpse of his uncle, they said: "By God, if Allah gives us victory over them someday, we will commit such mockery of them as none of the Arabs has ever done. . . ." The battle of Mount Uhud was a misfortune, a disaster. In it, Allah put Muslims to the test. He also tested hypocrites who believed in words, but hid their unbelief in their souls. (355–356, 363)

When fortune favored them, thorny questions regarding the division of the spoils arose:

> The Prophet moved through the oasis of Khaybar, captured one herd after another, one fortress after another, [as well as] property.... The Prophet took prisoners from them. Among the captives were Safiya.... and two nieces. The Prophet chose Safiya for himself. Dihya ibn Khalifa al-Kalbi asked the Prophet to give him Safiya, but the Prophet, choosing her for himself, gave him her two nieces. The prisoners from Khaybar were distributed among the Muslims. . . . Then the Prophet distributed the goods of al-Katiba . . . to his relatives, Muslims and Muslim women. The Messenger of Allah allocated two hundred bags to his daughter Fatima, Ali ibn Abu Talib—one hundred bags, Osama ibn Zayd—two hundred bags and

Figure 21.4. The Battle of Uhud. Miniature from "Siyar-i-Nabi" (The Life of the Prophet), 1595.

fifty bags of date stones, Aisha, the mother of believers—two hundred bags, Abu Bakr ibn Abu Kuhafa—one hundred bags, Akil ibn Abu Talib—one hundred and forty bags. . . . Muslims ate the meat of domestic donkeys residents of Khaybar. . . . The Prophet then forbade them four things: to take pregnant captives, to eat the meat of domestic donkeys, to eat the meat of any predatory beast, to sell the prey before its division. (448–449, 462)

21.9. Return to Mecca

When it was time to return to his homeland Mecca, the Prophet knew that he had to win the city back in battle against the pagan Quraish with their vile numerous idols (Fig. 21.5).

The Prophet ordered his leaders from among the Muslims, giving them the order to enter Mecca, to fight only with those who resist.But he gave the names of several people whom he ordered to be killed, even if they were found under the covers of the Kaaba. Among them was Abdallah ibn Saad. The Prophet ordered him to be killed because, having converted to Islam, he recorded revelations for the Messenger of Allah, but then, having renounced Islam, becoming a pagan, he returned to the Quraysh. O Quraish! Allah has removed from you the pride of paganism based on the greatness of your ancestors, because all people are from Adam, and Adam is made of clay. (490–493)

Figure 21.5. The Prophet enters Mecca (630), and his Muslim followers crush pagan idols there (https://www.pngfind.com/mpng/ixhTomo_15-minute-history-muhammad-conquest-of-mecca-hd /); figures on a horse and camel are hidden.

However, with numerous murders occurring in Mecca,

> The Messenger of Allah, addressing us, said: "O people! Allah made Mecca forbidden, sacred on the day He created the heavens and the earth, and it will be holy until the Day of Judgment (Figs. 21.6 and 21.7). And it is not allowed for a person who believes in Allah and on the Day of Judgment to shed blood in it and cut down a tree. No one is allowed—neither now nor after me—from this hour to show hatred to the residents of the city. Mecca is becoming a shrine, as before. Let the present inform the absent about it. If anyone tells you that the Messenger of Allah fought in it, answer: "Allah allowed it to His Messenger and did not allow it to you, the Khuzaites! Stop killing. A lot of murders!" (496). Curiously, blood is forbidden from being shed in the Shrine of Mecca, except by the Messenger of Allah himself. When Muhammed died, he discovered that the disease, as a result of which he died—Allah called him to himself, under his mercy. . . . The first manifestation of this disease, as I was told, was that he went to the cemetery called Baqia al-Gharkad in the dead of night (Figs. 21.8, 21.9) and asked Allah to absolve the sins of the people buried here. Then he returned to his family. In the morning he began to have severe pains. . . . Aisha said: "When the illness overcame the Messenger of Allah, he said: "Tell Abu Bakr: let him lead the prayer of the people!" I objected: "O Prophet of Allah! After all, Abu Bakr is an impressionable person, with a weak voice and often cries when he reads the Koran." But the Prophet repeated: "Tell him: let him lead the prayer of the people!" (625, 629)

21.10. The Prophet's Death and the "Caliph" Title Intrigues: The Heir of the Prophet

Assuming the role of leading the people in prayer meant becoming the Caliph, the successor of the Prophet, a position that immediately sparked ambitions for power, leading to intrigue and unrest among Muslims.

Figure 21.6. Mecca during the Hajj in 2009; the world's largest Haram mosque, in the courtyard of which is the main shrine of Islam: Kaaba.

Figure 21.7. In the worship of Allah ('ibadah) today.

Figure 21.8. The cemetery of Baqia al-Gharkad/al-Baki in Mecca before the destruction of its monuments by King Saud in 1925; then the tombstone of Khadija, the first wife of the Prophet, was preserved on it.

Figure 21.9. The cemetery of Baqia al-Gharkad/al-Baki in Medina in our time after the elimination of tombstones on it by the king of Saudi Arabia in 1925 (left) and the mosque of the Prophet, whose grave is located under the green dome of the mosque.

A man came to Abu Bakr and Omar and said: "The Ansars gathered around Saad ibn Ubada in the courtyard of the Banu Said: if you want to lead the people, then hurry to them before their business takes a serious turn!" And the Messenger of Allah was lying in his house—unwashed and untidy, his relatives locked the doors of the house. Omar said: "I then told Abu Bakr: "Let's go together to our Ansar brothers and see what they are up to!" (639)

. . . Ibn Hisham reported: "that most of the inhabitants of Mecca, after the death of the Prophet, planned to renounce Islam and wanted to do so already. Even Attab ibn Usaid (he was the ruler of Mecca when the Prophet died) was afraid of them and disappeared. Then Suhail ibn Amr stood up, praised Allah, mentioned the death of the Messenger of Allah and said: "This will only strengthen Islam. And whoever sows the seeds of doubt among us, we will cut off his head!" (648)

21.11. The Righteous Caliphs, Victories Islam Ascents, and the Schisms in Teaching

After the Prophet's death, Abu Bakr became the first righteous Caliph given his close and unwavering support of Muhammed throughout numerous bloody battles. The Prophet himself frequently mentioned Abu Bakr in his final moments, despite his initial impression of him as a sensitive man with a weak voice and prone to tears, details that were attested by Muhammed's wife. However, this assessment of Abu Bakr's character was incorrect, as evidenced in the map illustrating the victorious battles led by Muslim forces mobilized under Abu Bakr's leadership in Arabia (Fig. 21.10). Although his reign lasted only two years until he died in 634, intrigues regarding the caliphal throne persisted:

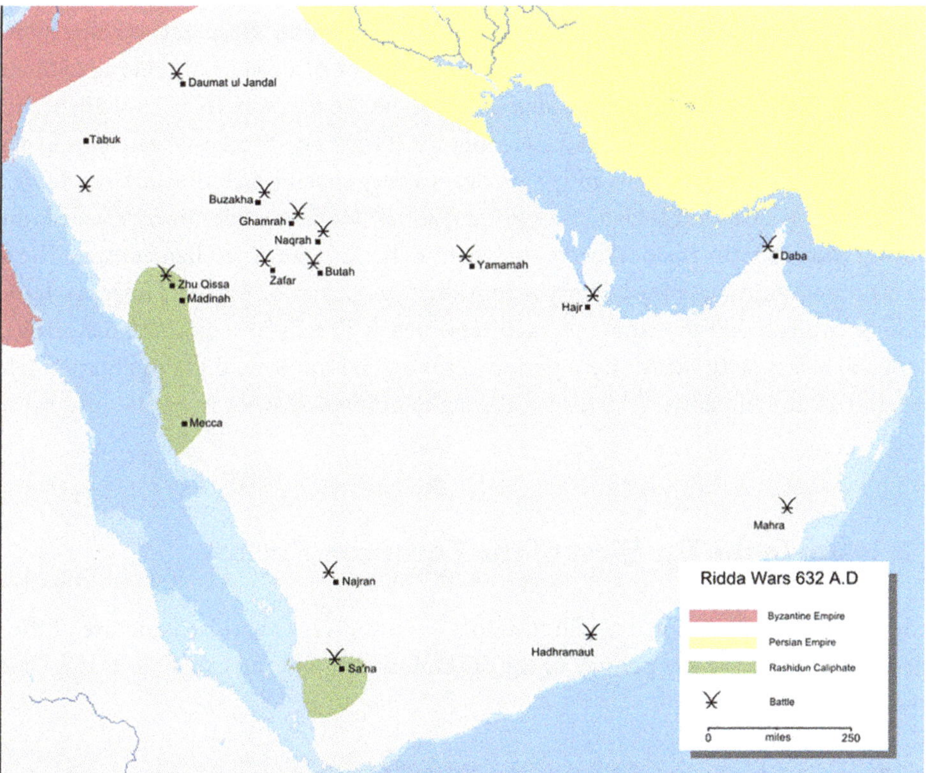

Figure 21.10. The original lands of Islam under the Prophet (green) and the subsequent battles in the spaces of Arabia during the period of the righteous Caliph Abu Bakr.

Sensing the approach of death, [Abu Bakr] summoned his old comrades for the last time and took an oath from them that they would obey the one whom he himself would choose as his successor. After pronouncing the oath, he appointed Omar, then soon closed his eyes calmly. (Muller 2004, 311)

Thus, Omar became the second righteous caliph.

Islam, like Christianity, has also been affected by schisms. Moreover, traces of this were observed already at the earliest stages of the formation of religion, and a certain completion of the processes of fragmentation was, as many believe, at the time of the Umayyad caliphate. There are usually three main currents in Muslim doctrine: Sunnis, Shiites and Ibadis. Sunnis firmly follow the laws and traditions of the Sunnah (path) established by the Prophet Muhammad, which was expressed, in particular, in the foundations of the caliphates, where the caliph or the successor of the Prophet was appointed by the predecessor and confirmed by society. Shiites recognize that the caliphs, and especially the first, righteous ones, should have been elected only by Allah himself and in no other way. Although the earliest of these righteous four—Abu Bakr—was named by the Prophet himself directly, or, in fact, by the mouth of Allah. In general, this election had to go back to the genetic line of the descendants of the marriage of the daughter of the Prophet Fatima. The Ibadis, on the other hand, recognize only the first two of the righteous caliphs, and the others, as it were, remain outside the law.

However, the goals of our book do not imply focusing special attention on such subtleties of Islamic doctrine. Therefore, we will focus very briefly on the actual estimates of the number of adherents of each of the directions, and at the present time, and those in this respect are very unequal. Sunnis completely dominate here—up to 84–85% of all Muslims belong to them. The Shiites are sharply inferior to them, numerically—there are no more than 14–15% of the total number of followers of Islam. The Ibadis account for at least one percent of Muslims in general. In addition, each of these currents of Islam covers a very specific area of southern Eurasia and North Africa (Fig. 21.11). It is also understandable that the Sunni communities cover most of the Muslim space. Shiites are localized mainly within the Iranian-Anatolian enclave. The Ibadis are concentrated on the relatively narrow southeastern corner of Arabia. However, let us pay attention once again that the whole picture presented is already based on the materials of the late, coupled, in fact, with the Modern period; although it is unlikely that the situation relating, for example, to the fifteenth to sixteenth centuries differed sharply from the one shown on our map.

21.12. Islam in the Far West of the European Enclave

At this chapter's conclusion, we will turn to a vivid and very expressive picture of the rapid Islamic conquests during the periods of the faithful caliphs and the caliphate of the Umayyad dynasty (Fig. 21.12).

Two mighty waves of Arab conquerors, like a huge surf, began to flood the neighboring lands, from the east and west. First, at the behest of the

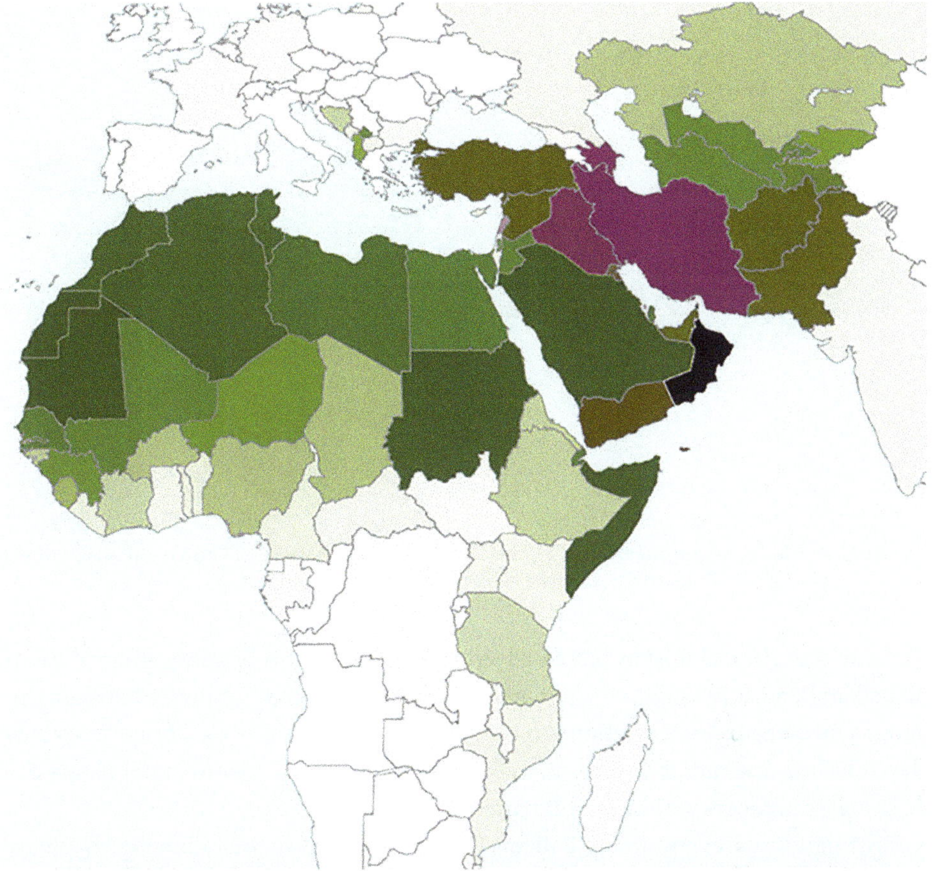

Figure 21.11. Areas of Muslim communities in our time; green—Sunnis, red—Shiites, black—Ibadis. The color differences indicated in the graph determine the relative share of adherents of these trends in each country (the share calculations are given according to the Pew Research Center, Religion and Public Life, 2014).

caliph, the first wave was sent with irresistible force: it flooded Persia to the Oxus, Syria, Mesopotamia, Armenia, some parts of Asia Minor as far as Constantinople, Egypt and the northern coast of Africa to Carthage inclusive. (Muller 2004, 313)

From 655–661, after the third righteous Caliph Uthman, the First Civil War erupted in Arabia, followed by the Second Civil War (680–699). These conflicts briefly tempered the expansionist ambitions of Muslims beyond their borders. However, the era of the four caliphs of the righteous path came to an end, and power transitioned to the Umayyad dynasty. Under the Umayyad caliphs, starting with al-Walid (705–715), Muslims experienced a resurgence in conquests. By 750, the Umayyads managed to expand their influence westward from Iberia, and eastward towards the Indus, once again confronting the legendary and seemingly insurmountable frontier of the Indus for many.

Both Roman Empires suffered the heaviest losses, particularly the east. For Christians of both Orthodox and Catholic denominations, Islam and its adherents became enemy number

Figure 21.12. Muslim conquests during the Umayyad Caliphate (661–750); the place of the battle with the Franks at Poitiers is marked.

one for many centuries, almost as sacred adversaries. Initially, this hostility affected Byzantium and the Balkan Peninsula, where constant battles and unquenchable conflicts between the three Abrahamic faiths ensued and continue to our present day. However, not only Christians were considered Islam's enemies, but Jews as well, as Judaism diverged from many tenets of Islam, despite being its predecessor in the Abrahamic triad.

The most significant events in the Balkans will be discussed in the following chapter. Briefly touching on the extreme West, battles occurred between Muslim and Catholic armies. In 711, Muslim forces of the Umayyad Caliphate, led by Tariq ibn Ziyad, crossed the Strait of Gibraltar and entered the foothills of the Iberian Peninsula for the first time. The local Visigoth rulers and their principalities did not resist them, and by 732 (one hundred years after Muhammad's death and 21 after entering the lands), the Muslim army, led by Abd-ar-Rahman, breached Frankish territory held by Martell (Fig. 21.13). This time, the Muslims suffered defeat, and the Battle of Poitiers ended with a decisive Frankish victory, which erected an insurmountable barrier to the further advancement of Islam in Europe.

Despite differences in ruling philosophy, Muslims have remained on the Iberian Peninsula for over 800 years. The most significant Islamic formation was the Emirate of Cordoba/Caliphate, dating from 756–1031 (Fig. 21.14). The Arabs referred to the entire territory of the Iberian Peninsula as al-Andalus or Andalusia. The final chapter of continuous confrontations was the Emirate of Granada, and later the Caliphate of Granada, whose narrow strip of possessions appeared closely pressed against the Mediterranean coast to the east of Gibraltar. However, in 1492, the army of the Spanish king Ferdinand of Aragon) and Isabella of Castile ended this lengthy Muslim rule in the far western region of Europe. Interestingly, this defeat was acutely felt even in the newly formed and relatively distant Ottoman Empire. Granada was portrayed in a curious manner in Ottoman nautical guidebooks and maps from the sixteenth century, though it was unlikely that Islamic ships could enter the harbors and bays of Granada even then (Fig. 21.15).

Figure 21.13. The Battle of Poitiers in 732 between the Franks of Charles Martell (in the center of the picture on a white horse) and Muslims led by Abdur-Rahman ibn Abdallah. The latter is also in the center, with a long gray beard and a sword; he died during the battle (K. Steuben, 1837, Versailles).

Figure 21.14. The territory of the Cordoba Caliphate around 1000; Muslims called this territory al-Andalus.

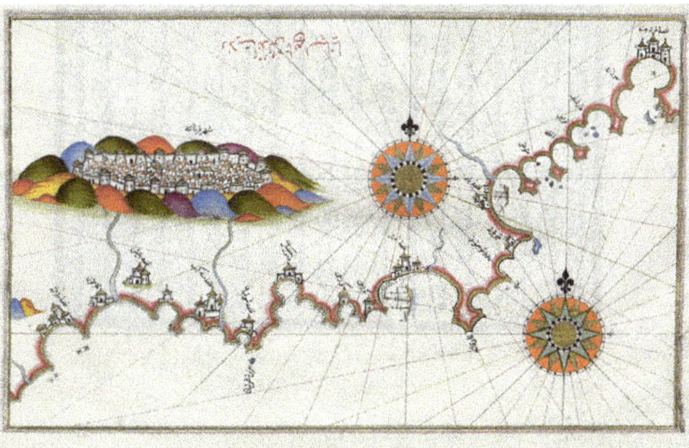

Figure 21.15. This was the idea of Granada and the adjacent Mediterranean harbors and bays on the sea guide of the Ottoman Empire of the sixteenth century.

Chapter 22

Christianity and Islam in Battle: Constantinople—Kazan—Moscow

22.1. Symbolic and Significant Dates: 1500, 1453, and 1552

The beginning of this chapter marks a leap of several hundred years to the date 1500 CE. This date is symbolic as it typically signifies the beginning of the New Age. We will continue our narrative by analyzing the correlation of doctrines of the Abrahamic triad during the centuries leading up to this symbolic year.

One striking aspect is the incredible speed at which Islam spread across the lands south of Europe (Fig. 22.1). Despite originating six centuries after Christianity, adherents of Islam managed to rapidly cover about 17–20 million square kilometers of Eurasian space, as well as the vast north of the African continent, identified as an enclave of the African Saharan or near-Mediterranean shield (see Section 2.1). By the time of the end of the so-called "eve" of the New Age, Christian communities managed to "overcome," or more precisely, to "preserve" 5 million square meters (Fig. 22.1), thus yielding to their now hated enemies, the Muslims, nearly four times the total spatial coverage.

All this occurred within a boom of rapid expansion between the years 1453 and 1552. The year 1453 saw the great success of Islam with the fall of Constantinople and the Orthodox Byzantine Empire. Conversely, 1552 reflected significant losses for the Muslims on their eastern flank, as Orthodox Russian Christian forces managed to crush both Kazan and the Kazan Khanate.

22.2. The Tragic Year 1453 for Orthodox Christians

After the collapse of the Great Roman Empire into a Western and Eastern part (known as Byzantine) following the Great Schism of 395 (see Section 12.4), the Byzantine Empire endured for more than a thousand years. Its end, however, was notably tragic against the backdrop of the rise of Islamic communities and the establishment of the Ottoman Empire in 1299. By the beginning of the fifteenth century, the Ottomans surrounded the constantly shrinking territory of Byzantium with the seemingly impregnable Constantinople at its center (Fig. 22.2).

Figure 22.1. Western Eurasia: areas of distribution and dominance of Christian (Chr) and Islamic (Isl) cultures by the turn of the fifteenth and sixteenth centuries.

A year before the capital was taken, the Ottoman Turks attempted to capture it from all sides. Perhaps the most important effort was the construction of the Rumeli-hisar or Bogaz-Kesen fortress, equipped with powerful bombards capable of blocking the Bosporus Strait which separated Europe from Asia (Fig. 22.3). The residents of Constantinople were also expecting an external assault and even managed to block the Golden Horn Bay with powerful chains to prevent Ottoman ships from penetrating the city's poorly defended areas (Fig. 22.4). However, the fate of Constantinople was sealed. The Ottoman forces were well-prepared for a heavy assault on the seemingly insurmountable walls of the city. Their weapons were also enriched with monstrous bombs suitable for crushing stone walls and hurling stone cores weighing up to 200–300 kg (Fig. 22.5).

Constantinople fell on May 29, 1453, marking the end of the thousand-year-old Byzantine Empire. The young Ottoman leader Sultan Mehmed II (only 21 years old at the time) accquired a significant addition to his title Fatih the Conqueror. The last monarch of Byzantium, Constantine IX Palaiologos, died protecting the city in its final battle. Monuments were erected for both monarchs in various places. The majestic monument to the victor Mehmed rises in the center of the Old City of Istanbul (Fig. 22.6, left), while an incomparably modest monument to the last Byzantine monarch Constantine was erected in Athens (Fig. 22.6, right).

To this day, Turkey celebrates May 29 as the Great Victory Day. By the end of Mehmed II's rule, the outlines of the Ottoman Empire had expanded significantly (Fig. 22.7, top). However, about eight decades later, by the end of the conquests of Sultan Suleiman the Magnificent, the Ottomans began to play the most prominent role in the vast block of Islamic cultures of Eurasia and North Africa (Fig. 22.7, bottom).

Figure 22.2. Constantinople. A bird's-eye view of the capital of Byzantium (reconstruction, Wikipedia).

Figure 22.3. The fireballs with which the Muslims crushed the walls of Constantinople and blocked the Bosphorus. https://rusnext.ru/recent_opinions/1507965176.

Figure 22.4. Fragments of iron chains that blocked the mouth of the Golden Horn Bay depicted in the painting (Istanbul Archaeological Museum).

Figure 22.5. The Siege of Constantinople, depicted in the Istanbul Top-Kapa Museum in the Panorama-1453 image cycle (Tarih Muzesi).

22.3. The Orthodox Crush Kazan and the Kazan Khanate: The Year 1552

Now we will shift our focus from Constantinople/Istanbul to Kazan and the Kazan Khanate in the far northeast (Fig. 22.8). Both events which will be discussed in this chapter are separated by almost a hundred years—from the landmark date May 29, 1453, to October 2, 1552. It is also curious that both heroes of these battles turned out to be surprisingly young: Mehmed II was only 21 years old and John IV (at that time not yet the Terrible) was only 22 years old. For the "Kazan" section of the chapter, we will turn to the revered Russian historian Nikolai

Figure 22.6. The monument to Mehmed II in the center of the Old City, Fatiha district, on the left; on the right is the figure of Constantine IX Palaiologos on the Metropolitan Square in Athens (the monument was built at the end of the twentieth century).

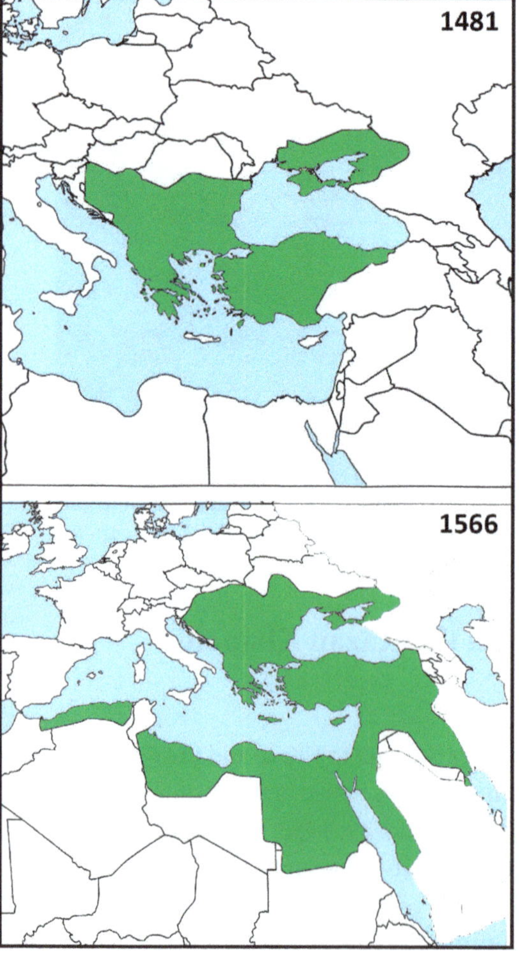

Figure 22.7. Above, the Ottoman Empire at the end of the reign of Mehmed II in 1481, and below, under Sultan Suleiman the Magnificent in 1586.

Figure 22.8. Schematic map of the Kazan Khanate and the formations surrounding it (electronic atlas; tatarhistory.ru).

Mikhailovich Karamzin (1766–1826), who is difficult to rival in terms of breadth and knowledge of historical materials. All the numerous excerpts given below are from N. M. Karamzin's work, "The History of the Russian State," Volume 8, chapters 3 and 4:

Seventeen-year-old John, burning with the zeal of glory, wanted to lead the army to Kazan himself and left Moscow in December [1549]; but fate tempted his firmness with bad luck. . . . Instead of snow, it rained incessantly; wagons and cannons sank in the mud. On February 2, when the Tsar, having spent the night in Yelnya, 15 versts from Nizhny Novgorod, arrived on the island of Robotka, the whole Volga was covered with water: the ice cracked; a firearm shell [cannon] failed, and many people died. For three days the Emperor lived on the island and waited in vain for the way: at last, as if terrified by a bad omen, he returned with sadness to Moscow . . .

In Nizhny Novegorod, the regiments joined together and on February 14 [1550] became near Kazan. . . . We made tours and started to the city. Hitherto, our Sovereigns have not been under the walls of this rebellious capital, sending only Voivodes to punish its treacherous inhabitants: here the young, cheerful, beloved Monarch himself drew his sword . . . with his voice and courage he called the soldiers to glory and easy victory. The tsar of Kazan was in swaddling clothes [the baby Utyamysh-Giray], her noblest nobles died in sedition or were transferred to us, surrounded John and through their secret friends persuaded fellow countrymen to submit to his generosity. 60,000 Russians aspired to a wooden fortress, crushed by the terrible thunder of battering guns. But the last hour has not yet come for Kazan . . . Russians killed a lot of people in the city . . . but they could not seize the fortress. In the following days there was a thaw; it rained heavily, the guns did not fire, the ice on the rivers broke, the roads deteriorated, and the army, having no supplies, was afraid of starvation. It was necessary to yield to necessity and go back with the greatest difficulty . . .

Having sent forward a large regiment and a heavy shell, the Sovereign himself followed them with light cavalry in order to save the guns and hold the pressure of the enemy; he showed firmness, did not lose heart and, engaged only in one thought, the overthrow of this malicious, hateful Kingdom for Russia . . .

And as if almost suddenly in front of the young tsar . . . an immeasurable view opened up in all directions: to Kazan, to Vyatka, to Nizhny Novgorod and to the deserts of the present Simbirsk Province. Surprised by the beauty of the place, John said: "There will be a Christian city here; we will embarrass Kazan: God will give it into our hands." Everyone praised his happy thought, and Shig-Aley and the Tatar Nobles described to him the wealth and fertility of the surrounding lands—and the Sovereign, hoping for future success, returned to Moscow with a cheerful face [March 25, 1550].

In the hated Kazan, however, there were many who were already quite willing to make peace with Moscow.

John said that he would announce the terms of peace if the Kazan people would send five or six of the noblest Nobles to Moscow—and, without wasting time, at the very beginning of spring—after many meetings with the Duma Boyars and with the Kazan exiles, after a solemn prayer service in churches, accepting the blessing from the Metropolitan, he released Shig-Aley with five hundred noble Kazan citizens and with a strong army to the mouth of the Sviyaga River, where they had to set up a city in the name of Ioannovo.

A dense forest overshadowed the mountain: leaving their swords, the soldiers took axes, and in a few hours its top was exposed (Fig. 22.9). They appointed, measured the place, walked around it with crosses, holy water, founded walls, a church in the name of the Nativity of the Mother of God and St. Sergius and in four weeks made the city of Sviyazhsk . . . (Figs. 22.10 and 22.11)

The whole Mountain side—the Chuvash, Mordvins, Cheremis—idolaters of the Finnish tribe, once conquered by the Tatars and not tied to them by either unity of Faith or unity of language, sent their noble people to Moscow, swore allegiance to Russia . . . were assigned to the new city of Sviyazhsky and released from the Yasaks for three years, or dani. To make sure of their sincerity, John ordered them to fight Kazan; they did not dare to disobey, gathered together and, transported in Russian ships to the Meadow side, in the presence of our officials, had a battle with the Kazan people in the middle of the Arsky field: although scattered by cannon shots, they fled in disorder, however, without proving their courage, they at least proved their loyalty. Their Princes, Murzas and centurions during this summer incessantly traveled to Moscow; dined at the palace and, rewarded with fur coats, fabrics, armor, horses, money, praised the mercy of the Tsar and boasted of the new fatherland. The sovereign then poured out silver and gold, sparing no treasury for the fulfillment of great intentions . . .

Shig-Aley constantly pestered him about the Mountain Side, wishing that he would return at least half or part of it . . . But having learned that some Nobles, according to the old custom, secretly commit sedition, are sent with Nogai, plotting to kill him and all Russians, Aley did not doubt to resort to cruel measures: he gave a feast in the palace and ordered to slaughter guests caught or only suspected of treason: some were killed in his dining room, others in the Royal courtyard, a total of seventy people, the most notable; the executioners were their own Aleev Princes and Moscow Archers. Blood flowed for two days: the people froze; the guilty and innocent fled from fear [1552]. This terrible incident revealed to John the need to look for new ways to pacify Kazan.

The only way turned out to be a dense and durable siege. Then came the year 1552.

Regiments surrounded Kazan. Tents and three canvas churches were set up . . . In the evening, the Sovereign, having gathered the Voivodes, verbally gave them all the necessary orders. The night was calm. The next day there was an unusually strong storm: it tore down the Royal and many tents; sank ships loaded with supplies, and terrified the army. They thought that everything was over; that there would be no siege; that we, having no bread, should retire in shame. John did not think so: he sent to Sviyazhsk, to Moscow

Figure 22.9. During the construction of the fortress on Sviyazhsk (Sviyazhsk, Front chronicle, book 20, 444).

Figure 22.10. View of the island-city of Sviyazhsk (engraving by M. I. Makhaev, middle of the eighteenth century).

Figure 22.11. Sviyazhsk today.

for food supplies, for warm clothes for soldiers, for silver and was preparing to spend the winter near Kazan . . .

In the movement, in the heat of heroism, the Kazan people did not feel their weakness; and as in the most desperate determination hope still lurks in the heart, they counted all our unsuccessful attacks on their capital and said to each other: "it's not the first time we'll see Muscovites under the walls; it's not the first time they'll run back home, and we'll laugh at them!" Such was the disposition of the Tsar and the people in Kazan; but John offered mercy in order to fulfill the measure of long-suffering, according to the Policy of his father and grandfather . . .

On August 27, Boyar Mikhail Yakovlevich Morozov, rolling a battering shell [cannon] to the tours, opened heavy fire from all our loopholes; and the squeakers fired into the city from the trenches . . . (Fig. 22.12)

Wishing to use all means to take Kazan with less bloodshed, the tsar ordered a skilled German engineer (that is, an engineer) serving in his army to dig a tunnel from the Bulaka River between the Atalakov and Tyumen gates. Murza Kamai informed the Sovereign that the besieged take water from a spring near the Kazanka River and go there by underground. . . . Our voivodes wanted to open this hiding place, but they could not. . . . For this purpose, razmysl sent his disciples, who, under the supervision of Prince Vasily Serebryany and Ioannov's favorite, Alexei Adashev, dug in the ground for ten days; they heard above them the voices of people walking behind a hiding place for water; they rolled 11 barrels of gunpowder into the tunnel and let the Sovereign know. On September 5, early, John left for the fortifications. Suddenly, in his eyes, the ground, the hiding place, part of the city wall, a lot of people exploded with a thunder, with a crash; logs, stones, soaring to a height, fell, crushed the inhabitants, who froze with horror, not understanding what had happened. At that moment, the Russians, seizing banners, rushed to the collapsed wall; they broke into the city itself, but could not stay in it.

Then followed a new excavation and a new assault on the city:

The Kazan people stood on the walls: the Russians in front of them, under the protection of fortifications, under the banner of banners, in silence, motionless; only tambourines and trumpets sounded, enemy and ours; neither arrows flew, nor guns rattled. We watched each other; everything was in anticipation. The camp was deserted: in its silence, the singing of the Priests who served Mass could be heard. The sovereign remained in the church with a few of his neighbors. The sun was already rising. The deacon was reading the Gospel and barely uttered the words: let there be one flock and one Shepherd! there was a strong thunder, the earth shook, the church shook . . . The sovereign went to the porch: he saw the terrible effect of the excavation and the thick darkness over the whole Kazan: blocks of earth, fragments of towers, walls of houses, people were rushing up in clouds of smoke and fell on the city . . .

But this victory has not yet been completely decided. Desperate Tatars, broken, overthrown from above the walls and towers, stood as a firm stronghold in the streets, were

Figure 22.12. Siege weapon of the sixteenth century. The original drawing was by R. Stein; and engraving of the drawing (here) by P. Dziedzitz (https://www.runivers.ru/upload/iblock/edf/page_182.jpg).

flogged with sabers, grabbed hands with Russians, cut themselves with knives in a terrible dump. They fought on fences, on the roofs of houses; everywhere they trampled heads and bodies with their feet . . . Ours were overpowered in all places and pushed the Tatars to the fortified court of the Tsar. Ediger himself, with the noblest nobles, slowly retreated from the breaches, stopped in the middle of the city, at the Tezitsky or Merchant moat, fought hard and suddenly noticed that our crowds were thinning: for the Russians, having seized half of the city, famous for the riches of Asian trade, were seduced by its treasures; leaving the sich, they began to smash houses, shops—and the officials themselves, whom the Sovereign ordered to follow the soldiers with drawn swords, so that none of them would be allowed to plunder, rushed to avarice. Here the cowards, lying on the field as if dead or wounded, also came to life; and servants, cooks, even merchants came running from the wagons: everyone was hungry for loot, grabbed silver, furs, fabrics; carried them to the camp and returned to the city again, not thinking to help their own in battle. The Kazan people took advantage of the fatigue of our soldiers, loyal to honor and valor: they hit hard and pushed them, to the horror of the robbers, who all immediately took flight, rushed over the wall and screamed: they are being flogged! flogged! . . .

The High Priest Kul Sharif [Kul-Sharif] met the Russians not with gifts, not with prayer, but with weapons: in a frenzy of anger they rushed to certain death and every single one fell under our swords. (Fig. 22.13)

The city was taken and burned in various places; the battle was over, but blood was pouring; irritated soldiers slaughtered everyone they found in mosques, in

Figure 22.13. "The last battle at the Kul-Sharif mosque" (F. Khalikov).

houses, in pits; captured wives and children or officials. The Tsar's courtyard, streets, walls, deep ditches were littered with the dead; from the fortress to the Kazanka, further on the meadows and in the forest, bodies still lay and rushed along the river. The firing ceased; only the blows of swords, the groans of the slain, the cry of the victors were heard in the smoke of the city . . .

So one of the famous Kingdoms founded by the Genghis Moguls within present-day Russia fell to the feet of Ioannov . . .

As we conclude this section, we will repeat the words of Karamzin, that the siege of Kazan, coupled with the Battle of Mamayev, lives in the memory of the people until our current times as the most glorious feat of antiquity, known to all Russians, in both low and high places.

22.4. Kazan Victory and a Monument in Her Honor

Not a single monument was erected in honor of the Russian victory over Kazan and the Kazan Khanate, at least not one comparable to those in Istanbul or Athens. However, the Orthodox victory over Islam at the behest of Ivan IV was reflected in another bright and impressive form—the famous Pokrovsky Cathedral on Moscow's Red Square, also known as the Cathedral of the Intercession of the Most Holy Theotokos on the Moat, or more popularly, the Cathedral of St. Basil the Blessed, named after the revered Moscow fool (Fig. 22.14). The cathedral is incredibly significant and is included in the list of UNESCO World Heritage Sites.

The cathedral was built over the course of six years from 1555 to 1561. Traditionally, it was claimed that Russian architects Barma and Postnik (Posnik) were the project's creators. There was even a rumor of Tsar Ivan being awed by their work and ordering them to be blinded in order to prevent them from creating something more beautiful (Yukhimenko, 23–27). The narrative then shifted from two architects to one, Barma Postnikov. Finally, there were doubts about whether it was built by Russian architects at all, implying instead that the masters must have been Italian.

However, it is not our concern to resolve these disputes, and it is thereby important to add that there is no memorial plaque on the Cathedral itself or on its elevated foundation to indicate that it was constructed in honor of the Kazan victory of 1552. There are only signs in Russian and English that read "Church-Museum Pokrovsky Cathedral" and "Intersession Cathedral (St. Basil's)." There is also a famous monument to Minin and Pozharsky nearby, although this dynamic sculpture, depicting the zemstvo elder Kozma Minin from Nizhny Novgorod warmlly appealing to Prince Dmitry Pozharsky to lead the fight against the Poles.

It is likely that such a kind of impersonality of this unique Temple is due to the extremely strange position of Ivan IV the Terrible, which occupies his figure in the long Russian history. After all, the personality of the destroyer of the Kazan Khanate was not even honored to receive a place among other 129 iconic figures on the famous monument "Millennium of Russia," built in 1862 in Veliky

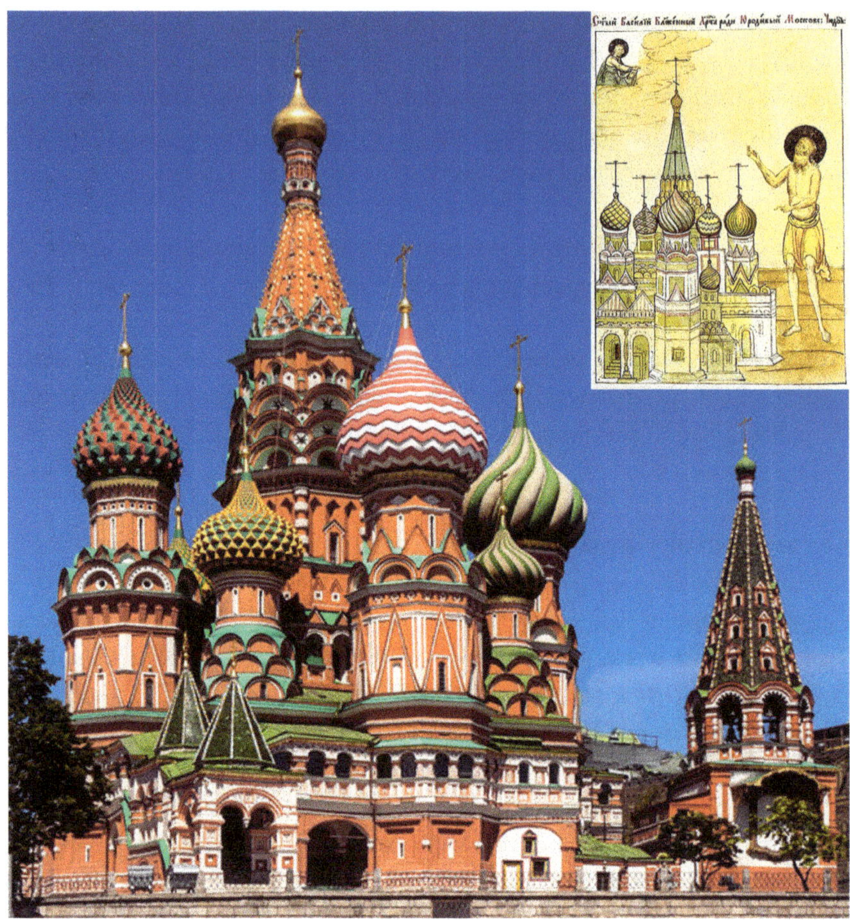

Figure 22.14. Cathedral of the Intercession: St. Basil's Cathedral in the very center of Moscow (above: the image of the holy fool Basil the Blessed, created before the nineteenth century).

Novgorod. At that time, both historians and politicians of the Romanov tsarist dynasty agreed that the harm inflicted on Russia by this tyrant-despot turned out to be incomparably greater than all his benefits. Maybe that's why the Cathedral of the Intercession on Red Square in Moscow is not marked in any way with this extremely controversial, although very iconic figure of Russian millennial history; in any event, the author of the book is inclined to this interpretation.

22.5. Kazan's Revenge: The Burning of Moscow—Ivan the Terrible, Devlet Giray, and the Year 1571

After the fall of the Kazan Khanate in 1552, the Astrakhan Khanate was next, succumbing two years later. Though perhaps the most inconspicuous of the Golden Horde's fragments, its fall was significant. With its collapse, Russia effectively bisected the western half of the Steppe

Belt from the Urals to the Caspian Sea, a crucial factor in the subsequent history of the vast Eurasian region.

Nineteen years after the victory in 1552, the Steppe Belt reasserted itself violently. This time, the threat came from the Crimean Khanate, which had often seemed peripheral and unimportant against the vast Steppe expanse. By this time, the Ottoman Empire had embraced the already Islamized Crimea, and Crimean Khan Devlet-Giray began making claims against Russia and Ivan IV on behalf of his Ottoman ruler, Sultan Selim. This intervention sounded far more threatening than that of a minor khanate.

> The author finds it much more expedient, as in the third section of this chapter, to give the floor to N. M. Karamzin again, rather than to express the thoughts of our outstanding historian in his own language.

> So, the next embassy of the Muslim Turks arrived to Tsar Ivan in 1571 with severe demands: . . . But this Embassy . . . did not have the desired effect, although the Tsar, to please Selim, agreed to destroy our new fortress in Kabarda. The proud Sultan wanted Astrakhan and Kazan, or that John, owning them, recognized himself as a tributary of the Ottoman Empire. A proposal so ridiculous remained unanswered. At the same time, the Tsar learned that Selim was asking Kiev from Sigismund for the most convenient confluence with Russia; that he had ordered bridges to be built on the Danube and grain to be stored in Moldavia; that the Khan, excited by the Turks, was preparing for war with us; that the Prince of Crimea had defeated the Sovereign's father-in-law, Temryuk, and captured his two sons. Already Devlet-Giray in direct relations with Moscow again began to threaten, demand tribute and the restoration of the Kingdoms of Batu, Kazan, Astrakhan . . .

> All the more he [John] was alarmed at the onset of spring (1571), although the Khan, having armed all his ulus, a hundred thousand or more, entered the southern borders of Russia with extraordinary speed, where he was met by some fugitives, our Boyar Children, expelled from the fatherland by the horror of the Moscow executions: these traitors said Devlet-To Giray, that famine, ulcer and incessant disgrace in two years destroyed most of John's army; that the rest is in Livonia and in the fortresses; that the way to Moscow is open; that John, only for glory, only for appearance, can go out into the field with a small oprichnina, but will not hesitate to flee to the Northern deserts; that they vouch for the truth of that with their heads, and will be faithful guides of the Crimeans. The traitors, unfortunately, told the truth: we already had much less courageous commanders and serviceable troops . . .

> Moscow remained without an army, without superiors, without any device: and the Khan was already standing thirty versts away! But the Royal commanders from the banks of the Oka, without resting, came to protect—and what did they

do? Instead of meeting and repelling the Khan in the field, they occupied the suburbs of Moscow, filled with countless fugitives from the surrounding villages; they wanted to defend themselves between cramped, perishable buildings . . .

The next day, May 24, on the feast of the Ascension, the Khan approached Moscow—and what happened was to be expected: he ordered the suburbs to be lit. The morning was calm and clear. The Russians courageously prepared for the battle, but saw themselves engulfed in flames: wooden houses and huts broke out in ten different places. The sky was clouded by smoke; a whirlwind arose and in a few minutes a fiery, stormy sea spread from end to end of the city with a terrible noise and roar. No human power could stop the destruction: no one thought to extinguish; the people, the soldiers in unconsciousness sought salvation and perished under the ruins of burning buildings, or pressed each other in close quarters, striving to the city, to China, but everywhere driven by the flames rushed into the river and drowned. The bosses no longer commanded, or did not obey them: they only managed to block up the Kremlin gates, not letting anyone into this last refuge of salvation, fenced with high walls. People burned, fell dead from the heat and smoke in stone churches. The Tatars wanted, but could not plunder in the suburbs: the fire drove them out, and the Khan himself, terrified by this hell, retired to the village of Kolomenskoye. At three o'clock Moscow was gone: neither Posadov nor China-the city; one Kremlin survived, where Metropolitan Kirill sat in the Church of the Assumption of the Mother of God with a shrine and a treasury; the beloved Arbat Palace of the Ioannovs collapsed. An incredible multitude of people died: more than one hundred and twenty thousand soldiers and citizens, except wives, babies and villagers who fled to Moscow from the enemy; and all about eight hundred thousand . . . "Who saw this spectacle," eyewitnesses write, "he always remembers it with new horror and prays to God not to see this is the second time." (Fig. 22.15)

Thus, the phial of the wrath of Heaven was poured out on Russia. What was missing from her disasters, after famine, plague, fire, sword, captivity and—tyrant? Now we will see how cowardly the tyrant was in this first, the most important misadventure of his Reign. On June 15, he approached Moscow and stopped in Bratovshchina, where he was presented with two messengers from Devlet Giray, who, leaving Russia, as a majestic winner, wanted to sincerely explain to him. The tsar was dressed in simple clothes: Boyars and Nobles also as a sign of sorrow or disrespect for the Khan. To John's question about the health of his brother, Devlet-Giray, the Khan's official replied: "Thus says our King to you, we were called friends; now we have become enemies. Brothers quarrel and reconcile. Give Kazan and Astrakhan: then I will diligently go to your enemies. . . . I burn and empty Russia (Khan wrote) solely for Kazan and Astrakhan; and I apply wealth and money to the ashes. I looked everywhere for you, in Serpukhov and in Moscow itself; I wanted a crown and your head: but

Figure 22.15. Devlet-kettlebell and burning at his behest Moscow in 1571 (https://eadaily.com/ru/news/2017/06/03/etot-den-v-istorii-1571-god-krymskiy-han-devlet-girey-szheg-moskvu).

you fled from Serpukhov, fled from Moscow—and you dare to boast of your Royal greatness, having neither courage nor shame! . . .

What did John do, so arrogant against the Christian, famous Crowned Kings of Europe? He beat the Khan with his forehead: he promised to cede Astrakhan to him at the solemn conclusion of peace; and until that time he begged him not to disturb Russia; he did not respond to abusive words and sarcastic ridicule . . .

Indeed, ready in the extreme to abandon his brilliant conquest, John wrote to the Naked Man in Taurida that we should at least, together with the Khan, confirm the future Kings of Astrakhan on their throne; that is, he wanted to preserve the shadow of power over this Power. Changing our state honor and benefit, he did not doubt to change the rules of the Church: to please Devlet Giray, he gave him at the same time a noble Crimean prisoner, the son of a Prince, who voluntarily accepted the Christian Faith in Moscow; he gave him out for torture or for changing the Law to an unheard-of temptation for Orthodoxy.

Our outstanding historian N. M. Karamzin very unflatteringly, although from my point of view, rightly, assessed the strange figure of Ivan IV, who played an ambiguous and extremely cruel role in Russian history. This was especially vividly reflected in the pages of his works—above all, to 1571, when the monarch, called the Terrible, shamefully ran away from the enemy, leaving his subordinates to the will of the victors. But, one way or another, both the defeat and the burning of Moscow were usually considered by the Muslim enemies of Russia as revenge for the crushing of Kazan (and Astrakhan).

Chapter 23

Eurasia Ushers in a New Era

23.1. 1453–1552: The Landmark Century

Maps of Eurasia, and indeed the entire globe, often depict two contrasting states. In borrowing marine terminology, the first is relatively calm—each ethnic group and its associated culture recognizes its place within the habitat without actively seeking changes, whether willingly or unwillingly. Consequently, the overall picture remains undisturbed. The second state, by contrast, is tumultuous, resembling a stormy sea. The general picture is fragmented, marred by deep fissures, making it difficult to discern the reasons behind the disappearances of old cultures or the emergence of new, distinct ones. While a calm sea is easier to study and depict, the turbulent waves, due to their complexity and mystery, hold a greater allure for researchers.

Typically, a calm map undergoes disruptions due to significant migrations of peoples, both intercontinental and intracontinental, which affects numerous cultural blocks. These migrations often result in notable transformations, as exemplified in Chapter 7. Each of the three eastern waves discussed—the Huns, Turks, and Mongols—spurred the conquest of vast, new territories inhabited by previously unknown groups. Such periods typically led to abrupt and significant changes in the fundamental structures of many Eurasian enclaves.

The dramatic historical events discussed in the previous chapter were likely stimuli or impulses that directly or indirectly generated multiple metamorphoses across the globe. The pivotal century of 1453–1552 marked the commencement of the Great Geographical Discoveries, which, while not directly related to battles such as at Constantinople or Kazan, were indirectly influenced by them. For example, the notable victories of Islam over Christianity in the Balkans were offset by the significant losses of Muslim positions on the Iberian Peninsula. By 1492, the Emirate/Calphate of Granada, the last Muslim stronghold closely confined to the Mediterranean coast, fell to the armies of Castile and Aragon (Fig. 23.1). Sixty years later, Islam faced another major loss thousands of miles from Iberia, this time in the Volga region.

Figure 23.1. The surrender of Granada by the Muslims to the Kings of Castile and Leon Ferdinand II and Isabella I, January 2, 1492 (Francisco Pradilla, 1882).

23.2. European Sailors Embark on Overseas Ventures

While we extensively covered these subjects in our previous book (Chernykh 2019, 227–278), in this chapter, we will provide a survey of the notable events of century.

In the overseas expansion and colonization of territories beyond Eurasia, we identify two primary waves: the early Spanish-Portuguese and the subsequent British eras. The initial wave of overseas discoveries largely coincided with the pivotal century of 1453–1552, while the British wave predominantly occurred in the years after this period.

The maritime crews of the Spanish-Portuguese wave were primarily composed of Catholics from communities within the southern-Mediterranean sub-enclave, particularly from the Iberian and Apennine peninsulas. In contrast, the British wave was primarily comprised of communities from the northern Baltic sub-enclave and the British Isles. English rulers played a significant role in organizing naval expeditions, although representatives from the southern sub-enclave and their Catholic communities also actively participated in this wave as well. This involvement was especially notable in the early phase of the second wave of discoveries. Additionally, the Dutch and individuals from other European ethnic groups were actively involved in long-distance voyages during this period.

At the same time, it is important to note that virtually all navigators aimed to reach India as their primary goal. Below, we will briefly outline the events, beginning with the Spanish-Portuguese Catholic wave of voyages.

Enrique the Navigator (1394–1460), tried to organize fifteen (!) Portuguese voyages around Africa, but was only successful in the sixteenth. In 1434, the Portuguese navigator Gil Eanesh successfully navigated the waters around the

Figure 23.2. A 100 escudo coin issued in Portugal in honor of Gilas Eanes (pictured), who overcame in 1434 The "barrier" of Cape Bohador in the extreme west of Africa.

Figure 23.3. Christopher Columbus sets foot for the first time on the land of America (however, he believes that it is still India); however, initially and certainly such land should be consecrated with a cross (Dioscoro Pueblo, 1862).

mysterious and mystical Cape Bohador in Africa, marking a turning point that heralded a series of fortunate events for the Portuguese (Fig. 23.2).

Christopher Columbus in his four voyages of 1492–1504 discovered America (Fig. 23.3), although he adamantly insisted that he discovered the Indies, a term that was then used in the plural for reasons not entirely clear. He also erroneously believed he could travel to Spain from the Indies by land. Nevertheless, it is important to acknowledge that this discovery was undeniably the central achievement of this cycle of maritime explorations.

Amerigo Vespucci, although he did not take part in the Columbian expeditions, decidedly asserted during Columbus's fourth voyage that he had discovered the New World as opposed to India. It seems unlikely that he could have imagined that, due to Columbus's misconceptions, this New World would come to be known as America rather than Columbia.

Vasco da Gama in 1498 managed to circumnavigate the whole of continent of Africa and reach the Hindustani enclave on the west coast of India. In this way, the European sailors' dream was finally realized, although the result was far from their century-old preconceived notions.

Pedro Alvarish Cabral in 1500 came upon South America, where the densely forested shores rose before him. This region would soon become the vast expanse of Portuguese-speaking Brazil.

Fernand Magellan in 1520 set off on the first world-wide trip, which was compelted in 1522 by Juan Sebastian Elcano, after Magellan was killed in battle by natives from the Philippines, never realizing his long-term dream.

Now we turn to the British wave of voyages, wherein representatives of the northwest Baltic sub-enclave played a prominent role, although the southerners from the Mediterranean enclave were also noticeable. The British authorities took the lead as organizers and financiers of these long voyages. However, it is important to note that these expeditions were not solely British undertakings, as an Italian and Frenchman were among those who initiated these journeys. Additionally, all the navigators of this wave also aimed to reach India and China, albeit it by circumventing America or even Eurasia from the north.

John Cabot was a Venetian known as Giovanni Caboto. It is believed that he was the first to reach North America in 1497. He is commemorated by monuments in Italy, San Marino, and on the island of Newfoundland. A coin with his likeness was issued on the 500th anniversary of his discovery (Fig. 23.4). His mysterious death remains a topic of debate, and his son Sebastian continued his legacy as a sailor and organizer of voyages.

Figure 23.4. 10,000 lira coin issued in 1997 in San Marino in honor of Giovanni Cabot and the 500th anniversary of his discovery of North America.

Jacques Cartier, a Frenchman, also reached North America in 1541–1543, announcing that he had discovered the country of Canada.

Francis Drake was perhaps the most prominent and famous among the British sailors of that time. In 1577–1580, he made the second trip around the world after Magellan and Elkano, which ended quite successfully.

Thomas Cavendish, a sailor and a semi-pirate, undertook the third circumnavigation of the world from1586 to1588, completing it in a record time of two years and fifty days.

Hugh Willoughby and Richard Chancellor in 1554 set their sights on the same dream of reaching India but wanted to circumnavigate Eurasia from the north. However, Willoughby soon died, and Chancellor ended up in the White Sea, in Russia where he reached Ivan IV the Terrible himself. The two discussed building a bridge between the Russian and British thrones, an effort which ended unsuccessfully.

Willem Barents, a Dutchman, in 1597 hoped to reach India via the Arctic Ocean by also skirting Eurasia from the north. However, he met his perilous end in Novaya Zemlya, where he is buried.

Henry Hudson, from 1607 to 1611 also tried to reach China and India by circumnavigating America from the north. It is believed that it was he who discovered the Hudson Bay, which is named after him. However, this bay likely became the grave for him and his young son, when they disappeared forever.

Wilem Janszon, a Dutchman, in 1606, reached the Torres Strait and the western shores of Australia, which he called New Holland.

Abel Tasman, also a Dutchman, later circumvented Australia, though his role will be discussed in more detail below.

This brief overview provides a snapshot of the explosive era of discoveries, which largely coincided with the pivotal century of 1453–1552. In the years leading up to this era, significant efforts were mainly undertaken by Portuguese sailors, with Gil Eanes's success being a notable example. Beyond this century, many more voyages were undertaken, all of which relied on and were forever connected to the discoveries of the main phase, as reflected in the cartographic materials of those centuries.

23.3. Voyages, Discoveries, and Their Depictions on World Maps

As with navigation, we recommend that readers refer to our previous book for a more detailed presentation and numerous illustrations on these topics (Black 2019). Here, we limit our discussion to mentioning only the key figures in cartography and their significant works.

Nicholas Germanus (fifteenth century) was a Benedictine monk and German cosmographer who reconceptualized the data from Claudius Ptolemy's "Geography," which was created approximately 1,300 years prior. In 1466 and 1477, he proposed his version of the earth's surface outlines.

The Treaty of Tordesillas (Castile), 1494. Prompted by the voyages and discoveries of the Portuguese and Spanish sailors, the authorities of these countries signed an agreement dividing the world (Fig. 23.5). Despite the rudimentary and uncertain contours of the globe at the time, this treaty was hastily ratified by Pope Julius II in a bull in 1506.

Martin Waldseemuller (c. 1470–1520) was a German cartographer whose 1507 map named the New World after Amerigo Vespucci, dubbing it "America." However, in a later map, he referred to the region as Terra Incognita.

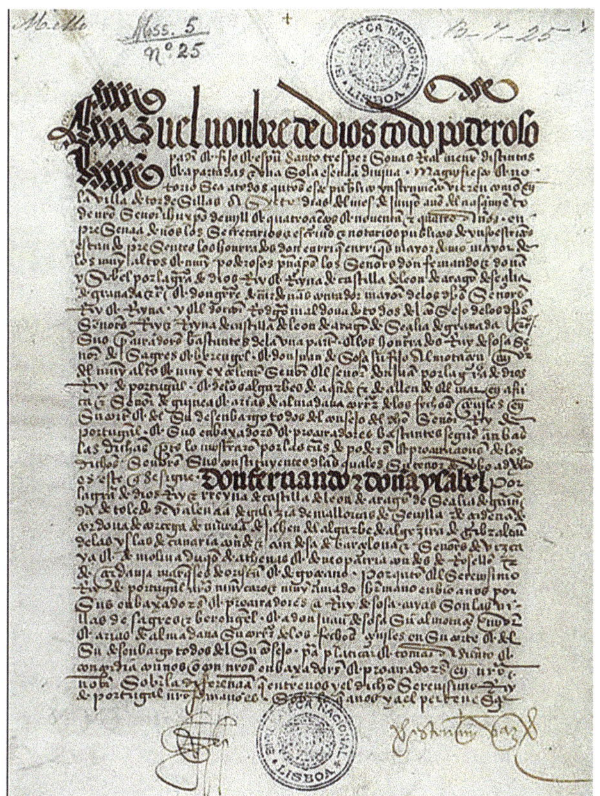

Figure 23.5. Title page of the agreement on the division of the Earth (Globe) between the Spaniards and the Portuguese, signed in 1494 in Tordesillas (Castile).

Abraham Ortelius (1527–1598) was a Dutch cartographer who published a series of maps from 1570 to 1587 which depict the continents' contours significantly different from Waldseemuller's.

Peter Plancius (1552–1622) was Dutch theologian, astronomer and cartographer, whose 1594 map, as developing from Abraham Ortelius' work, is distinguished by the sonorous Latin name: "Orbis terrarum typus de integro multis in locis emendates," which translated means "A wall image of the Earth, corrected anew and in many places."

Vincenzo Coronelli (1650–1718) was a Venetian historian and cosmographer who gained fame thanks to a series of large globes made from 1680 to 1690 commissioned by the French king Louis XIV. However, in addition to globes, he also created excellent maps of Europe on copper plates. (Fig. 23.6)

23.4. Giacomo Gastaldi and Sigismund Herberstein: What Is Muscovy?

Giacomo Gastaldi (c. 1500–1560) was an Italian-Venetian cartographer who produced a significant number of maps, with his most notable work published in 1548. This map Eurasia and America as a single continent. However, twenty-six years later in 1574, Gastaldi separated Asia from America on his new map (Fig. 23.7). During this period, he also published a number of local maps, of which the map of Muscovy in 1550 holds significant interest (Fig. 23.8).

For many—not only Russian, but also for European inquisitive readers—Gastaldi rose to fame through his association with Sigismund Herberstein (1486–1566; Fig. 23.9). Gastaldi granted Herberstein the right to use his map of Muscovy in Herberstein's book *Notes on Muscovy*, originally published in Latin in Vienna in 1549. This book quickly gained fame, leading to multiple translations and reprints (Fig. 23.10). Herberstein's firsthand account of the

Figure 23.6 Vincenzo Coronelli: A globe and a copper plate with a map of Europe printed on it.

Figure 23.7. Giacomo Gastaldi's map of Asia.

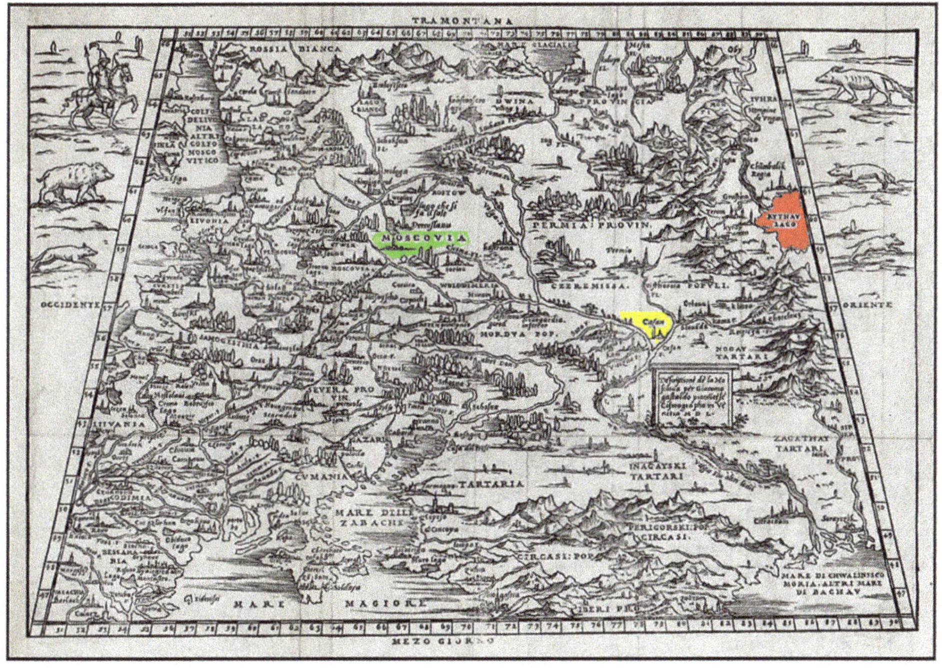

Figure 23.8. Giacomo Gastaldi's map of Muscovy included in the book
by Sigismund Herberstein.

337

Figure 23.9. Sigismund Herberstein in a boyar's garb, granted to him by Grand Duke Vasily III.

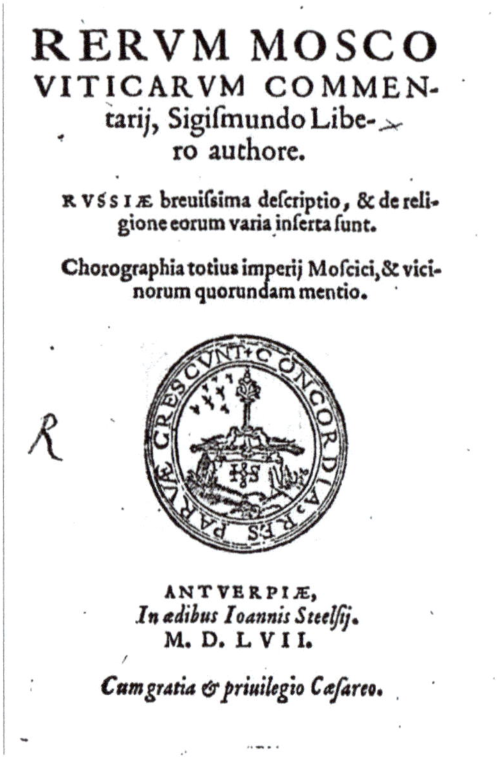

RERVM MOSCO

VITICARVM COMMEN-
tarij, Sigismundo Libe-
ro authore.

RVSSIÆ breuissima descriptio, & de religione eorum varia inserta sunt.

Chorographia totius imperij Moscici, & vicinorum quorundam mentio.

ANTVERPIÆ,
In ædibus Ioannis Steelsij.
M. D. L V I I.

Cum gratia & priuilegio Cæsareo.

Figure 23.10. The title page of the Dutch edition of 1557 of S. Herberstein's book "Notes on Muscovy."

remote and seemingly barbaric land of Russia/Muscovy was both engaging and accessible to European readers, significantly contributing to its popularity.

> For me, it is a constant mystery: how did the cosmographers of that distant time in their works, which have already been mentioned, manage to publish a fair number of different maps, including the most all-encompassing ones, that is, of the globe, without actually ever participating in any distant cognitive expeditions and therefore without having the slightest experience in concrete earthly estimates. A certain exception in this row was, say, Amerigo Vespucci. But after all, he did not make maps, but on the basis of his own observations and impressions, he was able to offer an extremely significant correction in the interpretation of discoveries, and this soon reflected on a number of crown details in understanding the global earth structure. Consequently, cosmographers—as they were called at that time—made maps based only on the stories and descriptions presented to them by persons who either visited these extremely remote places themselves, or used at least those rumors that seemed reliable enough to them.

In this context, the descriptions of the vast expanses of Muscovy in Herberstein's aforementioned book are particularly noteworthy. Before it's publication, he visited Muscovy twice—in 1517 and 1522—with his last visit lasting approximately nine months. However, it remains uncertain whether he managed to travel to Moscow. If he did, it is unclear which areas he visited and the length of his excursions there.

There is little doubt that the detailed accounts in his book, particularly those concerning specific routes or ethnographic details, are based on retellings of various informants. Nevertheless, it is commonly asserted that Herberstein carefully verified the information he received through repeated conversations with other sources. Here, we will focus on a single example from his narrative about the Urals and Siberia:

> The ascent of Kamen Mountain takes three days; descending from it you will get to the Artawische River, from there to the Sibut River, from it to the Lyapin fortress (Lepin), from Lyapin to the Sosva River (Sossa). Those who live on this river are called Vogulici (Wogulici). Leaving Sosva on the right, you will reach the Ob River, which originates from the Chinese (Kitaisko) lake. They barely crossed this river in a whole day, and even then when driving fast: is it so wide [long?], which stretches for almost eighty versts. . . . From the mouth of the Irtysh River to the Grustinka fortress is a two-month journey, from here to the Chinese Lake along the Ob River, which, as I said, originates from this lake, more than three months of travel. From this lake, a very large number of black people come who do not have a generally understandable speech, and bring with them a variety of goods that are bought by the peoples of the Sad and Serponians.

The map of Muscovy, even at a quick glance, leaves a rather curious impression on its viewers. Observers often point out features such as the Chinese Lake, which supposedly conceals the regions of China beyond it. Additionally, there are references to the Ob River, which is depicted as being only eighty kilometers long from this lake. This portrayal suggests Teletskoye Lake, the source of the Biya River, which merges with the Katun River to form the great Ob. However, Teletskoye Lake is located in Altai, and the placement of Altai on this map is questionable. These curiosities raise several questions.

It also remains unclear whether Herberstein included details about these unfamiliar routes and regions of Muscovy in his own writings, although the author is confident that Herberstein had no personal knowledge of these areas. Alternatively, the question remains—did he add these details later after examining Giacomo Gastaldi's map and integrating his own descriptions with it? This raises another challenging question: where did Gastaldi obtain the data to create his map? Perhaps he relied on Herberstein's accounts, given that Herberstein had returned to his homeland long before the map's publication.

However, for the purposes of our current discussion, these questions are not of significant importance. Thus, we will limit ourselves to what has been mentioned.

23.5. Abel Tasman and Semyon Dezhnev: The Invisible Pioneer Duo

The title of this section immediately raises several questions: firstly, why are these figures referred to as pioneers when they traversed lands and seas that had been known to true pioneers many thousands of years before their time? Additionally, why is the term "duo" used when these individuals neither met nor were aware of each other's existence? Furthermore, the two men appeared quite different from one another, particularly in terms of education: Dezhnev, for instance, was likely illiterate, relying on others to write reports for him.

During their early years, Tasman and Dezhnev lived worlds apart. Yet, for more than five decades, their challenging lives unfolded synchronously, sometimes even in a similar rhythm, and generally followed a west-to-east trajectory. Despite this synchronicity, they never crossed paths, with their routes separated by about 12,000 kilometers.

> Abel Tasman was born in 1603 in the small village of Lutegasta, in the north of the Netherlands; he died in 1659 and was buried in Batavia, now called Jakarta, the capital of Indonesia. Semyon Ivanovich Dezhnev was born two years later— in 1605 in the village of Esipovskaya, which is about two hundred kilometers southeast of Arkhangelsk. He died in Moscow, having barely reached the capital with a heavy load of walrus tusks at the beginning of 1673; however, the place of his burial is unknown to us.

Tasman Abel was hired around 1632 to serve in the Dutch East India Company, and by 1634, as already listed as the skipper of one of the ships. His career developed quite successfully: by

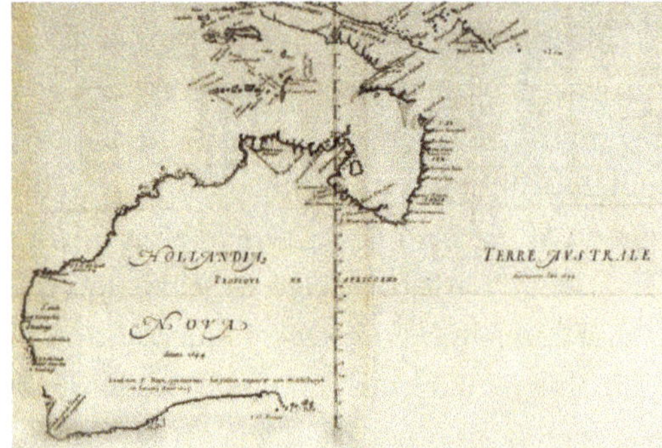

Figure 23.12. The map compiled by Abel Tasman after he completed his voyages around Australia.

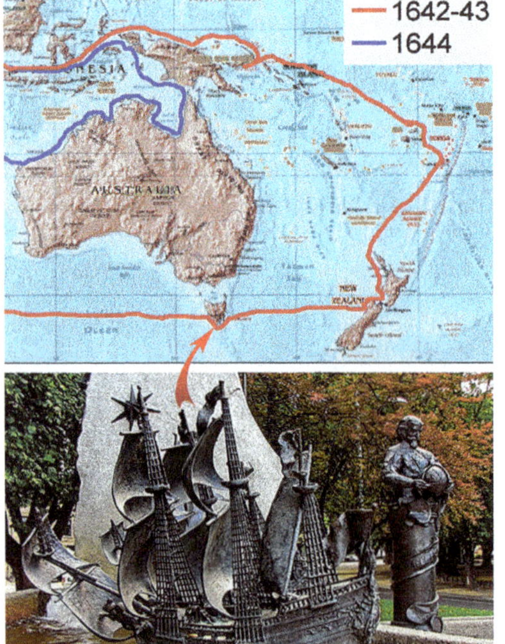

Figure 23.11. Routes of the Dutch sailor Abel Tasman 1642–1644 and a monument to him in Hobart on the island of Tasmania (the place of the city is indicated by an arrow).

1638, he had become captain of the ship. However, perhaps the most significant of his actions were his discoveries during two expeditions initiated by Van Diemen, the Governor-General of the East India Company's possessions from 1642–1644 (Fig. 23.11). Abel succeeding in presenting the continent of Australia (Hollandia Nova). Australia was revealed not to be the western terminus (Fig. 23.12) of the vast Southern continent (Terra Australis), which many at the time believed was necessary to balance the globe (see Abraham Ortelius's map of 1570, Fig. 23.13). During these same years, Tasman and his team discovered a pair of islands, which they named New Zealand, stretching almost two-thousand kilometers from the southwest to the northeast. Following these remarkable successes, Tasman transitioned into the trade sector, where he became very wealthy.

> In 1631, Semyon Dezhnev, in the city of Veliky Ustyug, on the Sukhona River, managed to hire himself into the sovereign's service for the most difficult to characterize and not always understandable jobs in Siberia.Having settled into this so remote region, he, in the detachment of Mikhail Stadukhin (by the way, his future detractor and opponent), went in 1641 to the remote and then virtually unknown Verkhoyansk Siberia, to the Indigirka Valley, to collect fur tribute from the natives. Six years later, he joined the company of the merchant Fedot Popov (Alekseev), and already on sailing small vessels (kochs), more suitable for sailing on rivers than on seas, they managed, as if, to circumnavigate the entire northern, as well as the eastern shores of Chukotka (Fig. 23.14). Having rounded its capes It became obvious that Asia and America are separated by a wide strait. In any case, the date of the opening of this strait, which received the name Bering much later, is considered to be 1648.

And here we are faced with the fact that many refuse to recognize Semyon Dezhnev's naval feat. The first to document Dezhnev's success was the explorer of Siberia G. F. Miller (1705-1783) in 1758. However, he was unable convince others of the worthiness of Dezhnev's discoveries. For example, the academician Johann Georg Gmelin (1709–1755), also a researcher of Siberia, rejected it. The prominent Russian historian P. A. Slovtsov (1767–1843) discussed

Figure 23.13. Abraham Ortelius map of 1570 with the giant continent *terra australis nonum cognita* (the southern land is still unknown), balancing the northern unknown land: terra septentrionalis incognita.

Figure 23.14. Monument to Semyon Dezhnev in Veliky Ustyug and lighthouse monument on Cape Dezhnev (Chukotka, the easternmost point of Asia).

this phenomenon in his remarkable two-volume work *Historical Review of Siberia* (book One, Chapter 6). The twists and turns of these discussions were also sufficiently covered by V. Bakhmutov in a special article (Bakhmutov 2018). However, the author of this book is not inclined to participate in these endless debates. He acknowledges that even if Dezhnev did not circumnavigate Chukotka by sea, he believes the Russians still reached Bolshoy Chukchi Cape (later renamed Cape Dezhnev).

Russians will only add that in the published and preserved "replies" of Semyon Dezhnev (Notes of Russian travelers) sent to steward and voivode Akinfiev, there are indeed no explicit references to sea voyages and the cape. The descriptions provided are rather vague, such as the following:

> And from the Kovyma [Kolyma] river to go by sea to the Anandyr [Anadyr] river there is a nose, went out to sea far away; and not the nose that lies from the Chukhocha River, Mikhailo Stadukhin did not reach that nose. And there are two islands opposite that nose, and on those islands live the Chukhchi, and their teeth are cut, their lips are cut, a fish tooth bone. And that nose lies between the siver on the half-moon, and on the Russian side of the nose there is a sign: a river has come out, the station here at the chukhoch is made like towers made of whale bone. And the nose will turn around to the Onandyr River, and I'll run from the nose to the Onandyr of the Troi River for a day, but no more. And it's a week to go from the shore to the river, because the Anadyr River has fallen into the lip.

The author does not dare to imagine exactly the route of Dezhnev with access to the sea around Chukotka. However, already in our time, quite a lot of maps have been drawn with its proposed routes, but they are unlikely to be drawn up correctly enough in reality. In any case, Dezhnev's "replies" contain incomparably more information about the bad behavior of, say, Mikhailo Stadukhin, or about the death, murders and the unthinkably hard life of polar explorers who were searching for the natives unexplained until then:

> Last in the 156 year of June, on the 20th day, I, the Family [Semyon], was sent from the Kovyma River to the new river on Anandyr for the search for new country people. And last year, in the 157th year of the month of September, on the 20th day, going with the Forged Rivers by the sea, at the shelter of the merchant man Fedot Alekseev, chukhoch people were wounded in a fight, and that Fedot with me, a family, went missing at sea. And it carried me, the Family, by sea after the Intercession of the Mother of God, everywhere involuntarily, and threw me ashore at the front end of the Anandyr River. And there were twenty-five of us on the koch [ship]. And we all went uphill, we don't know our own way, cold and hungry, naked and barefoot. And I, a poor Family, walked from tovarysh to the Anandyr River for exactly ten weeks and we got to the Anandyr River below the sea, and we couldn't get fish, there was no forest, and we, poor people, were starved apart. And twelve people went up the Anandyr. And walked twenty den, people and argishnits, roads foreign, not seen. And they turned back and, not having reached the camp for three days, spent the night, they dug holes in the snow.

Despite the differences between geographers and researchers, the easternmost cape of Chukotka began to be known as Cape Dezhnev much later, in 1898, commemorating the 250th anniversary of its supposed discovery (Fig. 23.14, right). Similar to Abel Talsman, the brave Russian pioneer also ventured in commercial pursuits, particularly in fishing and walrus tusk extraction, which proved to be more profitable than fur trading.

The question remains, what do Talsman and Dezhnev have in common? At first glance, they appear vastly different—one traveled by land, the other by sea, each in significantly different environments and seemingly in opposite directions. Strangely, however, the answer to this question is not so difficult to discern. Both navigated through unknown, vast spaces, spanning thousands of kilometers, moving synchronously from west to east, despite their routes being separated by great distances. While one sailed through largely uncharted seas and the other traversed harsh and unfamiliar taiga, forest tundra, and tundra, both arrived at points where they perceived their journey could go no further, almost at the exact same time—Talsman in 1643/44, and Dezhnev three years later in 1647/48.

Furthermore, there is an intriguing similarity in their voyages, as both concluded their expeditions near the iconic 180 ° longitude dividing the Eastern and Western hemispheres of the Globe (Fig. 23.15). Tasman turned west after discovering New Zealand and glimpsing Fiji, just shy of reaching this invisible edge. Dezhnev, whether circumventing Chukotka, crossed this line when rounding its extreme capes, entering the Western hemisphere. It is unlikely that Dezhnev suspected the Earth was round or acknowledged the concept of hemispheric division.

One way or another, however, both pioneers outlined paths—northern Eurasian and the southern oceanic—that would see fervent explorers and missionaries carrying Christian truths during subsequent intercontinental and intra-continental mitigations in the centuries that followed.

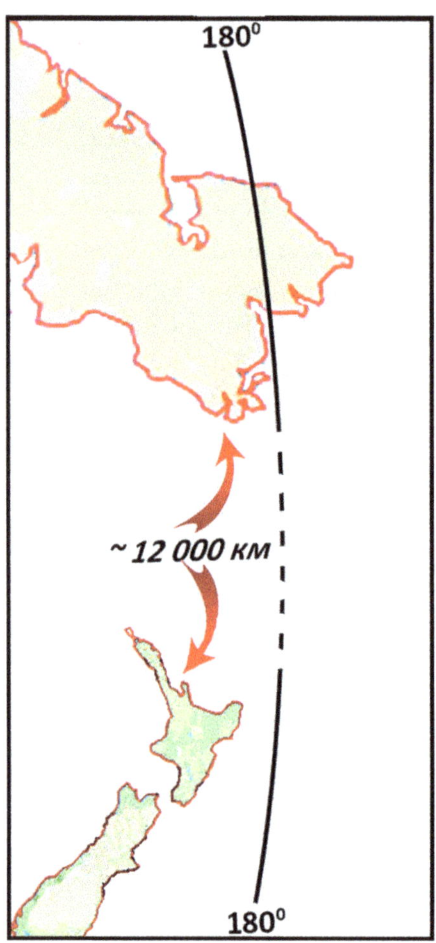

Figure 23.15. Abel Tasman's routes in New Zealand and Semyon Dezhnev in Chukotka in relation to the division of the globe into the western and eastern hemispheres.

23.6. Chukotka and Kamchatka: The Russian Breakthrough to the North and East

Russia/Muscovy and its expansion into the east is also a well-known topic, although not as widely discussed as are, for example, the Great Geographical Discoveries made by European sailors. After all, Russia is essentially eastern branch of the European root, with which it has not always maintained friendly or impartial relations—often quite the opposite. Additionally, the Russian ethnic identity differed significantly from the Steppe Belt and forest/tundra enclaves of the Urals and East Siberia towards which it expanded. Following the liberation from the Turkic-Mongol rule, Russia began its push eastward and northeastward. The first attempts to corss the Ural Mountains date back to the time of Ivan III.

It may seem strange, but the prophecies of the monk-elder Philotheus (1465–1542) of the Pskov Spaso-Eleazar Monastery (Fig. 23.16) made an unusually deep impression on the rulers of the Moscow kingdom of that time, and on many of the Grand Duke's entourage, when, for example, he addressed them with these words:

"The sovereign Grand Duke to the clerk, Mr. Mikhail Grigoryevich, your poor praying mantis elder Philotheus prays to God and beats his forehead. . . . So know, lover of God and lover of Christ, that all the Christian kingdoms have come to an end and have converged into a single kingdom of our sovereign, according to the prophetic books, this is the Roman kingdom: for two Romans have fallen, and the third stands, and the fourth will not happen. Many times the Apostle Paul mentions Rome in the epistles, in the interpretations it says: 'Rome is the whole world.'"

And here's what's curious: the idea expressed by the elder Philotheus—"Moscow is the Third Rome, and the fourth will not happen"—lasted for half a thousand years; and even today it can be repeated here, however, often in a mocking tone. In any case, there is no doubt that it could really stimulate the processes of Russian colonization of Eurasia.

When Ivan the Terrible managed to crush the Kazan Khanate, it largely opened the way to the desired, but completely foggy distances. The role of the most prominent, initial figure in this quest for the unknown east belongs, of course, to the legendary, but so adored in the folk epic Ermak, who penetrated to the wild brega of the Irtysh in 1579–1585. Finally, Russia penetrates further—already to Alaska, to the west coast of North America—although, in the end, American colonization ended in an insulting failure for Russia.

The final result of this rapid breakthrough to the east was the formation of a new—and unlike the globally transoceanic British—land Russian Empire. And its most extensive basic geoareas were two giant Eurasian northern forest and forest-tundra enclaves—the Ural and East Siberian (Figs. 23.17, 23.18). Russia has actually absorbed these enclaves entirely throughout the vast territory of Northern Eurasia. But those geoareas were in fact the eternal possessions of the peoples of the Stone Age, starting from the Pleistocene and Paleolithic. The crown ethnic groups of these enclaves often became, however, already in our days and gradually, only objects of ethnographic reserves.

23.7. Steppe Belt Nomads as Russia's Eternal Enemies

The Russian Empire's most prominent enemies continued to be the elusive and nomadic people of the Steppe Belt.

Figure 23.16. Moscow, the third Rome: the monk-elder Philotheus, offering supplications to God. Above the figure of Philotheus in an arched contour is Moscow with the Kremlin; behind the arch are the ruined Constantinople (left) and Rome.

Figure 23.17. The Steppe Belt and the borders of the Russian Empire, Qing China by the 80s of the nineteenth century. Territories: 1—Russia; 2—China, 3—the borders of the Steppe Belt.

Fighting against them turned out to be extremely difficult. The objective of their conquest once again became a paramount concern. For instance, consider the testimony of Kutlu-Muhammad Tevkelev, a Tatar murza who later became the Russian general Alexei Tevkelev:

> Peter the Great [in 1722] was pleased to have a desire for the whole fatherland of the Russian Empire to have a useful intention in bringing the vast Kirghiz-Kaysak hordes, heard from time immemorial and at that time almost unknown, into Russian citizenship by his high monarchical person, I, the lowest, had the intention to use it so that if this horde did not wish to become an exact citizen, then try to keep me, despite the great costs, at least to melion, but tokmo so that only one sheet under the protection of the Russian Empire is obliged to be. For how . . . Peter the Great in 722, being in the Persian campaign and in Astrakhan, through many deigned to be notified of this horde, although this Kirghiz-Kaysatskaya steppe and frivolous people, tokmo-de to all Asian countries and lands this horde is the key and the gate; and for the sake of this reason-de horde needs to be under Russian protection, so that only through them in all Asian countries to have comonicacy and to take useful and capable measures to the Russian side. (Kazakh-Russian relations, 31; Suleimenov, Basin, 28, 29)

While text in this message is quite convoluted, Peter the Great's intent is clear: the Steppe Belt, with all its diverse peoples, should belong to the Russian Empire. However, achieving this goal took about a century and a half, as evidenced by the "General Map of the Russian Empire," compiled in 1745 (Fig. 23.19), and even then, it required significant effort. It was

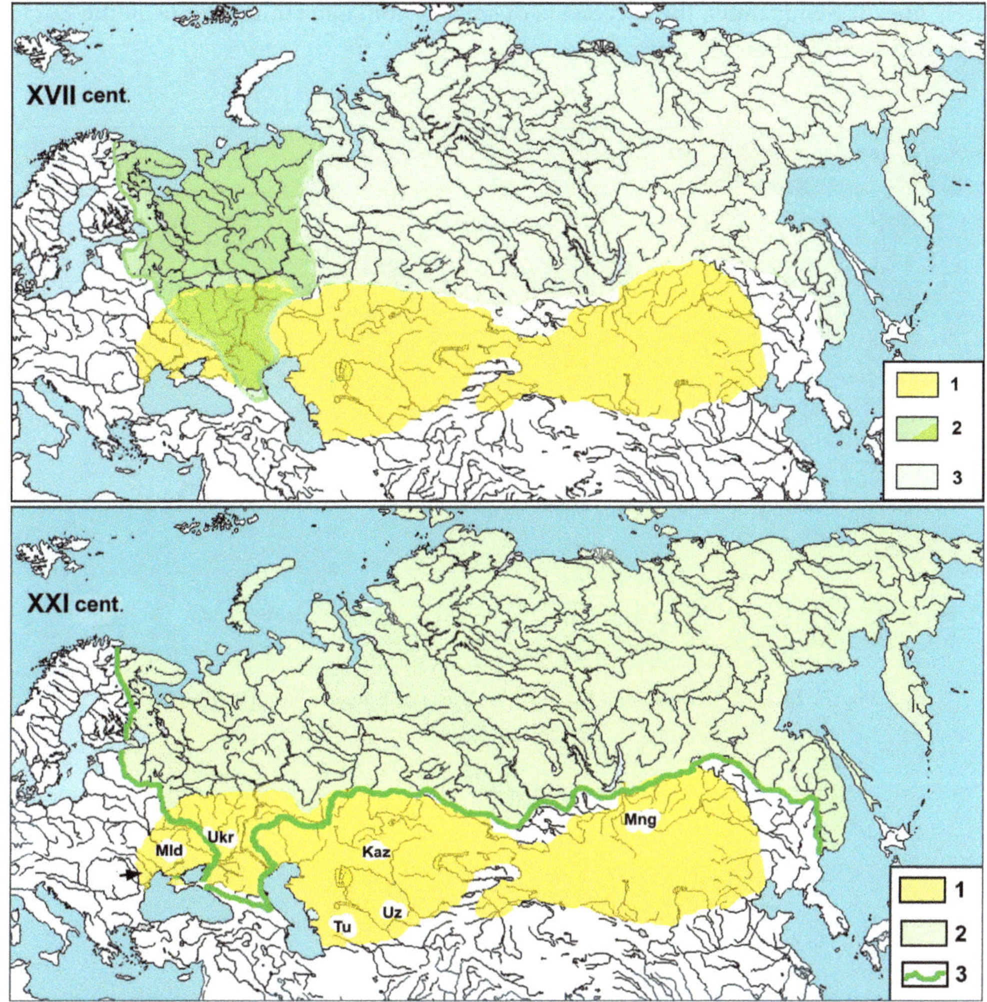

Figure 23.18. Russia at the turn of the seventeenth to eighteenth centuries (above) and at the beginning of the twenty-first century. Territories in the upper figure: 1—the Steppe Belt, 2—the Moscow Kingdom, 3—the North Asian colonies of the Moscow Kingdom; in the lower figure: 1—the Steppe Belt, 2—the Russian Federation, 3—the borders of the Russian Federation. Symbols: Ml—Moldova, Uk—Ukraine, Tu—Turkmenistan, Uz—Uzbekistan, Ka—Kazakhstan, Mn—Mongolia.

not until 1886 that the entire Asian part of the Steppe Belt was designated as the Turkestan Region. Interestingly, its borders coincided with the northwestern boundaries of the expansive Chinese Qing Empire (Fig. 23.17). At that time, it seemed possible to permanently conclude the millennia-long history of nomadic ethnic groups and cultures in the Steppe Belt, much to the satisfaction of sedentary neighbors to the north and south—Russians and Chinese alike.

However, this hope was short-lived. Approximately a century later, during the rapid collapse of the Soviet-Russian Empire, almost all states within the Steppe Belt swiftly declared full independence, stretching from Mongolia in the east to Moldova in the west. This chain of newly formed nations spanned almost seven thousand kilometers (Fig. 23.18). Suddenly and in an entirely unexpected manner, the successors of ancient nomadic ethnic groups of the Eurasian Steppe Belt asserted their identities.[1]

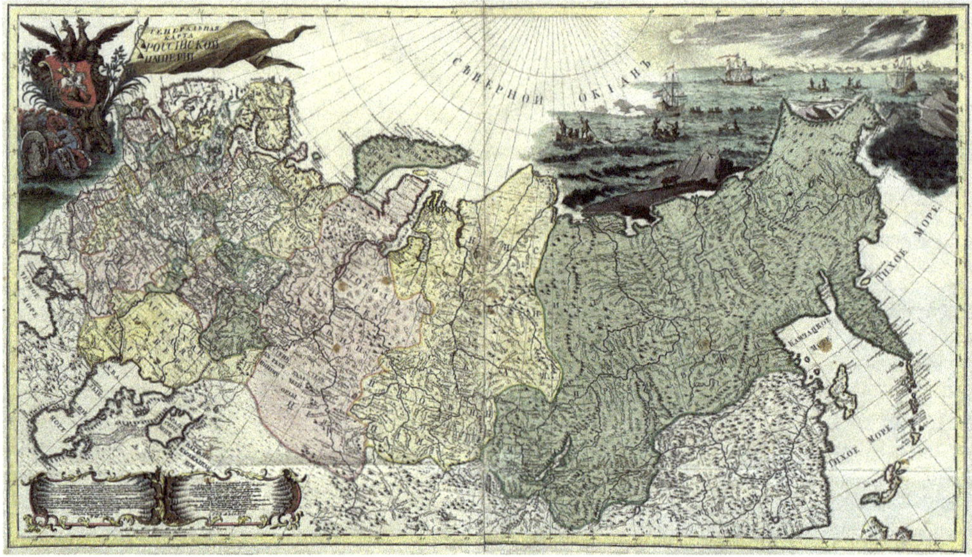

Figure 23.19. The "General Map of the Russian Empire" compiled in 1745.

---

[1] This topic has already attracted the author's attention in much more detail (See: Chernykh 2019, 318–320, 324–329, 333–339, 342–345, Figs. 11.39–11.41, as well as Chernykh 2013; Chernykh 2017).

Chapter 24

The Abrahamic "Triad" against Other World Doctrines

The previous chapter began with the statement that geographical maps of the whole world (not only Eurasia) reflect to polar states described in marine terminology as calm and stormy. Our exploration of this phenomenon, however, highlights mainly the stormy situation. The reasons for this are quite clear: while this book is small in volume, it extensively covers a great breadth of topics, both chronological and spatial. Hence, there is a clear need to focus on significant historical moments in the history of Homo communities—not only Eurasian, other continents as well. Throughout history, Eurasia has been intricately linked with the narratives of other continents, despite the varying stability in the nature and pace of intercontinental interactions. However, it was during the Era of Modern Times that Eurasian ethnic groups and their cultures emerged as catalysts for nearly all profound transformations in the global human experience.

The type of attention to the critical and stormy periods of human history have prompted this author to resort quite often to such concepts as jump and leap based on the need to emphasize the rapid transition from one critical moment to its replacement. It is quite possible that this may cause consternation among a number of readers, since my presentation cannot satisfy interest in the calm periods of existence of ethnic groups and their cultures; after all, those periods are hidden by emphasizing such jumps. However, it seems to the author that it was precisely this style of presentation of materials that has allowed us to concisely and more logically reconstruct not only the general canvas of living space, but also the dynamics of the development of the painting itself. So now we will make a deliberate time jump from the turn of the new era, quite clearly associated with the landmark century of 1453–1552, directly to our own days. If the beginning of global processes was connected with that century, then what happened five hundred years later? In all likelihood, we will be able to get a clearer assessment of the consequences of those turning points that originated at the turn of the new era.

24.1. Explosive Activity and Spatial Coverage of Christian Cultures

In the previous chapter, we began with a sense of awe, observing the rapid spread of Islam across Eurasian and North African regions—a testament to its status as the youngest branch of the Abrahamic faiths (see Fig. 21.11). However, as the sixteenth and seventeenth centuries unfolded, profound changes were not confined to Eurasia alone but rather extended across the broader canvas of the Earth. Firstly, the cultures of the Euro-enclave surged to prominence, directing their sailors towards uncharted overseas horizons. Secondly, Orthodox Russians, by the mid-seventeenth century, reached the farthest and hitherto unknown eastern extremity of the Eurasian continent.

Among the most remarkable outcomes of the shifts in the past five centuries, one undoubtedly includes the global spatial leaps made by adherents of the Abrahamic doctrines. Christianity and Islam, in particular, stand out for their striking and expansive trajectories, which are vividly depicted in corresponding graphs and maps (Figs. 24.1 and 24.2)

Today, there are approximately 2.3 billion adherents of the Christian faith across the planets, spanning various denominations such as Catholic, Protestant, and Orthodox and across the globe (Fig. 24.2B). In contrast, Islam, with around 1.8 billion adherents, is noticeably smaller in both number and geographic distribution (Fig. 24.2C). Unlike Christianity, which has extended primarily overseas from its territorial origins, Islam has firmly established itself as a continental presence rooted in Iranian-Anatolian and Arabian enclaves. Outside of southern Eurasia and northern Africa, its presence is much more sparse, with notable exceptions in regions like Malaysia, Kalimantan, and New Guinea.

One of the essential reasons behind such a rapid, explosive spread of Christianity and Islam often attributed to their assertive nature. Historically, adherents of these doctrines demanded—often through forceful means—that politically subjugated peoples either voluntarily convert or face coercion, sometimes resulting in enslavement. This approach is exemplified by Pope Urban II to the Crusaders:

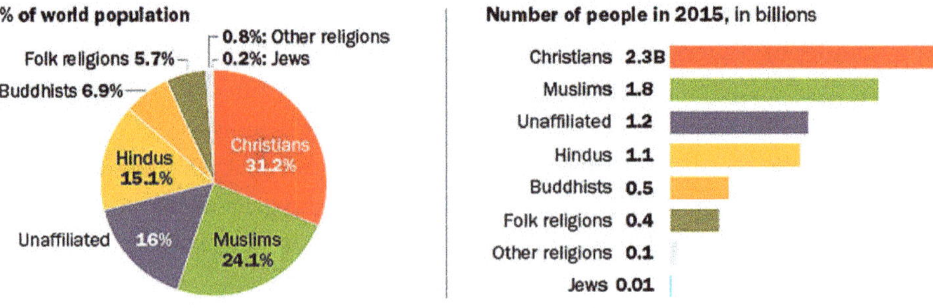

Figure 24.1. Distribution around the globe in percentages and absolute values of groups—adherents of various religious doctrines, as well as outside of affiliation with the intended groups (Pew Research Center demographic projection[1]).

---

[1] The number of adherents of different religions of the globe is most thoroughly and successfully studied by the analytical Pew Research Center by their extensive program "The Changing Global Religious."

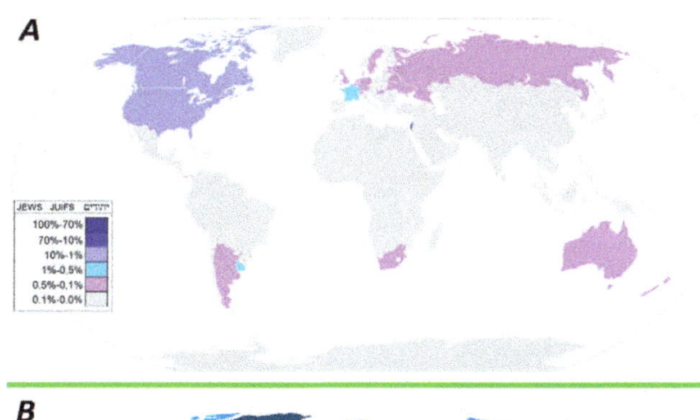

Figure 24.2. Schematic maps of the distribution of adherents of the Abrahamic religions by countries: A—Judaism, B—Christianity, C—Islam.

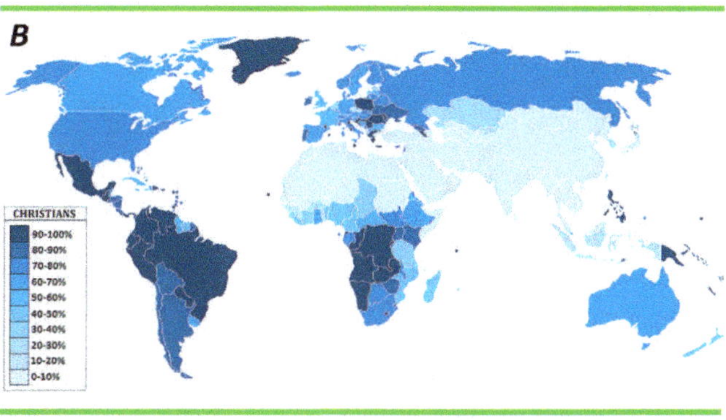

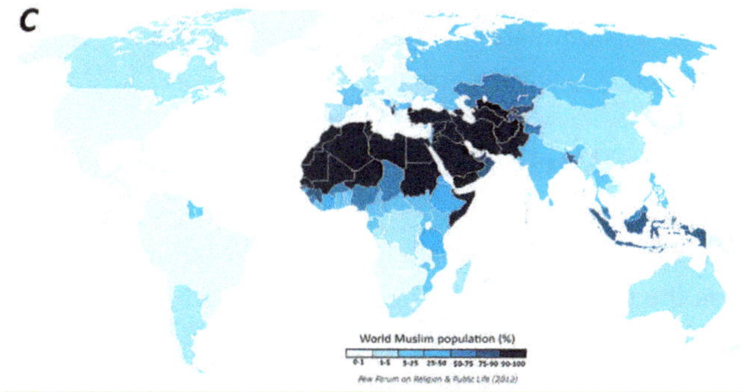

Enter the road to the Holy Sepulchre; snatch this land from the wicked race and subjugate it to yourself. . . . God has rewarded you over all nations with great glory in weapons. Accordingly, undertake this journey for the forgiveness of your sins, with confidence in the imperishable glory of the Kingdom of Heaven.

Hinduism, with over a billion worshippers, predominantly resides within the Indian subcontinent, emphasizing its deep historical and cultural roots in Hindustan (Fig. 24.3).

Hinduism's concentration in the Indian Peninsula, which has puzzled European explorers for centuries, highlights its unique geographical and cultural context (Fig. 24.3). Similarly, Buddhism, with approximately 520 million followers, holds sway over extensive territories particularly in East Asia, notably connected with the Han enclave and China (Fig. 24.3).

On the other hand, Christianity stands out for its global spread, enveloping central areas across a wide span of the Eastern Hemisphere, forming a robust chain from the Pacific to the Atlantic, while also extending influence from the north and west, including into the Americas

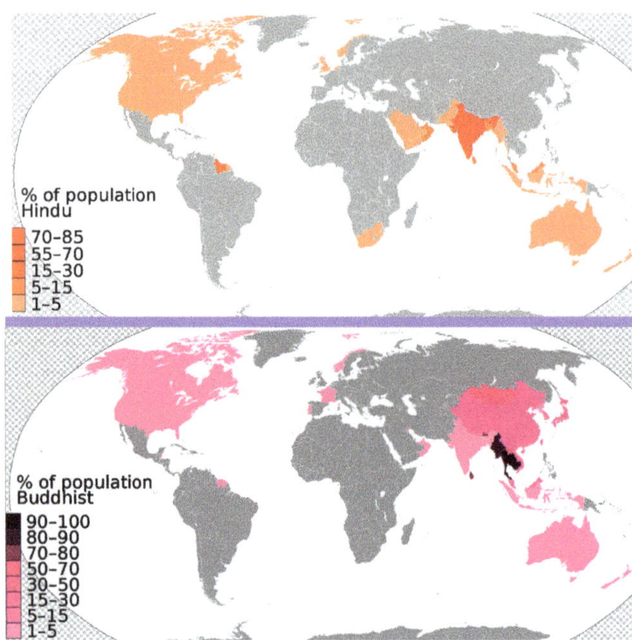

Figure 24.3. Schematic maps of the spread of adherents of Hinduism and Buddhism.

(Fig. 24.2B, C). This spatial distribution underscores Christianity's expansive reach compared to other major religions, which predominantly dominate specific regions of Eurasia.

24.2. Anomalies and Paradoxes in World Doctrines

Against the background of these other world religions, attention should be drawn to the seemingly mysterious fate of Judaism as the true patriarch-progenitor of the Abrahamic triad of monotheism. It is striking due to the small number of adherents, approximately 14 million, which is a mere fraction of both Christianity and Islam (only 0.2%!). More than half of the Jewish population resides in Israel, highlighting the enclave's significance as the historic cradle of the entire Abrahamic tradition, despite its small geographic size compared to other religious centers (Fig. 24.2A).

Another notable anomaly lies in the category of unaffiliated individuals, comprising approximately 1.2 billion worldwide (Fig. 24.1). This group surpasses even the number of Hindu followers and is structure into three segments—atheists (12%), agnostics (17%) and 71% who either could not clearly define their religious beliefs or chose not to specify (Fig. 24.4).

The rise of this unaffiliated group reflects broader societal changes influenced by modernity and the spread of scientific thought, which provided individuals with more avenues for expressing skepticism or uncertainty about religious affiliations. This shift underscores the evolving dynamics of religious identity and belief systems in the contemporary worlds.

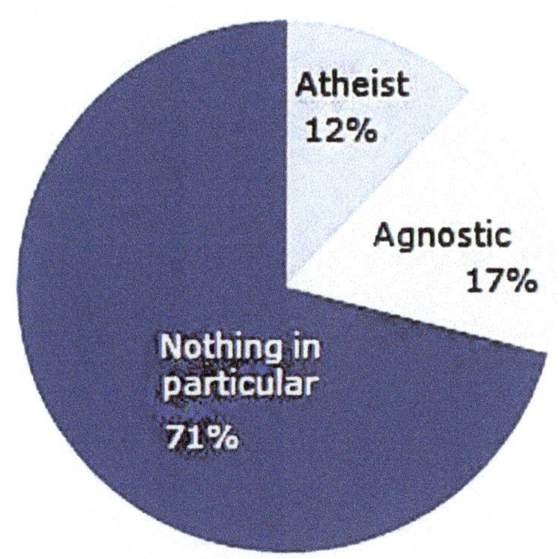

Figure 24.4. A diagram of the ratios of respondents in the group who are not affiliated with religious movements and who have recognized themselves as a) atheists, b) agnostics, and c) undecided with the answer to the questions received.

24.3. Visions of Unity and a Spectrum of Fractured Christian Beliefs

One of humanity's most enduring yet potentially unattainable dreams for any human ethnic groups or collective of *Homo* has been an insatiable thirst for unity of mind and connection. Without it, life would seem unbearable, perhaps even bleak and pessimistic. However, achieving this unanimity has always remained a utopian ideal, never succeeding in practice. Even when the utopian ideal seemed to be attained, victory often bathed a distorted chimera of consensus that proved difficult to unravel.

It seems to the author that it would be appropriate to offer several epigraphs to this section on like-mindedness, offering deeper insight into this profoundly important topic. Numerous candidates vied for the role of epigraphs, prompting the author to select just three. The first is a quote from the eminent Dutch philosopher Baruch Spinoza (1632–1677) (Fig. 24.5). In 1670, the philosopher published the *book Theological and Political Treatise*, reflecting on this very theme:

> I have often been surprised that people who boast of professing the Christian religion, i.e., professing love, joy, peace, abstinence and trust in everyone, more than unfairly argue among themselves and show the most bitter hatred towards each other every day. (Spinoza, 11)

However, exactly two hundred years after Spinoza, our unforgettable classic, Kozma Prutkov—behind whom Mikhail Yevgrafovich Saltykov-Shchedrin hid—offered his own playful take on like-mindedness and its understanding. His proposition, imbued with the characteristic humor of Kozma Prutkov's writings, is presented here as the second epigraph, attributed to the mayor of Sztupovszky, Basilisk Borodavkin:

> It is necessary that unanimity reigns among the mayors. So that they speak all over the face of the earth with one mouth. I will briefly discuss the harm of the mayor's polytheism...

Figure 24.5. Benedict (Baruch) Spinoza (right: Barend Graat, 1666; left: monument in Amsterdam, Nicolas Dings, 2008).

The mayor is obliged to plant sciences. This is true. But this time it is necessary to give yourself an account: What sciences? There are different sciences; some treat of fertilizing fields, of building human and animal dwellings, of military valor and invincible firmness—these are useful; others, on the contrary, treat about harmful Freemason and Jacobin freethinking, about some supposedly natural concepts and rights to man, and even the structures of the world are harmful. . . . Having thus established itself in the very center, the mayor's unanimity will inevitably entail universal unanimity. . . . For they will have nowhere to go from such unanimity. Consequently, there will be no quarrel, no discord, but there will be orders and shooting everywhere. In conclusion, I will say a few words about the mayoral autocracy and other things. This is also necessary, because without the mayor's autocracy, the mayor's unanimity is also impossible. But there are different opinions on this. Some, for example, say that the mayoral autocracy consists in the conquest of the elements. One mayor personally told me: "What kind of mayors are we, brother! I have the sun rising every day in the east, and I can't order it to rise in the west!" (Saltykov-Shchedrin 1969, 427–428)

However, in arriving at the present day, thoughts on the criteria of like-mindedness and unity as well as the approach to understanding this desired state begin to evolve. Such is the view of Archbishop Theophylact of Pyatigorsk and Circassian whose perspective resonates with earlier recommendations by Kozma Prutkov, thereby meriting inclusion as another quotation here:

Like-mindedness has a proper name—this is Christ. In Christ is the essence of our unity. The point of contact is God. And if we are looking for some other reasons for our similarity, for our like-mindedness, then we will get lost rather than find it. In Christ we can forgive, in Christ we can love, in Christ we can be friends and we can hope. In Christ we can be grateful. In Christ we can stop hatred and malice. Everything is in Christ. And if this is not the case, then our humanity turns on, and it is already so inventive, so complex that it leads not to unity, but to the only correct opinion that only I am right and others are wrong. And the end is for all unity. . . . St. John Chrysostom has wonderful words: it is only necessary, he says, to put Christ in the main place, as everything in your life will fall into place. This is probably the recipe for like-mindedness and unity. If Christ is at the center of your life, everything becomes organized, unified, integral. (Pravoslavie.ru)

In this case, the question remains—how should the words of Jesus Christ in the Gospel according to Matthew (10: 32–38) be understood and interpreted through the prism of like-mindedness?

> Therefore, whoever confesses Me before men, I will also confess him before my heavenly Father; but whoever denies me before men, I will also deny him before my Heavenly Father.

> Do not think that I have come to bring peace to the earth; I have not come to bring peace, but a sword, for I have come to divide a man with his father, and a daughter with her mother, and a daughter-in-law with her mother-in-law. And man's enemies are his household. Whoever loves the father or mother more than Me is not worthy of Me; and whoever loves the son or daughter more than Me is not worthy of Me; and whoever does not take up his cross and follows Me is not worthy of Me.

Is the primary idea here universal love or a punishing sword?

Spinoza was deeply surprised only by the dissonance between Christianity's Ten Commandments and the realities of life surrounding him. The similarity between the last two thinkers—Kozma Prutkov with his deep humor, and Archbishop Theophylact, without any hint of it—is that there can be only one opinion. And yet, the juxtaposition of these two figures in our chosen epigraphs—where a faceless boss and Jesus Christ appear side by side—highlights a profound dichotomy. For many, what matters most is their perceived supremacy, compelling people to follow without room for the inventive, human nature that often disrupts such mandated unanimity.

However, how do we reconcile the stark contrast between the Evangelist Matthew's depiction (Fig. 24.6) and the viewpoint of Archbishop Theophylact of Circassian? According to Matthew, the message brought by God's Son was not one of reconciliation, but rather of a sword that brought punishment to nearly everyone. Thus, Spinoza's puzzlement over witnessing Christians exhibiting intense hatred towards each other on a daily basis appears contradictory.

Figure 24.6. The angel who inspires the revelations of the Evangelist Matthew
(Michelangelo Caravaggio, 1599–1602).

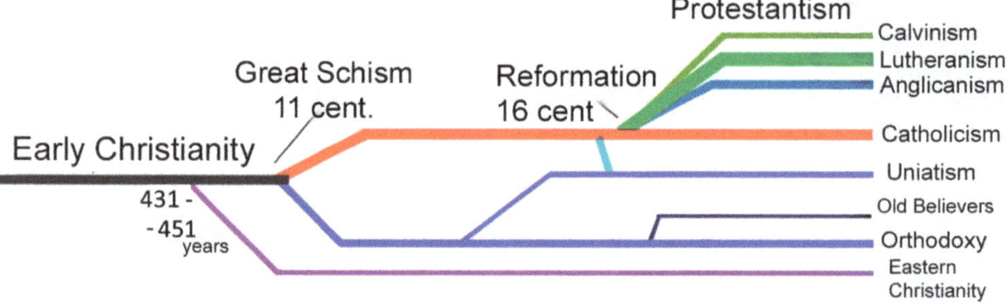

Figure 24.7. Chronological diagram of the most important ideological "breaks" in the global
Christian doctrine (Wikipedia).

24.4. Christianity's New Crisis: Protestants and Catholics

Within the history of Christianity, the noble pursuit of unanimity and connection within
the faith was marred by significant ruptures, notably during the division of the Roman Empire
into Eastern and Western halves, which culminated in the Great Schism discussed in Chapter
20. However, even more damaging to the reputation of the Roman Catholic Church was the
schism that coincided with the period from 1453 to 1552. This period marked the dawn of the
early modern era that era that witnessed a proliferation of heresies and challenged the authority
of papal Rome and spawned movements such as Protestantism and reformism.

Protestantism emerged with the rise of Lutheranism, Calvinism, and Anglicanism (Fig. 24.7).
Each denomination branched off from Roman Catholic doctrine and was led by influential
figures. Martin Luther (1483–1546; Fig. 24.8), the German theologian, for example, ignited

the Reformation with his famed 95 theses, which he posted on the Castle Church door in Wittenberg on October 31, 1517. Luther's uncompromising stance and theological reforms marked a significant departure from Roman Catholic orthodoxy, thereby earning him reverence and recognition as a pivotal figure in Christian history.

Before Luther, in 1412, the Czech theologian and preacher Jan Huss (~1370–1415) challenged the sale of indulgences at the University of Prague. Huss's outspoken critique of church practices led to his condemnation and execution as a heretic in 1415, inspiring the Hussite movement (Fig. 24.9). Jean Calvin (1509–1564; Fig. 24.10), a French theologian, emerged as another prominent figure in the Protestant Reformation. His intellectual prowess and theological insights, which he began developing from a young age at the University of

Figure 24.8. Martin Luther (Lucas Cranach the Elder, 1532).

Figure 24.9. The burning of Jan Hus in 1415 (Konstanzer Richental Chronik, 1464).

Figure 24.10. The image of Jean Calvin (stained glass in the American Reformed Church, Paris) and the sculpture of Miguel Servet (Canton of Annmas, France, but 4 km from Geneva).

Figure 24.11. The English King Henry VIII (Hans Holbein the Younger, 1539–1540).

Paris, culminated in his push for reform within Catholicism. Calvin's theological influence was most profoundly felt in Geneva, where he played a pivotal role in shaping religious and civic life from 1541 to 1553. His involvement in the trial and execution of the Spanish physician and anti-Trinitarian thinker Miguel Servetus complicated Calvin's legacy and earned him both admiration and condemnation within Protestant circles.

The English monarch Henry VIII (Fig. 24.11) decided to take a significantly different approach to challenging papal authority. His desire for autonomy from Rome and his quest to establish supremacy over the English Church led to the enactment of the "Supremacy Act" in 1534. This legislative move solidified Henry's position as the head of the Church of England and eventually contributed to Anglicanism branching off as a distinct doctrine within Western Christianity.

These movements and figures illustrate the fragmentation and diversity within Christianity during the early modern era, reflecting ongoing struggles for doctrinal unity amid profound theological and political transformations.

Chapter 25

The Modern Era: Brewing Disagreements

Martin Luther's 95 Theses (Figs. 25.1, 25.2) are widely considered to be the catalyst for a profound conflict within the heart of papal Catholic Rome. The authority of the Papal Throne, at least in the European enclave, had lasted for nearly 1,500 years. This declared war, characterized by periods of strife and temporary lulls, had persisted over the course of five centuries.

Martin Luther's complex personality and his swift and decisive actions had a significant impact on historical developments, not just within European, but on a broader, global scale. Therefore, it is fitting to devote several key sections of this chapter to exploring this influential figure and his significant deeds.

25.1. Martin Luther as Prophet of the New Doctrine: His Monologues and Dialogues

Martin Luther's prolific pen produced a wealth of writings, seldom characterized by a calm tone. It is easy for researchers to uncover pieces of his thoughts that are explicitly abusive, denouncing, and stigmatizing. As a rule, Luther typically initiated these confrontations, and in the event of a response, we might term them dialogues, despite their vehement nature. In the absence of responses, they remain monologues.

The first and perhaps the most famous of his monologues is the 95 Theses addressed to Pope Leo X and the entire Catholic Church. From Luther's point of view, the sale of indulgences was egregious and reprehensible, as the Pope often granted immediate forgiveness for past sins in exchange for money (Fig. 25.3a, b). Luther's critique echoed Jan Huss's protest from a century earlier. In 1517, Luther had not yet fully distanced himself from the Papist camp, and he expressed his initial monologue in relatively mild terms comparably. However, some theses stand out for their primarily harsh tone:

> 69. Bishops and priests are obliged to receive the commissioners of these indulgences with every favor.

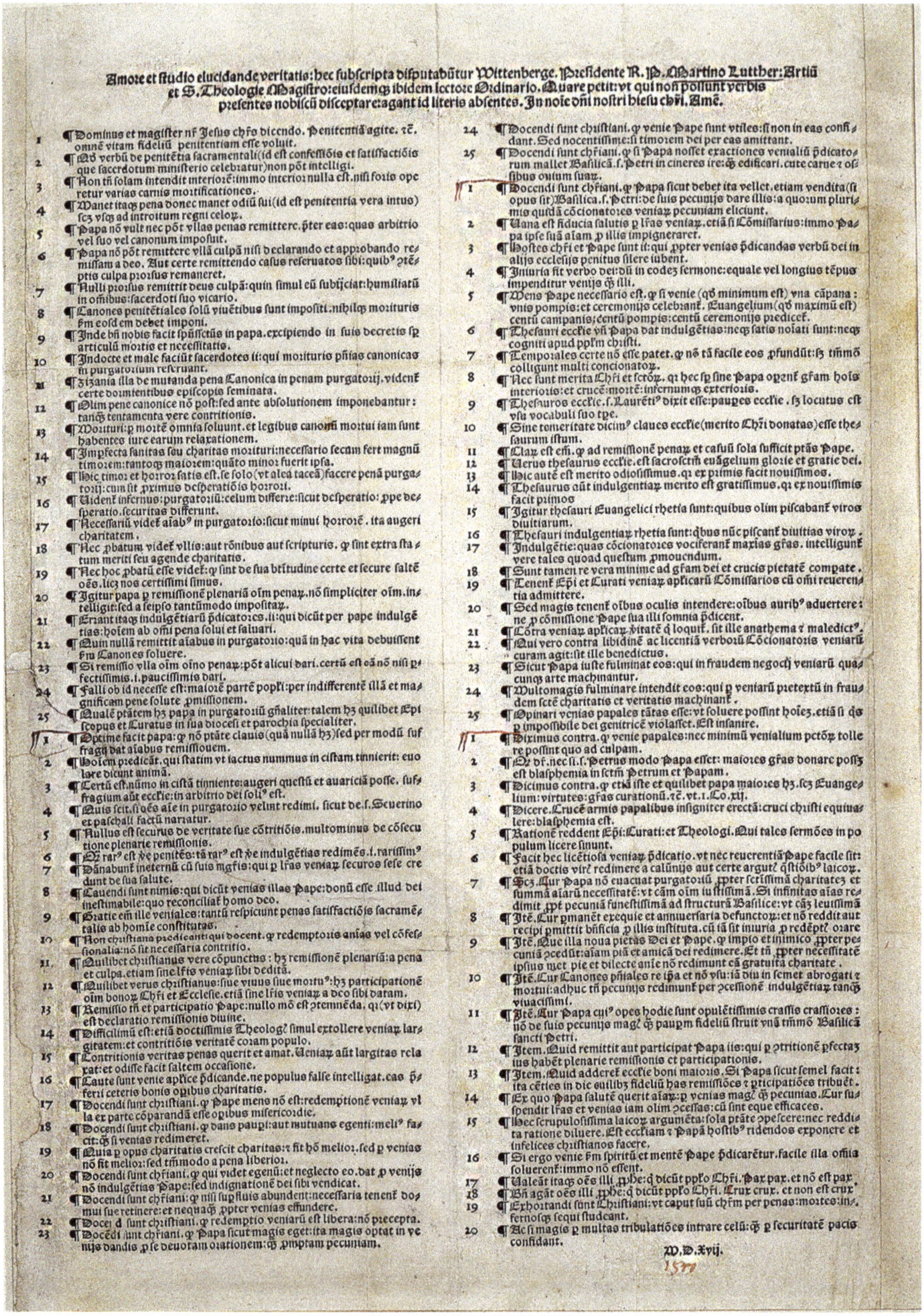

Amore et studio elucidande veritatis: hec subscripta disputabuntur Wittenberge. Presidente R. P. Martino Luther: Artiu[m] et S. Theologie Magistro: eiusdemq[ue] ibidem lectore Ordinario. Quare petit: vt qui non possunt verbis presentes nobiscu[m] disceptare: agant id literis absentes. In noie dni nostri hiesu chri. Ame[n].

M.D.Xvij.

Figure 25.1. An excerpt from Luther's 95 theses, 1517.

Figure 25.2. An engraving depicting Luther writing his theses with a giant pen on the door of the Castle Church in Wittenberg; the back end of the pen penetrates into the head of a lion symbolizing Pope Leo X.

Figure 25.3a. A brisk sale of indulgences in the hall where the Pope sits on the throne (an engraving of the sixteenth century).

Figure 25.3b. Satan distributing indulgences (from a Czech manuscript, 1490s).

Figure 25.4. Lithograph "The Life of Martin Luther and the heroes of the Reformation." It depicts Luther throwing a papal bull into the fire with his excommunication from the church; there are also figures of Jan Huss, Girolamo Savonarola, John Wycliffe, Kaspar Kreuziger, Philip Melanchthon and Johann Bugenhagen and others (the guardian).

70. However, all of them, much more, are obliged to look with all their eyes and listen with all their ears, so that these commissars, under the guise of a papal commission, do not preach their own inventions.

71. Let anyone who denies the truth of these papal indulgences be anathematized and thereby cursed!

While it is true that these words are less clear, the message of the seventy-first thesis remains particularly ambiguous: could there be some truth in the papal indulgences that it is considered criminal to deny them?

Luther received an answer, but after a long delay. Two and a half years later, Pope Leo responded to the accusations in the bull *Exsurge Domine* (Arise, Lord), dated May 15, 1520, which effectively excommunicated Luther:

If this Martin, his supporters, adherents and accomplices, to Our great regret, stubbornly do not submit to the above-mentioned conditions within the specified time, then We, following the teaching of the Holy Apostle Paul . . . will curse this Martin, his supporters, adherents and accomplices, as barren vines that do not abide in Christ, preaching offensive doctrines contrary to the Christian faith and insulting the Divine majesty, to the detriment and shame of the entire Christian Church, and defaming the keys of the Church, as stubborn and well-known heretics.

362

Luther's answer to Pope Leo X from Wittenberg was almost immediate. On June 6, 1520, he responded with more inflammatory language (See Fig. 25.4).

> A certain Augustinian monk, Martin Luther, rushed like a madman at the Holy Mother, the Church, wanted to strangle her with blasphemous books. . . . This is not a man, but Satan in human form. . . . Combining into one stinking pile . . . all the heresies that had been before him, he added new ones to them. . . . We forbid anyone from now on to dare to receive, defend, or support the said Martin Luther with words or deeds. On the contrary, we want him to be detained and punished as the notorious heretic he deserves, to be brought before us personally or to be securely guarded until those who captured him inform us, after which we will order properly, based on what Luther said. Those who help in his capture will be generously rewarded for their good work. (Wikipedia, Diet of Worms)

Luther's capture did not succeed—by that time, he had garnered a considerable number of supporters who also desired the humiliation of Rome and the Pope. Consequently, he was able to continue his work unabated. His primary efforts then shifted to translating the Bible into German, which many, including even his detractors, consider one of his greatest achievements.

> Then he sometimes convulsively seized on work; he spent whole days and nights reading, writing, studying Greek and Hebrew, preparing to translate the Holy Scriptures into German. (Merezhkovsky, 75)

He disguised himself with the pseudonym Junker Jörg and hid in the Wartburg castle, where in just eleven weeks he translated the New Testament from Greek into German, thereby allowing it to be accessed by ordinary people, which was an invaluable treasure in the history of Protestantism (Fig. 25.5).

> The richest fruit of Luther's leisure in Wartburg and the most important and useful work of his whole life is the translation of the New Testament, with which he brought the teaching and example of Christ and the Apostles into the mind

Figure 25.5. Martin Luther with the "New Testament" translated by him into German (Wittenberg).

and heart of the German in a living reproduction. It was a reissue of the Gospel. He made the Bible the people's book in church, school and home. If he had done nothing else, he would have been one of the greatest benefactors of the German-speaking race.

His version was followed by Protestant versions in other languages, especially in French, Dutch and English. The Bible ceased to be a foreign book in a foreign language and became naturalized, and therefore became more understandable and expensive for ordinary people. In the future, the Reformation no longer depended on the deeds of the reformers, but on the book of God, which everyone could read for themselves as their daily guide in spiritual life. This invaluable blessing of an open Bible for all, without the permission or intervention of the Pope and priest, marks a huge progress in the history of the church and can never be lost. (Shaff, vol. 7, § 62)

However, the differences between Papism and the emerging Lutheranism-Protestantism continued to widen, and this was vividly reflected in Luther's relentless monologues and dialogues. According to the historian Gobry, "Insults seemed to give him strength, the presence of enemies filled him with a sense of contentment. Surrendering to the heat of the controversy, which required him to endlessly reread the Scripture, analyze each of its commas, search for weaknesses in the opponent's argument, he temporarily forgot about his personal experiences." Nevertheless, the influence of Protestantism in Germany was growing, a fact that Luther himself reported with a certain sense of pride: "Here with me are 30 men, including 12 night guards and two sentries on the towers" (Gobry, 383).

In 1530, the Augsburg Reichstag took place, which Luther did not attend because he had been anathematized at the previous Worms Reichstag (Fig. 25.6). However, while staying in Coburg, which is not particularly close to Augsburg, he sent streams of letters about the Reichstag to ensure that the actions of his Protestant adherents were not left to chance. At the conclusion of the Reichstag, Luther published a pamphlet titled "Reflections on the so-called Imperial Edict." Its solemn theological tone was interspersed with crude vernacular expressions:

I, Dr. Luther, an unworthy Evangelist of our Lord Jesus Christ, assure you that neither the Roman, nor the Turkish, nor the Tatar emperor, nor the pope, nor the cardinals, nor the bishops, nor the churchmen, nor the princes, nor the knights will be able to do anything against this work, even if the whole world and all its demons. They only bring deadly hatred on their head. This is stated by me, Dr. Martin Luther, speaking in the name of the Holy Spirit ... papists are incorrigible, and, despairing of destroying their infernal plans with good, I break with them from now on and will pursue them with my malice until death. They won't hear another kind word from me.

Catholics most likely were confused about what kind words, if any, they had heard from Luther during these ten tumultuous years. They were no longer in doubt about the future, especially after Luther declared:

I can't pray anymore without cursing!
When I say, "Hallowed be Thy name!,"
I must add, "Cursed be the names of
all papists and blasphemers!" When I
say, "Thy Kingdom come!," I must add,
"Cursed be the papacy!" When I say,
"May Your will be done!," I must finish
with, "May all the plans of the papists
be cursed!" Such a prayer I now passionately recite every day. . . . However,
I maintain the friendly disposition of a
peaceful Christian towards everyone.
Everyone knows this, even my worst
enemies. (Gobry, 386, 387)

His worst enemies, the Catholics, were
clearly unaware of this supposed friendly disposition. Their response came through the words
and pen of Duke George of Saxony:

Figure 25.6. Luther's monument in Worms,
where he was excommunicated from the
Church at the Worms Reichstag in 1521.

There are as many lies in your "Reasoning" as there are words in it. You vile apos-
tate, you allow your mouth to spew only lies. You shamelessly claim that Papism
forbids marriage; you lie when you say that anyone who uses unsanctified salt
will die. You are lying when you claim that the Duke of Saxony and the Elector
of Brandenburg provided the emperor with five thousand horsemen for the
war against the Lutherans. You call all papists, including the emperor, traitors,
scoundrels, thieves of God and rabid dogs. . . . Your thunder and lightning do
not frighten me and will not prevent me from telling the truth. Do you want to
yap at me? I'll bark back, too, so much so that you, a hellhound, will spit fire! . . .
For princes who disagree with you, you have only two words: thieves of God and
madmen. So know this: the thief of the Lord God and the madman is you! The
incarnate demon! . . . You are all the evil, all the abomination, all the sinfulness and
all the shame of the world. Here it is, the pinnacle of villainy! (Gobry, 389)

As new developments appeared, discussions—if indeed they can be defined as such—multiplied and increased in frantic tone.

25.2. The Swift Rise of the Protestant Church and the Words of Its Prophet

The growth in Martin Luther's followers facilitated the emergence of a Protestant church
within the context of Christian doctrine. Simultaneously, the author contemplates a recurring

phenomenon wherein the endorsement of new doctrines invariably prompts and is exacerbated by intense mutual reproaches among conflicting parties engaged in theological disputes.

> Luther founded a new, true Church that returned to its evangelical origins and set out to replace the old Church, which in turn considered itself to be true. In 1533, he stated, contrasting both Churches: One of them, a false one, although it calls itself the Church and God's people, is not really such; the other, although it does not bear such a big name, is in fact the true church. The first one has been engaged for centuries in deceiving believers who did not suspect that they trusted the "Church of the Antichrist," the Pig Church, the new Church of adulterers and the devil. This church is possessed by many demons, the most dangerous of which are the demon of pilgrimage, the demon of indulgences, the demon of bulls, the demon of fraternities, the demon of mass, the demon of purgatory, the demon of monasteries, the demon of the priesthood and the demon of the papacy. (Gobry, 353–354)

Among the adherents of Protestantism, familiarity with all such expressions may understandably not be enjoyable. However, it is impossible to do without such acquaintance, as they reveal the authentic essence of that turbulent era. This period is not just reflected on geographical maps, but profoundly impresses with its linguistic intricacies.

The preacher from Zurich, Bellinger, referred to Luther as the "German Prophet and apostle," adding the following caveat:

> Unfortunately, it is obvious and beyond doubt that no one except Luther has ever written about faith and other serious things in a more rude, obscene, indecent and shameless style, contrary to all Christian decency! And further: "Most people are terribly fond of Luther's cynical, dirty and obscene eloquence. . . . Thomas More, the author of a Latin work containing an analysis of Luther's language, came to the conclusion that in this stream he saw nothing but latrinae, merdae, stercora (latrine, shit, manure) . . .

> Perhaps it is more interesting just to give a number of definitions that will give us the opportunity to feel with what hurricane force Luther tried to free his soul from what hindered him. Dad?—A pile of shit that the devil piled on the Church. . . . Or: The head of the Christian world is the front and back hole through which the devil has piled a bunch of rubbish into this world, such as mass, monks and other outrages. . . . Does the Pope demand repentance from sinners? Let him kiss our ass!

> Or—An appeal to the devil: Dear devil! I have committed many other sins that are not in your registry, namely, I dirty my pants! You can take them for yourself, wrap them around your neck and run your nose into them! (Gobry, 406–407)

Perhaps what is most intriguing here is that Martin Luther was a perceived prophet who illuminated the truths of the Holy Scriptures for the Germans by translating the texts from Greek and Hebrew into the common vernacular.

25.3. The Prophet's Image through the Eyes of His Supporters

Luther was also renowned for his remarkable talent and energy, celebrated by the famous German artist Lucas Cranach the Elder (1472–1553; Fig. 25.7), who created a series of portraits of Luther and his relatives. Of particularly interest are the painting where Luther is depicted in close proximity to Jesus Christ, such as in scenes depicting the resurrection of Lazarus (Fig. 25.8), and the Last Supper (Fig. 25.9), where it almost appears as if the Last Supper took place in Luther's own home. However, following the supper, Christ's agonizing torment is vividly depicted (Fig. 25.10). This prompts the questions: did the artist himself believe in these accounts? Were they his own creation or influenced by others?

Nevertheless, what is important for our discussion is the likelihood that perception of Luther as prophet which underscored the feeling many had at the time. Eventually, the decision was made to commemorate the pivotal events of 1517 by renovating the famous Castle Church in Wittenberg (Fig. 25.11). It was not until 1858, however, that the legendary doors, to which Luther had famously nailed his Theses, were recreated in bronze, thereby immortalizing the text of the Theses themselves (Fig. 25.12). Thus, the image of the prophet was firmly solidified in the city, where he was often referred to as the Wittenberg Pope.

Figure 25.7. Portrait of Lucas Cranach the Elder (Albrecht Durer, 1524).

Figure 25.8. Lucas Cranach the Elder: Martin Luther (the figure is indicated by a white triangle) is present at the resurrection of Lazarus by Christ (1555).

Figure 25.9. Lucas Cranach the Elder: The Last Supper,
Jesus Christ with 12 disciples-apostles. Martin Luther also
participates in this famous meeting, giving a vessel
to one of the Apostles (1547).

Figure 25.10. Lucas Cranach the Elder: the Tortures
of Jesus Christ (1558).

Figure 25.11. Castle Church in Wittenberg.

Figure 25.12. The gates of the Castle Church with 95 Theses
of Luther cast in bronze plates.

25.4. The Prophet and His Colleagues against Catholicism

Luther's esteemed colleagues in the fight against Catholicism were not those who surrounded him directly in Germany and whom he regarded as spiritually subordinate to him, such as Philip Melanchthon, whom many considered to be Luther's closest friend (Fig. 25.13). Rather, by esteemed colleagues, we refer to those who were called leaders of the Reformation—Jean Calvin (Fig. 25.14) and the English King Henry VIII. Luther's relationships with them were notably distinctive.

The English monarch composed and published his work "Protection of the Holy Sacraments," for which he was bestowed the title of "Guardian of the Catholic Faith" by the Pope. Luther's response to this work was an acerbic pamphlet titled "The Book of King Henry VIII of England," in which the argued that the king had dishonored himself by defending the "Roman harlot." Luther referred to Henry as "a madman, pig shit, donkey shit, he dirties the crown of the Heavenly King with his dirt. This swindler and liar, this carnival buffoon, should be pelted with mud and manure. The king, in Luther's opinion, behaves like a loose slut himself. . . . It is immediately clear that there is not a drop of royal blood in his veins" (Gobry, 223).

The reaction of Luther's allies to this pamphlet varied in their scope, but were largely conditional:

> Luther's original sin is an insufficient sense of proportion—the will to immensity, and the demon of "Teutonic rage" that possesses Luther, generated by this will. . . . This evil is all the more dangerous for him because he considered it good. "I speak and write best in anger. . . . To write well, to pray, to preach, I have to get angry. . . . I am deeply unbridled, violent and very belligerent; I was born to fight with countless monsters and devils" . . .

Figure 25.13. Sculptures of Luther and his associate Melanchthon on a bench in the park of Bretten, Melanchthon's hometown.

Figure 25.14. Sculptural images of Luther and Calvin on the pulpit of the Protestant church in the city of Mikolow (South Poland).

His worst enemies could not harm him more than he harms himself with such words as these: "Satan himself has spewed his huge uncleanness—the Pope on blood-soaked Rome." Or these: "Good Christians should wash their hands in the blood of papal henchmen . . . and pull their blasphemous tongues to the back of their heads." This is not human speech, but an animal howl or the cry of a demoniac. . . . "Oh, if Luther could be silent!" Melanchthon almost cries; and Calvin prays: "Oh, if Luther could control himself! May God grant him to calm the raging storms so violently in his heart!" (Merezhkovsky, 117–118)

25.5. The Prophet and the Peasant War

In 1524, the initial sparks of what would become the famous Peasant War in Germany began, reaching its apex in 1525 before subsiding the following year. Many believed that Martin Luther's actions played a significant role in inciting peasant protests against their landlords, in which they demanded freedoms and relief from harsh forms of exploitation. This sentiment was encapsulated in the renowned "Twelve Articles" of 1525 (Figs. 25.15 and 25.16). However, Martin Luther and his associates Philipp Melanchthon and Ulrich Zwingli's responses portrayed a contrasting perspective. Herein lie Luther's true teachings:

> "Make sure that our sovereign does not defile his hands with the blood of these people"—the leaders of the Peasant uprising. Here is one end of the rope, and here is the other: "Strike, if you can, souls, cut, who can!" "I, Martin Luther, exterminated the rebellious peasants; I ordered them to be executed; may their blood fall on me, but I will lift it up to God, because He commanded me to speak And do what I said and did. . . . Until now (before the Peasant Uprising) I thought that it was possible to rule people according to the Gospel. . . . But now (after the uprising) I realized that people despise the gospel; to manage them, you need a state law, a sword and force. . . . Most sovereigns are the greatest madmen or the worst scoundrels in the world, but still they must be obeyed, because they are executioners and jailers in the service of God. . . . The power of sovereigns should be autocratic. . . . The sovereigns are gods, the people are Satan."
>
> In a good war (to defend the motherland), it is the duty of Christian love to destroy enemies mercilessly, rob and burn them according to the laws of war . . . because God himself helps the strong. (Merezhkovsky, 98–99)

The peasants resorted to force, and now it is pointless to call them to peace. All their actions have always been guided by the desire for looting and bloodshed. Their main fault is that they rebelled against the authority established by God; and only secondarily they are guilty of fires, robberies and murders. Everyone has the right to attack them with weapons, as robbers are attacked. Let everyone who has a burning desire to beat them act. They must be torn to pieces, strangled and cut, in open combat and from around the corner, as mad dogs are cut.

Let the authorities remember their duty! Wherever the peasants do not want to obey the authorities, they must be overtaken by its punishing sword! Every prince is a servant of God on earth. The time of mercy has passed. The time has come for war and anger. (Gobry, 300)

It is quite possible that commentators are correct in their assessments of Luther's unexpectedly ferocious reaction to the instigators of the Peasant War. While it is often observed that many rulers are regarded as the worst of scoundrels, every aspect of existence has long been predetermined by the Almighty. Therefore, there can be no room for doubts or deviations, as this is the Absolute.

Everything that is created by God, the one God, and is moved, everything is put into action by His omnipotence. No creature can escape His power, no one can change anything. The will itself depends on God, and when a person wants one or the other, it means that it is pleasing to God. (Gobry, 321)

Luther's ideology presents an infinite contradiction: on the one hand, he suggests that even human desires are predetermined by God, akin to fulfilling divine mandates ingrained from birth—a concept paralleling modern genetics. On the other hand, how should one interpret the countless bloody conflicts in this framework? Are all such conflicts also predetermined by God's will? In this scenario, the desires of warring parties are attributed not to humanity but to the Almighty Himself, a notion that is unsettling in its implications.

Nevertheless, the author refrains from delving into these intricate and seemingly endless debates within the ideological landscape of cultures. Instead, it appears more prudent to acknowledge these contradictions—whose abundance is seemingly boundless, even within the philosophy of a singular thinker like Martin Luther.

25.6. The Prophet on Human Intellect and Advanced Knowledge

We again convey the words of D. S. Merezhkovsky:

"Our great prophet," "our holy Apostle, Martin Luther," his disciples, who are partial to the teacher, will say. But the former dubious friend, the future enemy, the great humanist, Erasmus of Rotterdam, can least of all be suspected of partiality when he is forced to say about Luther: "This new prophet Elijah is Hercules, sent into the world to purify the Augean stables (of the Roman Church)." And then he said: "The world will be saved by reasonable doubt." (Merezhkovsky, 4)

Luther's sharp rebuke follows:

The world will be saved by mad faith. "God's worst enemy is the human mind: it is the main source of all evil. . . . Reason is the devil's greatest harlot. . . . Reason is

Figure 25.15. Peasant War in Germany, leaflets and small publications with the requirements of 12 articles-conditions of the rebels (1525).

Figure 25.16. Peasant war in Germany; on the left—Luther's pamphlet "Against . . ." (1525).

the most opposite of faith . . . the believer must kill it and bury it. . . . Destroy your mind—otherwise you will not be saved." Luther is rampaging and exaggerating here again. However, he himself knows that reason is "as necessary for a person here on earth as lust." But Luther is raging, maybe not for nothing. He seems to have a presentiment that Erasmus, the evangelist of reason, is followed by an infinitely stronger and most dangerous enemy; we now know his name—Kant. (Merezhkovsky, 104)

And here, Luther's assessment of science and, in particular, of the great achievements of Nicolaus Copernicus becomes completely clear to us (Fig. 25.17):

> They are talking about some new astrologer who proves that the Earth is moving, and the sky, Sun and Moon are motionless; as if the same thing happens here as when moving in a cart or on a ship, when it seems to the rider that he is sitting motionless, and the earth and trees are running past him. Well, but now anyone who wants to be known as a smart guy is trying to invent something special. So this fool is going to turn the whole astronomy upside down . . .

> Astrology is framed by the devil, until the end people may be afraid of entering into the state of matrimony and from any divine and human service and vocation; for stargazers portend nothing good on the planets; they offend the conscience of people regarding the misfortunes to come, which all stand in the hand of God, and through such crafty and useless reflections to annoy and torment all my life. . . . The interpreters of the stars cause great harm to God's creations. God created and placed the stars in the firmament so that they could shine on the kingdoms of the earth to the end; to create people. Rejoicing and rejoicing in the Lord, and being good signs of the years and seasons. But stargazers pretend that these creatures created by God darken, disturb the earth, and they are harmful; however, everything created by God is created only for good, although humanity, abusing this, turns them into evil. Overshadowing all this, indeed, are monsters, and they love strange and untimely births. Finally, believing in the stars, or something similar, or being afraid of them is idolatry, and the first commandment was directed against this.

This is how Luther articulated his views during the so-called free table conversations (*Tischreden*)[1] with his interlocutors, who attempted to record his thoughts, which were later published two decades after his death (Kobe 1998).

In 1530, reflecting on his brief period of leniency, Luther expresses regret over his "foolish softness," stating, "Now I am of the opinion that authorities should punish not only rebellion but even mere blasphemous ideas with the death penalty" (Gobry, 354). However, it was the government that ultimately decided how to define and punish blasphemous thoughts, often resulting in executions.

---

[1] Martin Luther's "Table conversations" ("Tischreden") were collected by his student Aurifaber (Aurifaber) in "Tischreden oder Colloquia Doct. Mart. Luthers" (Eisleben, 1566; critical edition by K. E. Forstemann U. H. E. Bindseil, 1845). Currently, all of Luther's "conversations" recorded by his students and contemporaries are combined in an academic ("Weimar") edition of his works: Die Werke Martin Luther. Historisch-kritische Ausgabe. Department of table discussions. Bd. 1–4. Weimar, 1912–1916.

Figure 25.17. Sculpture of Nicolaus Copernicus
in Olsztyn.

25.7. Antisemitism and Christianity: Back to Luther

Merezhkovsky argued:

> Anti-Semitism and Christianity are an eternal question, insoluble in the plane
> where it is almost always solved—nationally or even internationally-politi-
> cally,—solvable only in the religious plane, where it arose. . . . Oh, of course,
> this is not only the Russian question of our days, but also a worldwide, eternal
> one! Let it be distorted, perverse—yet it is deeply reflected in the decisive fate
> of mankind, the struggle of two tribes, two bloodlines, Semitic and Aryan. The
> law of religious polarity—attraction-repulsion, enmity-falling in love—has
> been operating between them forever and worldwide. . . . The struggle of
> Christianity with anti-Christianity has been going on for two thousand years
> and their last, worst fight is now taking place on the body of Russia. The essence
> of "Christianity" is the absolute denial of Judaism. (Merezhkovsky 1928)

Martin Luther upheld a similar position:

> Jews are real liars and bloodhounds who, with their false interpretations from
> the beginning to the present day, not only constantly distort and falsify the entire
> Scripture. . . . The sun has never shone on a more bloodthirsty and vengeful

people than this one, who imagined themselves to be God's people, who were commanded to kill and strike the Gentiles. In fact, the most important thing they expect from their Messiah is that he will slay and kill the whole world with his sword . . .

What should we Christians do with this rejected and cursed people, the Jews? Since they live among us, we dare not tolerate their behavior now that we are aware of their lies, and swearing, and blasphemy. If we tolerate them, we will become accomplices in their lies, vilification and blasphemy . . .

First of all, their synagogues or schools should be burned, and what does not burn should be buried and covered with dirt so that no one can ever see either the stone or the ashes left from them.

Secondly, I advise you to level and destroy their houses. For in them they pursue the same goals as in synagogues. Instead of (houses), they can be settled under a roof or in a barn, like gypsies. This will help them finally understand that they are not masters in our country, as they boast, but that they live in exile and in captivity, as they constantly groan and cry out for us before God.

Thirdly, I advise you to take away from them all prayer books and Talmuds in which they teach such idolatry, lies, cursing and blasphemy.

Fourthly, I advise from now on to forbid their rabbis to teach on pain of death.

Fifthly, I advise that Jews should be deprived of the right to a security certificate when traveling . . .

Sixth, I advise them to forbid usury, and take away all their cash, as well as silver and gold, and postpone all this.

Seventh, I recommend giving young, strong Jews and Jewish women an axe, a hoe, a shovel, a spinning wheel and a spindle, and let them earn bread by the sweat of their faces, as was said to the children of Adam . . .

However, if the authorities are reluctant to use force and do not curb the Jewish diabolical rampage, the latter should, as we said, be expelled from the country. And they should be told to return to their land and to their possessions in Jerusalem, where they can lie, curse, blaspheme, revile, kill, steal, steal, engage in usury, mock and condone these shameful abominations that they are engaged in between us. And leave our government, our country, our life and our property, and most of all leave our Lord the Messiah, our faith and our church undefiled and untainted by their diabolical tyranny and malice. (Merezhkovsky 1928: Gruber, appendix to the article "Theologian of the Holocaust")

The fact that Kristallnacht on November 9, 1938 was on the eve of Luther's birthday was quickly picked up by Hitler's loyalists. In particular, Protestant Bishop Martin Sasse in the preface to the pamphlet "Down with them!," wrote the following about Jews:

"On November 10, Luther's birthday, synagogues are burning in Germany. . . . During these hours, the voice of the German prophet of the XVI century should be heard, who began out of ignorance as a friend of the Jews, but then, driven by his knowledge, experience and prompted by reality, became the greatest anti-Semite of his time, who warned his people against the Jews." (Sasse 1938, WKP)

Indeed, Martin Sasse was correct: as early as 1519, Luther spoke unequivocally against anti-semitism and its abhorrent manifestations. However, in 1543, he published his pamphlet titled "On the Jews and Their Lies" (Fig. 25.18), from which several excerpts are cited.

25.8. Rumors of Luther's Demise

In 1545, a false rumor about Luther's death spread like wildfire. A book was published in Naples, informing this joyful news to all the faithful sons of the Catholic Church. The book was profoundly misguided, yet even individuals of such immense intelligence such as St. Ignatius of Loyola and St. Teresa of Avila judged Luther little better than those who authored it. This

Figure 25.18. Luther's pamphlet "Against the Jews and their Lies" (1553 version on the left, and its later translation into English).

Figure 25.19. Posthumous image of Martin Luther.

illustrates that responsibility for the events that unfolded in Christian humanity following the Second Division of the Churches will not rest solely on Protestants, but also on Catholics.

The Lord performed a terrible miracle for the glory of Christ and for the consolation of all the sons of the Church. The great heretic, Luther, has passed away. Before his death, he received communion and ordered to put his body on the altar, so that he would be revered as the body of a saint. But when, contrary to this order, he was buried in the ground, there was such a terrible noise in it, as if all the forces of hell collided with each other. A particle of the Holy Gifts appeared in the air, received by him unworthily. But just as they put it back in the bowl, that underground noise subsided. And the next night he got up again, with even greater strength. When, finally, the grave was dug, they saw that the body had disappeared without a trace, and such suffocating gray smoke came out of the ground that all those present became ill, and a great many witnesses of this miracle returned to the bosom of our holy mother, the Catholic Church. (Merezhkovsky, 110)

Nevertheless, the Catholics were not mistaken in their exultation. Luther was seriously ill at this point and died on February 18, 1546 (Fig. 25.19). In all cities and villages from Eisleben to Wittenberg, crowds of people joined Luther's funeral procession. The bells in the sky and the people on the ground were crying: "The great prophet of God, Martin Luther, has died!" (Merezhkovsky, 116)

25.9. A Return to the Dream of a Single Universal

The Protestants or Reformers precipitated a significant and deeply unsettling division within Western Catholicism, the most influential doctrinal split in terms of global adherents. Yet, it is equally noteworthy that Martin Luther, who was perhaps the most proactive and vehement opponent of Roman Church distortions during those years, was simultaneously engrossed in fervent aspirations for a Single Universal Christian Church. He asserted that "there cannot be two Churches, Eastern and Western; there is only one Universal Church; or one, or there is none." Clearly, Luther envisioned this one Church to be precisely the one he tried to establish—not the Roman Church with which he initially engaged in his relentless conflicts.

However, the great reformer's dream continues to be a utopia (and let us note in parentheses: did he himself realize that this was a utopia? After all, he was striving to crush the monolith of Catholicism that seemed to be ever present in Europe. However, exactly four centuries after the beginning of the Reformation, in 1937, in Edinburgh, adherents primarily of Protestantism and other diverse Christian traditions from 43 ethnic groups, representing approximately 120 Christian denominations, excluding Catholics and Orthodox, gathered with the same dream of unity. This assembly marked the Second World Conference of Protestants. There, they declared in unison: "We are all one in faith in our Lord Jesus Christ. . . . We humbly acknowledge that our divisions are contrary to the will of Christ, and we pray that the Lord, in His mercy, will lead us all by the Holy Spirit into the fullness of Unity."

Then, in a remarkable moment of agreement, all participants at the meeting joined together in a prayer of thanksgiving on August 18, 1937—a historic occasion marking the first such unified prayer in many centuries!

And these are the words with which the organizer of the Conference concludes it: "Our division is the greatest of the temptations of the world. . . . We should all be ashamed and repent that we have brought the Church to such impotence with our sins. . . . We deeply grieve that our work lacks the great Roman Church, which was better able than all other churches to speak to the peoples so that they could hear it." (The Second Conference 1938; see also Merezhkovsky, 18–19)

However, the Roman Church, in the writings of Pope Pius XI titled *Mortalium animos*, unequivocally denounced the first Conference of a similar nature held in Lausanne a decade earlier as "heresy," arguing, "The Apostolic See can in no way take part in their meetings, and Catholics have no right to support or participate in such gatherings; for if they do so, they will be promoting a false Christianity, entirely foreign to the one Church of Christ."

Eight decades have passed since that thanksgiving prayer in Edinburgh in 1938, but the divisions within Christian doctrine persist, albeit obscured beneath surface fractures. New denominations will undoubtedly continue to emerge, of which we are at present unaware. It appears that dreams of Christian unity, like those of any other ideology, remain in a familiar and elusive eternity, both for them and for us.

25.10. Postscript: Insights on Commentators

"Thousands of books have been written about Luther. And this is not surprising." Thus begins A. P. Levandovsky in his introductory article to Ivan Gobry's book *Luther*, translated from French into Russian. The range of assessments of Luther across numerous scholarly works is indeed remarkable, spanning attitudes from near deification to profound disdain.

Due to this vast array of perspectives, the author of this book chose to primarily reference the works of just two researchers, who hold contrasting views on Luther's significance—Dmitry

Figure 25.20. Dmitry Sergeyevich Merezhkovsky (on the left) and Ivan Gobri.

Sergeyevich Merezhkovsky (1865–1941) and the aforementioned French writer and philosopher Ivan Gobry (1927–2017). These commentators differ markedly from each other (see Fig. 25.20), despite Merezhkovsky having lived and worked in France for over two decades, where he was eventually laid to rest.

Books about Luther were published in France during different eras: Merezhkovsky's in 1941, the year of his death, and Gobry's exactly fifty years later in 1991. Both authors approached Luther's legacy in distinct ways. Gobry, who was a staunch Catholic and professor at a Catholic university, viewed Luther's reformist actions as fundamentally alien. On the other hand, Merezhkovsky's religious beliefs were ambiguous, as noted by N. A. Berdyaev, who described Merezhkovsky as detached from Orthodox Christianity and largely influenced by Rozanov's negative criticism. Despite this, the author of this book found Merezhkovsky's approach to Luther's works and his skilled presentation of extracted materials compelling. Similarly, Gobry's works, despite their differing assessments of Luther, also captivated the author with their stylistic presentation.

Some readers may perceive that the author has allocated an excessive amount of space to Martin Luther in his book and argue that such a deep exploration of his character is unwarranted. Indeed, Luther himself may not inherently merit such attention. However, it is relevant to the overall themes of this book as many continue to consider Luther to be the fourth prophet in the extensive lineage of Abrahamic religions, spanning approximately two and a half millennia. Following Baruch Spinoza's definition, Luther is seen as a mouthpiece of prophecy, akin to the prophets preceding him, rather than an apostle like Christ's disciples who convey divine truths.

From this perspective, it becomes crucial to trace the development of prophetic figures across Abrahamic traditions—from Moses, Jesus Christ, and Muhammad, and finally, to Martin Luther. Luther's figure, amidst these predecessors, exhibits a spectrum of human qualities—from the undeniably sublime to those that may evoke disdain. This complex portrayal brings Luther, as the last Abrahamic prophet, closer to our understanding. It is this nuanced perspective that has motivated the author to delve deeper into the extraordinary personality of Martin Luther.

Questions

26.1. Handcrafted and Divine Artworks: Questions and Reflections

The epilogue in any creative work serves as a farewell to the explored topic, a departure from its mission. Every author grapples, to varying degrees, with the questions: What did I achieve? What fell short? What completely failed? The author of this book views himself as a fairly typical writer, not markedly different from most others, and is similarly preoccupied with such questions. In wishing to share his reflection with his readers, he intends to shift from the author's position to that of the reader and, as a reader, attempt to evaluate the book's content. Should the reader's assessments be less than enthusiastic, the author will strive to explain his shortcomings. However, this is not a necessity.

Thus, we present a dialogue between the reader (R) and the author (A), beginning with a relatively simple question.

R. Was it justified to divide the analysis of materials into two discussions: a first which covered the earthly origins of the Eurasian world, and a second which addressed the endless whirl of assessments by *Homo* cultures of the world around them?

A. Probably yes, and if there was a need to deal with similar problems again, I would certainly repeat this kind of order. The author considers the basis of the first part of the book to focus on those realities that he calls geocreative or not man-made. Part two addresses the realities of *Homo*-creative or man-made culture. Moreover, we note that man-made realities are necessarily and very closely dependent on non-man-made ones. The existence of human formations and societies is inherently tied to the nurturing Earth, as we are inextricably its children. Understanding the authentic history of our planet, rather than a distorted or superficial representation, is therefore of paramount importance.

R. You have divided the whole of Eurasia and the entire Eurasian world along the lines of North-South and West-East. What happened here, and what failed?

A. The explanation provided may not sound particularly impressive, but it is necessary to address the complexity of the topic. The Alpine-Himalayan geosyncline has historically served as a definitive separator of the Eurasian continent along the North-South axis, due to the immense tectonic forces exerted by the colliding continental littoral plates. However, the structural complexity of the northern half of Eurasia—the Eurasian littoral plate—and its division into West-East blocks presents a more intricate challenge.

The Ural mountain system, traditionally considered a vertical separator delineating Europe from Asia, proves inadequate in this role (see Chapter 4). The historical and contemporary alignment of Eurasia into two continents, Europe and Asia, is based on the Ural division, as reflected in the term "Eurasia" itself. Nevertheless, the author, aligning with several leading geomorphologists, disputes the Ural-based continental division, advocating for the view that Eurasia constitutes a single continent unified by a massive littoral plate. This perspective, the author contends, should not be met with significant doubt. He posits that the Ural region is one of the largest enclaves in Northern Eurasia, with the Ural Mountain system dividing it into two sub-enclaves.

R. There is, however, another related issue to consider: Will the people themselves and our official authorities abandon such a division?

A. It is almost certain that the response will be a resounding no—they will not abandon this familiar separation. This division has been the foundation of our mental and spatial constructs for nearly two millennia, so why renounce it? For instance, the age-old and nearly sacred question for Russians has been: Is Russia part of Europe or Asia? This question continues to intrigue many and, consequently, remains unresolved to this day. Such intrigue, in general, can even be somewhat gratifying.

R. And finally, one more of the seemingly most complex points: megaenclaves, enclaves, sub-enclaves . . . Is this really both a complex and an extremely important issue?

A. Yes, indeed. However, we are currently only at the beginning of addressing this issue. The foundation of any enclave is a specific geoarea, which is distinct from its neighbors in essential geomorphological features, such as the richness and availability of ore deposits. Once a certain geoarea is inhabited by a specific ethnic group or groups, it can be classified as an enclave.

In the author's analysis, the Eurasian world—divided at the base of sixteen geoareas (see Chapter 5, Fig. 5.1) into 16 enclaves, which the author believes is an appropriate categorization for the time being. However, with more detailed exploration of these areas, further subdivision is inevitable: enclaves will "demand" fragmentation into sub-enclaves. Several such subdivisions are already outlined in the book, with the Balkan Peninsula and Palestine being among the most significant. These sub-enclaves have given rise to ethnic groups whose cultures and activities have had a profound and impactful influence on the history of the entire Eurasian world and beyond.

R. And another question: the second part of the book is noticeably larger than the first, although part one deals with events stretching over millions of years, and this is not counting the geological depths of billions of years. In the second part, the exposition concerns only the Holocene epoch, and not even the entire Holocene, but its final periods. How to explain such a dissonance?

A. The second part is primarily connected with the third, or ideological, facet of the foundational triangle that serves as the indispensable basis of any *Homo* culture (see Chapter 8). Deciphering the ideology and worldview of ancient societies is arguably the most challenging task for researchers, largely due to the antiquity that archaeology addresses. This field delves into the deepest, mostly unwritten past of humanity by studying a wide variety of monuments.

These monuments, left behind by ethnic groups that are often obscure or long extinct, are not always easily interpreted. In the majority of cases, these groups are unknown to us.

Although numerous, most monuments are voiceless and cannot reveal the secrets of the social or spiritual hierarchy of societies. Many correspond to traces of settlements with indistinct cultural layers and indistinct signs of structures, such as terrestrial dwellings or semi-subterranean houses. Alternatively, they might be burial grounds containing human remains without any distinctive indicators of the deceased's societal status.

However, even in these cases, we can occasionally extract hints to decipher the ideological aspects of cultures. These hints can deepen our understanding of the worldview structures of non-literate ancient cultures. Such structures are fundamentally based on an almost eternal principle of absolute faith in their unshakable truth and sanctity of their world understanding. Changes are primarily reflected in the specific forms and manifestations of these views, not in the principle itself. For instance, a culture must always believe in the truth of its worldview. If this steadfast faith becomes shaky or uncertain, it leads to the disintegration of the culture's most essential components and sometimes even the collapse of the ethnic group that gave birth to it.

This situation applies not only to religious systems. For example, Marxist-Leninist philosophy, while not a religion, served as the official ideology of Soviet society. Any deviation from its core dogmas was considered heresy and demanded severe punishment. The widespread rejection of this doctrine coincided with the catastrophic collapse of the Soviet Empire, after which the doctrine, once seen as the absolute truth, quickly faded. The author has analyzed these topics in more detail in his earlier works (see Chernykh 2013b, 311–327; Chernykh 2017, 562–581).

26.2. Eurasia through the Greek-Jewish Dichotomy

The debate persists, extending to the correlation issues between the ideological systems of the ancient Greeks and Israelites. In the history of Eurasia, particularly within the period spanning close to two to two and a half millennia, the Greek-Jewish dichotomy has played a significant role.

R. Could you explain how could these micro-ethnic groups inhabiting truly tiny tracts of land on our planet play a special role?

A. Indeed, the genesis of two disparate ideological realms stemmed from proximate micro-ethnoses (Fig. 26.1). The phenomenon where significant ideological innovations emerged from the milieu of relatively small human populations is not unprecedented. Large human collectives often prove less conducive to groundbreaking discoveries. As the doctrines of Hellenic and Eretz Israel peoples came into being, they entered into explicit, covert, and intense competition. Initially acting as self-appointed adjudicators, they were soon joined by neighboring groups, and subsequently by numerous Eurasian ethnicities. In the Modern Era, the resolution of conflicts has been governed by global arbitration, which can either be voluntary or coercive. Coercive arbitration involves the imposition of beliefs through the threat of severe consequences, thereby clouding the subordinate's full understanding of the supposed "truth."

R. What did the dissimilarity of these worlds affect?

A. Perhaps in this scenario, transitioning from scientific discourse to fiction would offer greater clarity. Consider this: before the onset of significant Hellenization in Eretz Israel, a Greek individual arrives unexpectedly. Fluent in Hebrew, he is driven by a profound curiosity to comprehend the essence of the Land of Israel—and here I invite the reader to plunge into a certain fantasy:

> Most likely, the Greek visitor was immediately struck by the external squalor of local life. Here he did not notice any squares surrounded by beautiful buildings with colonnades, similar to Hellenic temples. No one draws anything, because it is forbidden. There are no sculptors here, because their work is strictly prohibited. But the most important thing is that they believe in a certain singular deity who lives in heaven, but where exactly is unknown. And no one saw him. The name of God is pronounced extremely, extremely rarely; and it sounds like Yahweh, but it is the deepest mystery and it is a great sin to pronounce it. He is the creator of absolutely everything and decides absolutely everything for everyone. But how to understand this? And how to present his appearance, since portraying him in a drawing is recognized as grave blasphemy, which should be punished in the most cruel way. It is said that the daily existence of Jews is subject to many hundreds of regulations—prohibitions and permits for all occasions, large, small, even quite insignificant. Is this possible? And they read only one certain sacred book, called the Torah. Actually, they do not read, only their spiritual pastors read to them. They immediately explain to people how all this should be understood. They have not heard about our sciences and do not want to hear about them. They say, what sciences! We have a Torah, and you are talking about some sciences. Well, tell me, how do you want us Greeks to perceive these nightmares?

Now imagine a picture completely opposite: a Jew ended up in Athens, and there he somehow mastered the Greek language:

Figure 26.1. Enclaves of the Balkans and Palestine, as well as the area of Hellas in Eurasia.

It's impossible! How do these Greeks live? They have built pompous stone buildings with columns, and consider them temples dedicated, as they think, to some gods. And they have no account of these gods—is this possible? They not only endlessly draw these gods in different poses, but also carve their figures out of marble or sculpt them out of clay—and how can this also be explained? After all, these are the gods! But it's impossible to even imagine. And these countless gods live on some not-very-distant mountain called Olympus, and this mountain rises in their own country. The gods quarrel among themselves, easily communicate with ordinary people, and some male deities even get mistresses among the common people and produce children from them. And how can a Jew survive all such terrible pictures if he does not renounce his faith? And what they don't read: some fairy tales, songs, even jokes about the gods. What is this? And they haven't even heard of our sacred Torah. But if they had read it, then immediately, I think, they would have changed, corrected themselves.

And yet: they say that they have a lot of people whom they call scientists— and some even call the fathers of geography, history, astronomy and some other sciences unknown to us. They say that these scientists have established that our earth is supposedly round and all sorts of celestial bodies rotate around it, and most importantly, even the sun turns. This is absolutely terrible! After all, everything is said about this in the Torah, and all this is not true at all. Let them at least read about the six days from the Torah. What a poor people!

R. But this is your fantasy!

A. Yes, it is a fantasy, of course, albeit one that reflects reality. Despite presenting an extremely polar contrast in nearly every aspect, these two worlds were inevitably compelled to converge over centuries. They endeavored, perhaps with efforts deemed unimaginable, to bridge the gap and seek mutual understanding.

26.3. The Chosen People and Their Opponents

Adherents of the Christian and Jewish worlds also tried to somehow understand one another with disastrous results. Followers of Judaism argue as follows:

> If the church severed all ties with Jewish origins and completely rejected the Old Testament and the "Jewish God," . . . then Christianity would be, although hostile, but essentially a separate religion in relation to Judaism. However, the Church has persistently emphasized that Christianity serves as a direct continuation of that Divine action in history in which God's election of Israel played a primary role. If the Jews had disappeared from the historical arena, Christianity could, without feuding with them, treat them only as a preparatory stage of the "Kingdom of God." But the Jews continued to exist, they insisted on their exclusive right to the Bible and its interpretation and believed that its Christian interpretation was

heresy, falsification and idolatry. So there was an atmosphere of mutual hostility and denial, in which the relationship between Christianity and Judaism became more contradictory, complex and tragic than any other relationship of this kind in history. (Electronic Jewish Encyclopedia: "Christianity")

A. I believe this assessment is accurate. Initially, Jews considered Christianity to be heresy and idolatry. In response, their Christian counterparts waged a two-millennia-long war against Judaism. This reminds us of Baruch Spinoza, who was profoundly struck by the intense and enduring hatred exhibited daily among people who worship the same God, as previously noted.

R. What I find particularly interesting here is that these two worlds developed near each other. While they are not immediate neighbors, their occupied spaces are closely intertwined. How, then, can we explain the profound polar dissonance in their worldviews? What could have led to the emergence of such striking differences?

A. Your question is very appropriate, although it will be difficult to provide a satisfactory answer. Indeed, Hellas and Eretz Israel were located not far from each other. However, their enclaves were completely different from one another. Hellas is essentially a block of related societies. Perhaps it was less the continental territories that were the basis of their possession and more likely that the Aegean Sea, with its myriad islands spanning from the Balkan Peninsula to the western coast of Anatolia, played a crucial role. Here, over two thousand islands dotted the sea, ranging from larger ones like Crete, Cyprus, and Euboea to tiny ones barely discernible on maps. Amidst these islands and islets, life in the Aegean was characterized by constant movement, starkly contrasting with the environment in Eretz Israel.

Eretz Israel was intricately linked to the sub-enclave of Palestine, centered around the tectonic depression known as El Gor or the Jordan Rift Valley according to geomorphologists. Iconic geographical features such as the Dead Sea, Sea of Galilee (Lake Tiberias), and the Jordan River Valley define this sub-enclave. Geomorphological and geoecological characteristics distinctly set the Palestinian sub-enclave apart from neighboring regions: Arabia, Mesopotamia, and the broader Iran-Anatolia enclave, of which it is a part. A notable aspect of Eretz Israel's geoecology is its geographical isolation from the sea, surrounded predominantly by desert areas.

R. But how to understand isolation from the sea distances? After all, almost the entire southern half of the eastern coast of the Mediterranean Sea is subordinated to the current Israel?

A. You're correct; this perplexing paradox is challenging to comprehend, yet it remains a reality. At this point, I defer to A. I. Tyumenev (2003, 21–23), who extensively explored the nuances of Israeli attitudes towards maritime activities and the practicalities of sea voyages.

Navigation remained completely alien to the Israelites, and not only for the entire time of their independent historical existence, but equally in all later times up to the present. We see confirmation of this fact for the biblical era both in the helplessness of Solomon, who tried to establish external maritime relations and was forced to turn to the assistance of the Phoenicians for the construction of ships and for their subsequent use, and in the complete failure that befell a similar attempt to equip a naval expedition by one of his successors, an attempt that ended in the death of the ships equipped by him almost immediately after their exit from the harbor.

As for the favorable position at sea, this benefit was destroyed by the fact that really convenient harbors on this inaccessible seashore were in the hands of the Phoenicians. The best proof of this is the efforts that Solomon and some of his successors used to acquire the port city of Elath on the Red Sea, although this city was separated by a difficult and dangerous road through the desert . . .

Most likely, A. I. Tyumenev is correct in assessing the centuries-old failures of the Israelites in mastering the maritime business. However, why were there failures? It appears that the Jews did not particularly endeavor to excel in this realm. Interestingly, western Semites known as the Phoenicians resided along the entire eastern stretch of the Mediterranean coast, sharing linguistic similarities with the Jews. The Phoenicians were renowned for their exceptional seafaring skills. Notably, the name "Palestine" can be traced back to the Philistines, highlighting their historical presence in the region. This maritime way of life starkly contrasted with the inhabitants of Eretz Israel, whose migrations across Eurasia predominantly occurred via overland routes.

R. I want, however, to return our conversation to the issues of worldview. Is it true that the revealed polar dissonance between the Hellenic and Israelite worlds should be explained through the prism of cardinal differences in the sub-enclave geoecology? Is it perhaps possible, for example, to believe that the inhabitants of desert or semi-desert places *necessarily* had to explain the worlds through monotheism, or that the population would understand the areopagus of the world's rulers like the Olympic one?

A. Certainly not. We won't ever find any direct dependencies here. For example, the nomadic and semi-nomadic ideologies prevalent in the vast Steppe Belt of Eurasia did not align with any monotheistic dogmas. Similarly, the ideological canons of the numerous islands in the Torres Strait or the adjacent islands of the Malay Archipelago from the north bore little resemblance to Hellenic ideologies.

R. Let's shift the emphasis of our discussion to the situation of victorious explosions of Abrahamic monotheism, which has now engulfed most adherents of various doctrines of beliefs on Earth. How did it happen that a people living in relative isolation, who toiled in what was far from a paradise in one corner of the Earth, managed to persuade a vast number of peoples to embrace the truths they steadfastly professed for nearly three millennia?

A. If we trace the trajectory of influence, we will find that it was not the Jewish people themselves but their followers—Christians and Muslims—who were convinced. Initially, these followers lived near the Jews, yet their monotheistic beliefs differed notably. Christianity, for instance, developed the trinitarian doctrine—God the Father, God the Son (Jesus Christ), and the Holy Spirit—where each entity held equivalent sacred authority. This trinitarian concept subsequently led to the emergence of significant dissent among Protestants who contested what they saw as a departure from pure monotheism.

On the other hand, Islam adhered more closely to a strict monotheistic formula, where Allah corresponded directly to the lineage of Yahweh.

R. Still, Eretz Israel was at the heart of such a shift.

A. Indeed. The Jews began to suspect long ago that the one and only omnipotent God was much more attractive than a cluster of gods, deities, and other obscure characters. They concluded with the principle of monotheism, and this ensured the success of their doctrine.

Additionally, however, the most vital point of this argument is that a firm belief in a brief earthly life helped strengthen the idea of an afterlife. Earthly life, then, was viewed as a kind of exam for the afterlife. If one were to fail this test, they were guaranteed an eternity of suffering in hell, and vice-versa with heavenly pleasure. This dogma exhibited its extraordinary power, affecting the life and behavior of a very considerable number of individuals and cultures.

R. The question of why Judaism, as the pioneer of monotheism, did not become the primary engine and distributor of monotheistic ideas despite its foundational role is indeed perplexing. Despite Judaism's foundational contribution, its share in the vast array of monotheistic doctrines appears minuscule, accounting for only 0.2% (see Chapter 16; Fig. 26.1). How can this disparity be explained?

A. This, incidentally, is another paradox that appears frequently in various narratives of world history. Concerning the Jews, it echoes a longstanding and reiterated assertion: that they are a chosen people of God. Simultaneously, this belief is often accompanied by straightforward references to passages such as Deuteronomy 7:12–16:

> And if you will listen to the laws, to keep and fulfill them, then the LORD your God will keep a covenant and mercy to you, as He swore to your fathers, and will love you, and bless you, and multiply you, and bless the fruit of your womb and the fruit of your land, and your bread, and your wine, and your oil that is born of your cattle and of your flock of sheep, in the land that He swore to your fathers to give you; blessed will you be above all nations; there will be neither barren nor barren, neither with you nor in your cattle; and the LORD will remove from you all infirmity, and he will not bring any of the terrible diseases of Egypt that you know on you, but he will bring them on all those who hate you; and you will destroy all the nations that the LORD your God gives you: let not your eye spare them; and do not serve their gods, for this is a snare for you.

Reality, however, often did not align with these promises. The concept of being the chosen people stands as the most significant hallmark of Jewish national-religious consciousness. According to their belief, this status stems from God's special love for them, contingent largely upon the adherence to divine Jewish covenants and moral conduct. Conversely, disobedience could lead to just punishments, a fate Eretz Israel has experienced frequently. Nevertheless, the prophets among the people steadfastly believed in ultimate redemption, which would bring salvation not just to them but also to the entire world. This divine selection, they asserted, is based on the genetic inheritance of all the twelve tribes of Israel, excluding others with different genetic backgrounds from this privilege.

However, this steadfast belief in their chosen status has perhaps contributed animosity towards them from others. Why so? If God is one for the entire world, why do only Jews proclaim their superiority? Is this justified? Here, a significant contributing factor, identified by psychologists as narcissism or self-absorption, has played a crucial role.

This factor, combined with an unwavering belief in their moral righteousness, is essential for the peaceful coexistence of any society. The well-known psychological dichotomy of "us versus them," as discussed in Chapter 15, reinforces the notion that "we" are always right, while "they" are inherently inferior compared to us. Jews, as the pioneers of monotheism were convinced of this, and similar sentiments often prevailed among their followers, though directed towards their own ethnic groups rather than Jews:

> If the church severed all ties with Jewish origins and completely rejected the Old Testament and the "Jewish God," . . . then Christianity would be, although hostile, but essentially a separate religion in relation to Judaism. However, the Church has persistently emphasized that Christianity serves as a direct continuation of that Divine action in history in which God's election of Israel played a primary role. If the Jews had disappeared from the historical arena, Christianity could, without feuding with them, treat them only as a preparatory stage of the "Kingdom of God." But the Jews continued to exist, they insisted on their exclusive right to the Bible and its interpretation and believed that its Christian interpretation was heresy, falsification and idolatry. So there was an atmosphere of mutual hostility and denial, in which the relationship between Christianity and Judaism became more contradictory, complex and tragic than any other relationship of this kind in history. (Electronic Jewish Encyclopedia: "Christianity")

R. Therefore, only individuals tracing their origins to the tribes of Israel were considered eligible for inclusion in Judaism as the God-chosen people. Others were not permitted to join. This largely accounts for the comparatively smaller number of adherents of Judaism within the Abrahamic trio.

26.4. The Sacred Commandments and the 613 Jewish *Mitzvot*

A. Yes, it appears that this is primarily the reason. However, another significant factor contributing to this reduced number of adherents relates to Jews who do not observe the comprehensive set of 613 commandments in Judaism.

According to Jewish norms, every follower of the faith is obligated to adhere to 613 commandments, known in Hebrew as mitzvot. Among these, 248 are categorized as positive commandments (corresponding to the number of organs in the human body, though the method of counting them is a matter of interest), and 365 are negative commandments (corresponding to the number of days in a year).

It is impractical to list all 613 commandments here, so we will focus on highlighting a few that are both significant and, conversely, may seem commonplace or trivial, yet still merit attention. So, the commandments are prescriptive (their original numbering is preserved):

1. To know that there is a God

2. To know that He is one

3. To love Him

4. To fear Him

5. To serve Him

6. To strive to communicate with the sages of the Torah

7. To swear oaths in His name

8. To be like the Almighty in His actions

9. To sanctify Him name

10. To read the "Shema" every day in the evening and in the morning

11. To teach the Torah to others

156. To eliminate chametz from our possessions on the day of the Fourteenth Nisan

157. To tell about the exodus from Egypt on the night of the Fifteenth Nisan

158. To eat matzah on the night of the Fifteenth of Nisan

159. Rest from work on the first day of Passover

160. Rest on the seventh day of Passover

209. Respect the sages and stand up to them

210. Respect one's parents

211. Tremble before one's parents

212. Be fruitful and multiply, and strive for procreation

214. The husband must stay together with his wife during the first year of married life

215. Circumcision

216. Perform yibbum (levirate marriage)

218. The rapist is obliged to marry the girl he raped . . .

Now the commandments which are prohibitive:

1. Not to allow the thought that there are other gods besides the Almighty

2. Not to create statues for worship

3. Not to make idols even for others

4. Not to make sculptural images of a person

5. Not to worship idols

6. Not to serve idols in any way

7. Not to pass on their children to serve Molech

8. Not to perform the ritual of evoking the souls of the dead

9. Not to perform the ritual of divination

10. Not to turn to idolatry and not to study it

30. Not to follow the ways of other peoples

31. Not to cast spells

32. Not to be interested in astrology

33. Not to believe in omens

34. Not to practice magic

35. Do not cast spells

44. Do not shave with a blade

45. Do not scratch your body

46. Do not settle in Egypt

47. Do not allow ideas that contradict the worldview of the Torah

218. Do not use different types of animals together for work

219. Do not prevent working cattle from eating foot food

220. Do not cultivate the land in the seventh year

221. Do not take care of fruit trees in the seventh year

222. Do not harvest in the seventh year

301. Do not spread gossip and slander

302. Do not hate each other

303. Do not shame each other

304. Do not take revenge on each other

305. Do not show vindictiveness

307. Do not shave off the hair on an ulcer

308. Do not tear off the signs of leprosy from the skin

310. Do not leave the sorcerer alive

311. Do not tear the young husband away from home in the first year after the wedding . . .

Out of curiosity, I attempted to discover if any adherent of Judaism could recall all 613 commandments they are meant to observe. Unfortunately, I did not find anyone who could. This led me to ponder whether this extensive list of prohibitive and prescriptive commandments might discourage individuals who aspire to embrace Judaism but are deterred by the sheer breadth of these obligations. It appears unlikely, in my opinion, that anyone could definitively answer this question.

26.5. Ancient Wisdom Regains Power on Earth, a Millennium Later

A. But now, please, ask me the final question, as our readers are already tired of endless disputes. It might be better to explore a topic related to antiquity, as our discussion of the Greek-Jewish dichotomy seems somewhat one-sided towards Israel.

R. Indeed, you are correct. Did antiquity lose out to Judaism in every aspect?

A. Certainly not. Antiquity, in a sense, lost its heavenly standing, relinquishing its revered Olympic pantheon that once inspired deep sympathy and admiration. Perhaps the most significant blow came with the official prohibition by Roman Empire rulers in the late fourth century, which ended the worship of revered pairs such as Zeus-Jupiter, Aphrodite-Venus, Athena-Minerva, Poseidon-Neptune, and others that had endured in the collective consciousness until then. However, on earthly matters, the ancient sages emerged as undisputed victors. The scientific knowledge cultivated by Greco-Roman communities served as an indispensable foundation for modern science. Virtually all European scientists developed their intellectual frameworks on this basis, which later spread gloablly. This conclusion is widely accepted, as it is challenging to envision the development of modern society without such a profound ancient legacy. Indeed, Greek and ancient roots resonate in the names of nearly all scientific disciplines. Let's briefly list them here in a way that won't complicate our discussion:

Geology, Geography, Geomorphology, Geodesy...
Archaeology, History, Biology, Genetics, Physics...
Astronomy, Astrology, Astronautics, Dialectics...
Dichotomy, Diaspora, Propaedeutics, Pedagogy, Speleology...
Psychology, Cybernetics, Synthesis, Space, Parallel, Arctic/Antarctic...
It is quite possible that there are more such roots in medicine:
Bacteria, Bacillus Vaccine, Virus, Gonorrhea, Immunity...
Leukocytes, Lymph, Mycology, Mutation, Pathogenicity...
Toxins, Enzymes, Venereal diseases, Leprosarium...
Aesculapius/Asclepius...

It is impossible not to recall another phenomenon: the global spread of the Latin alphabet. Of course, this was especially affected after the internet essentially "enveloped" our entire planet. Here again we recall that the alphabet is basically a Greek word. Additionally, a system of Cyrillic writing was also built on the basis of Greek, developed by the Byzantine Greek pair Cyril and Methodius in the fourth century specifically for adherents of the Orthodox faith. These men were later canonized as saints as enlighteners of the Slavs on the same level of the Apostles.

Ancient roots are found in completely different words:
Poetry, Rhyme, Poet/Poetess, Hypnosis...
Angel, Heracles/Hercules, Sisyphus—Sisyphean works...
Achilles' heel...
Even in our names:
Peter, Pavel, Nikolai, Elena, Evgeny/Evgenia, Arkady...
Anna, Alexander/Alexandra, Vasily, Nikanor...

In fact, there are many more words with roots tracing back to Greek and ancient origins, although the author does not intend to list them all here. The unmistakable triumph of sciences rooted in ancient knowledge became most evident during the Renaissance or in modern times. However, achieving this "purification" took at least a millennium, overcoming stringent prohibitions rooted in the Old Testament and associated sacred texts. It is noteworthy that science has since been enriched by numerous talented individuals from Jewish backgrounds, which not long ago posed one of the most formidable barriers to the dissemination of views conflicting with Torah dogma. I would also note that the Russian philosopher G. N. Volkov extensively addresses this issue in his book "At the Cradle of Science."

The way to understanding the future of science lies through understanding its present and past. . . . Science [now] enters into all the pores of the life of our society, has an impact on all its spheres. And therefore, it is natural that interest in ancient philosophy as the source of modern science is now reviving with renewed vigor. And here's what else is curious. In the last century, interest in antiquity was almost entirely the lot of humanitarians. In our time, natural science itself has become seriously interested in ancient thought, primarily in its leading fields—in physics and mathematics. (Volkov, 6–7)

Indeed, our discussion has briefly touched upon a diverse array of sciences, not exclusively focused on foundational innovations. Let's also consider architecture. The enduring legacy of ancient Greek models, including temples with colonnades and other iconic architectural features (Fig. 17.2), is ubiquitous today, found in nearly every corner of the globe. At times, even in regions originally opposed to the ancient world like Palestine, these once deemed unacceptable architectural patterns have become irresistible.

Comprehending the Present and the Future

Each time, in reaching the end of the writing process of books devoted to the crowning insights of humanity, whether about the discovery of new lands, or the laws of natural development, or the development of fundamentally new canons of interpretation of the world surrounding *Homo* cultures, the author was seized with palpable sadness. Throughout the centuries and millennia, amidst the joys and innovations of civilization, violent and bloody conflicts have perpetually plagued human history. These conflicts often centered around the conquest of new lands and coveted wealth, and these lands frequently became places of exile for the unwanted—heretics, dissenters, and those who opposed the victors' imposed truths and light. The conquerors ardently believed that they alone possessed and upheld the ultimate truth, viewing the lives of the conquered as darkness and squalor before their arrival. The conquered, however, harbored deep resentment and hatred towards their oppressors, yearning for vengeance. This cycle of conquest, oppression, and resistance has perpetuated suffering, death, and bloodshed throughout history. Indeed, no understanding of history seems complete without acknowledging this reality.

The author feels inclined to view certain aspects of human history not through the lens of sorrow and sadness, but with humor and a wry smile. Perhaps adopting this perspective, depicted in the illustration at the beginning of Epilogue 2 (Fig. 27.1), can offer a more engaging and intriguing perspective. Let's attempt to explore this.

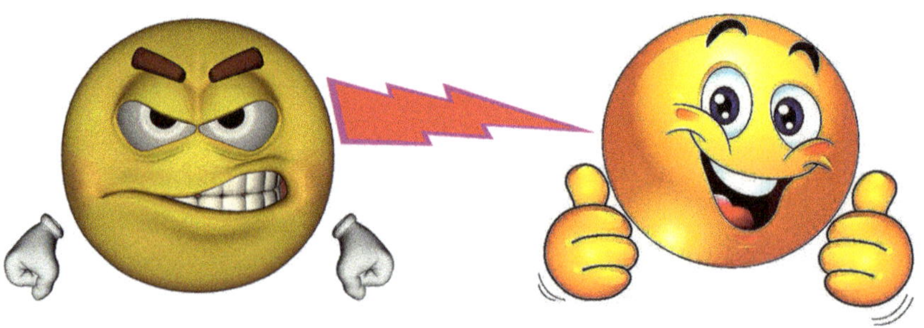

Figure 27.1 From bitterness and anger to humor and a mocking smile.

27.1. Inquiry into the Origins of Existence: The Quest of Glupov's Inhabitants

At least two and a half millennia have passed since the creation of the Torah, wherein the creation and history of the world was revealed to the people. There is a strange echo in the famous satirical book by nineteenth-century Russian Mikhail Yevgrafovich Saltykov-Shchedrin (Fig. 27.2) titled *The History of a City*, published in 1870. The publisher, in "rummaging in the Stupovo city archive, accidentally attacked a rather voluminous bunch of notebooks bearing the general name of the "Stupovo Chronicler." The "chronicle" primarily consisted of biographical accounts focusing on the mayors who governed the city of Stupovo for nearly a century. Despite its limited scope, these narratives offered glimpses into the city's character and illustrated how various historical changes influenced its trajectory. For instance, mayors during Biron's tenure were identified for their daring approach, those under Potemkin for their administrative efficiency, and Razumovsky's mayor is characterized by his mysterious back-story and knightly valor. Each governer's leadership styles were varied: while some ruled with strict discipline, others justified their methods by citing civilizational imperatives; still others encouraged resilience among the populace.

Before Surly-Burcheev took office, Sadilov believed "that [the city] would only be safe when the stupid people would all go to the vigil and when the inspector-observer of all the Stupid schools would be appointed [the famous fool] Paramosha. This last condition was especially important, and the poor people presented it very persistently." However, "moral corruption had reached a point where fools attempted to penetrate the mysteries of cosmology and openly applauded the calligraphy teacher Linkin, who, departing from his specialty, preached from the pulpit that the world could not have been created in six days. The poor meticulously calculated that confirming this opinion would precipitate the collapse of the entire narrow-minded world-view, so interconnected that disturbing one part risked unraveling the whole" (390).

In response, the impoverished residents took matters into their own hands. They forcibly entered Linkin's apartment, discovered a book titled *Means for the Extermination of Fleas, Bedbugs, and Other Insects,* and jubilantly paraded Linkin to the mayor's yard. Initially taken aback, Sadilov examined the book and explained it contained nothing against religion, morality, or public peace. However, the beggars disregarded his explanation. This incident solidified Sadilov's resolve; he later that day appointed Paramosha as inspector of the Stupovo schools and established a philosophy department at the county school for another fool, Yashenka.

In a brief period, the city's character underwent a radical transformation. Riotous revelry was replaced by solemn silence punctuated only by the tolling of bells, signifying a stark departure from its previous atmosphere. Soon after, the distinctive era of Mayor Ugryum-Burcheev commenced in Glupov. It began abruptly with Surly-Burcheev's intense gaze from the main exit, an impactful moment captured in the evocative description: "Surly-Burcheev stood at the very main exit and stared into the crowd with a numbing gaze. . . . But what a look it was . . . Oh, my God! what a look it was!" (Fig. 27.3).

Below follows an abstract to the publication in the Electronic Library RoyalLib.com:

Figure 27.2. M. E. Saltykov-Shchedrin
(by N. A. Yaroshenko, 1886).

Figure 27. 3. Ugryum-Burcheev
(by N. Remizov, 1907).

> This book, as you know, is passed in high school (it is "passed"), and completely
> in vain. You need to read it as an adult, otherwise a lot of it will seem uninterest-
> ing or incomprehensible. Read it, you won't regret it. This is our story, written by
> a person who is not indifferent to Russia. Contemporaries called it a "lampoon
> on the history of the Russian state," and in some ways they were right—it was
> written evil and ruthlessly. In fact, this book is relevant to this day.

The author completely agrees with the quotation provided here. He himself, however, long
ago, read this book in school, when there was a merciless struggle with the legacy of tsarism. He
didn't understand much then. But something about the narrative, evidently, remained—otherwise he would not have reread it and quoted it extensively in this book.

However, it probably makes sense to get closer to our days, to the thinkers of our time.

27.2. A Dialogue Between an Orthodox Archpriest and Charles Darwin at His Gravesite

Another century and a half had passed since the publication of a book about the history
of the city Glupov. Along the way, it became apparent that Charles Darwin had also deeply
unsettled adherents of Christianity, akin to the effect of the calligraphy teacher Linkin on the
inhabitants of Glupov. Recently, a peculiar and ancient discussion, revived on the internet,
appeared to continue unabated (Fig. 27.4), spanning millennia in its implications.

Figure 27. 4. Orthodox Archpriest Artemy Vladimirov and Charles Darwin.

Archpriest and confessor of the Alekseevsky Stavropol Convent in Moscow Artemy Vladimirov talked with the scientist Charles [Darwin], who lived in the nineteenth century. The priest told about this on August 5, 2019 on the air of the TV channel "Spas."

According to him, the dialogue took place at the grave of the scientist in Westminster Abbey in the UK. "I am not a fan of the monkey theory, so it was interesting for me as a priest to talk. 'Charles, how do you like it there?' I turned to the tombstone. 'What do you say now, do you think there are any intermediate links between species—between a zebra and a giraffe?' the archpriest quoted the conversation.

At the same time, as Vladimirov noted, he did not have to wait long for an answer from the scientist—the tombstone burst into a penitential speech.

"Father, do not be tempted by my theory, it was not from a great mind that I put forward this hypothesis, which I now repent of. You have nothing to do with a bear or a pig personally," the priest quoted the words allegedly uttered by Darwin.

The archpriest concluded his heartfelt story by expressing regret for the Anglo-Saxons and Danes: "Sorry for the Anglo-Saxons, sorry for the Danes. Hooray, we are Russians, what a delight! And let's bear witness to the triumph and reason of Orthodoxy throughout the universe," he concluded and suggested that the Spas TV channel broadcast its programs in English and Chinese."

Charles Darwin's book *On the Origin of Species by Means of Natural Selection, or the Preservation of Favored Races in the Struggle for Life* appeared out of print in its first edition in 1859. The first broad public discussion took place seven months after its release, when fierce battles immediately broke out over the basics of its content.

And for 160 years, up to the present day, these, to the surprise of the author, do not reduce the intensity of the violent disputes between supporters of the divine origin of *Homo* and their opponents. And for 160 years now, persistent adherents of the divine version of the creation of mankind have not been able to convince or defeat their stupid haters in any way. But everything turned out to be surprisingly simple! So why in 160 years no one has guessed to approach the grave of this perverter of the supreme truth Charles Darwin and directly ask his ashes. So here you are! Therefore, a warm applause to our Archpriest Artemy Vladimirov!

In his most recent book, *Homo Cultures*, the author has included a short section entitled "Science and Faith: Which is Stronger?" The section ended with the answer: "Of course faith: after all, science exists and lives in constant doubts and checks, but faith never!" (Chernykh 2019, 365–366). Well, Christian archpriests have once again underscored their stance: disbelief in sacred truths is inconceivable to them. It's been two and a half millennia, from the era when dissidents were burned for their lack of faith to today's demands for punishment over perceived insults to believers' sentiments.

Not everyone may agree with the author's take on the interplay between science and faith, and many might criticize his viewpoint. Yet, it wasn't the author who made light of it, but rather Archpriest Artemiy Vladimirov at Darwin's grave, albeit in a serious context. Can such occurrences go unnoticed in these times?

What remains exceptionally noteworthy is that this ongoing debate has spanned about two and a half millennia, tracing back to the latter half of the first millennium BCE. It was during this epoch that the foundational dogmas of future dominant Abrahamic religions were formulated, alongside ancient philosophers contemplating the structure of the world around us. These ancient musings continue to be referenced by adherents of atheism or agnosticism today.

27.3. Are We the Owners of the Earth?

In one of the songs that was almost a mandatory source of joy, though now decades old, we, the youth, sang, keeping pace and repeating the words of the poet Vasily Lebedev-Kumach:

> Step forward, Komsomol tribe,
> Joke and sing so that smiles bloom.
> We conquer space and time,
> We are the young masters of the earth.

In Russia, this song was brought to life by posters displayed everywhere—on buildings and fences—featuring diverse artistic talents. These posters depicted those who had conquered both space and time, symbolizing the young and joyous inheritors of the Earth (Fig. 27.5), who ardently believed in its vision.

Figure 27.5. Poster from Soviet times.

Figure 27.6. Tyrannosaurs, huge predatory animals that died in the Cretaceous period as a result of an explosion from a giant asteroid crashing into the Earth.

However, there is a significant caveat: what if the Earth starts to shake violently, causing the tectonic plates to shift dramatically? Or, more terrifyingly, what if, due to some unknown cosmic laws, the oxygen in our atmosphere suddenly disappears—are we capable of predicting such an event? Cosmic forces remain largely unpredictable to us. For instance, around 66 million years ago, during the late Cretaceous period, a giant asteroid approximately 15 kilometers in diameter, struck the Earth on the Yucatan Peninsula. This impact created the Chicxulub crater, measuring about 150–180 kilometers in diameter and up to twenty kilometers deep (Bralower et al. 1998). The energy released was estimated at 5×10^{23} joules, equivalent to 100 teratons of TNT. In comparison, the largest thermonuclear device ever detonated had a yield of about 0.00005 teratons, which is 2 million times less. It is believed that around 15 trillion tons of ash and soot were ejected into the atmosphere, plunging the Earth into darkness for one to two years (Bardeen 2017). This darkness significantly hindered photosynthesis and potentially reduced atmospheric oxygen levels during that period. Temperatures plummeted by 28 °C on land and 11 °C in the oceans. The collapse of phytoplankton, a crucial component of the oceanic food chain, led to the extinction of zooplankton and other marine animals. Among terrestrial animals, the tyrannosaurs—gigantic and awe-inspiring creatures—became extinct (Fig. 27.6).

This raises a necessary question: what will become of us, the so-called owners of the Earth, in such scenarios? How will we manage to survive? As the species *Homo*, we appear as a mere, almost microscopic appendage to our planet (see, for example, Fig. 1.2). Nothing more.

Nevertheless, *Homo sapiens* possess immense power, with nuclear weapons capable of destroying much of life on Earth, including ourselves. Therefore, it might not be wise to continue composing and singing triumphant songs about us as the masters of the earth. Perhaps it is better to stay grounded, even though it may not seem as joyous and inspiring as the call to new heroic deeds once was. Yet, can we truly do without this aspiration? At the very least, this remains highly doubtful.

27.4. COVID-19

Homo, of course, is a powerful species, and we possess nuclear weapons that can be unleashed to cause unimaginable devastation. Suddenly, however, in the first few months of 2020, the insidious coronavirus COVID-19 made waves throughout our planet. All the earthly civilizations of *Homo* trembled—this unexpected and invisible enemy managed to interject and disrupt our every aspect of our lives, from health to the economy, even our mentality. The world has undergone significant changes since then, and even outwardly almost all *Homo*s have adopted a defensive appearance: muzzled masks, gloves, a decrease in physical affections, separation from one another by one and a half meters. Giant megacities even began to resemble outwardly deserted and abandoned settlements.

But COVID-19, while a formidable biological adversary to humanity, was far from the most powerful. Perhaps it is better not to provoke fate, lest something cosmic and incomparably more powerful befall us. It might be wiser to refrain from entertaining ourselves with a cheerful song about the young masters of the Earth.

Yet, can we truly do without it? Can we?

Selected Bibliography

Adkins, Lesley, and Roy Adkins. *Ancient Greece: An Encyclopedia.* Moscow: Veche, 2008.

Aeschylus. *Prometheus Bound.* Translated by S. Apt. In *Ancient Drama.* Moscow: Artistic Literature, 1970.

Amirhanov, Kh. A. *Study of Oldowan Monuments in the North-Eastern Caucasus: Preliminary Results.* Moscow: Taus, 2007.

Ammianus Marcellinus. *Roman History.* Translated by Yu. A. Kulakovsky and A. I. Sonni. St. Petersburg: Aletheia, 1994.

Apollodorus. *The Library of Mythology, Books 1–3.* Edited by A. G. Borukhovich. Leningrad: Nauka, 1972.

Apollonius of Rhodes. *Argonautica.* Translated by N. A. Chistyakova. Moscow: Ladomir, 2001.

Aridan, N., and G. Sheffer, eds. *Israel and the Diaspora: New Perspectives.* Bloomington: Indiana University Press, 2005.

Arrian. *The Campaigns of Alexander.* Translated by M. E. Sergeyenko. Introduction by O. O. Kruger. Moscow-Leningrad: Academy of Sciences Publishing House, 1962.

Bar-Ilan, M. "Prester John: Fiction and History." *History of European Ideas* 20, no. 1 (1995): 291–298.

Bardeen, Charles G., Robert R. Garcia, O. B. Toon, and A. J. Conley. "On Transient Climate Change at the Cretaceous-Paleogene Boundary Due to Atmospheric Soot Injections." *Proceedings of the National Academy of Sciences* 114, no. 36 (2017): 6635–6644.

Bird, Peter. "An Updated Digital Model of Plate Boundaries." *Geochemistry, Geophysics, Geosystems* 4, no. 3 (2003): 1–52.

Bodnarsky, M. S., ed. *Ancient Geography: A Reader.* Moscow: Geografgiz, 1953.

Bongard-Levin, G. M., and E. A. Grantovsky. *From Scythia to India: Ancient Aryans, Myths, and History.* 3rd ed. St. Petersburg: Aletheia, 2001.

Bouvier, A., and W. Meenakshi. "The Age of the Solar System Redefined by the Oldest Pb-Pb Age of a Meteoritic Inclusion." *Nature Geoscience* 3, no. 9 (2010): 637–641.

Bralower, T. J., C. K. Pauli, Chap. Hill, and R. M. Leckie. "The Cretaceous-Tertiary Boundary Cocktail: Chicxulub Impact Triggers Margin Collapse and Extensive Sediment Gravity Flows." *Geology* 26, no. 4 (1998): 331–334.

Bromley, Y. V. *Essays on Ethnos Theory.* Moscow: Nauka, 1983.

Bronstein, V. A. *Claudius Ptolemy.* Moscow: Nauka, 1988.

Brown, L. A. *The History of Cartography.* Moscow: Litres, 2013.

——— *The Story of Maps.* New York: Dover, 1980.

Carpini, Plano. *History of the Mongols, Known as the Tatars.* In *Travels to Eastern Countries by Plano Carpini and Rubruck,* edited by N. P. Shastina, 23–83. Moscow: Geografizdat, 1957.

Carter, R. W. G. *Coastal Environments: An Introduction to the Physical, Ecological and Cultural Systems of Coastline.* London: Academic Press, 1988.

Carter, R. W. G., and C. D. Woodroffe, eds. *Coastal Evolution: Late Quaternary Shoreline Morphodynamics.* Cambridge: Cambridge University Press, 1994.

Chambers, O. *Biblical Psychology.* Translated by V. N. Gavrilova. Moscow: Wisshon, 2011.

Chekin, L. S. *Cartography of Christian Medieval Period (8th–13th Centuries): Texts, Translations, Commentary.* Moscow: Oriental Literature, 1999.

Chernykh, E. N. "Homo Cultures: Key Episodes in Million-Year History. Holocene: A Firework of Cultures and Their Paradoxes." *Nature* 5 (2018): 43–56.

——— *Metal, Man, Time.* Moscow: Nauka, 1972.

——— "On the European Zone of the Circum-Pontic Metallurgical Province." *Acta Archaeologica Carpathica* 17. Krakow (1977): 29–33.

——— *Mining and Metallurgy in Ancient Bulgaria.* Sofia: Bulgarian Academy of Sciences, 1978.

——— *Ancient Metallurgy in the USSR: The Early Metal Age.* Cambridge: Cambridge University Press, 1992.

———, ed. *Kargaly.* Vols. 1–5. Moscow: YASK, 2002–2007.

——— "Radiocarbon Chronology in Light of Systematic Analysis of Large Series of Dates (Expected Results and Paradoxical Results)." In *Interdisciplinary Integration in Archaeology,* edited by E. N. Chernykh and T. N. Mishina, 30–51. Berlin: MIA RAS, 2004.

——— *The Eurasian Steppe Belt: The Phenomenon of Nomadic Cultures.* Moscow: Manuscript Monuments of Ancient Rus, 2009.

——— *Nomadic Cultures in the Megastructure of the Eurasian World.* Vol. 1. Moscow: YASK, 2013.

——— *Nomadic Cultures in the Megastructure of the Eurasian World.* Vol. 2. Moscow: YASK, 2013.

——— *Nomadic Cultures in the Mega-Structure of the Eurasian World.* Translated by I. Savinetskaya and P. N. Hommel. Brighton: Academic Studies Press; Moscow: LRC Publishing House, 2017.

——— *Homo Cultures: Key Aspects of Million-Year History. Problematic Essays.* Moscow: YASK, 2019.

——— *The Megastructure of the Eurasian World: Through the Lens of Geology, Archaeology, and History.* Moscow: Taus, 2020.

Chernykh, E. N., and S. V. Kuzminykh. *Ancient Metallurgy of Northern Eurasia (The Seimin-Turbin Phenomenon).* Moscow: Nauka, 1989.

Chernykh, E. N., and L. B. Orlovskaya. "Databases of Radiocarbon Dates and Corrections to the Relative Chronology of the Early Metal Age." *Analytical Studies of the Laboratory of Natural Science Methods* 1 (2009): 26–40.

Chernykh, E. N., and L. B. Orlovskaya. "Radiocarbon Chronology of the Cultures of the Eurasian Steppe Belt and Its Surprises." In *Multidisciplinary Methods in Archaeology: Recent Results and Perspectives,* 370–381. Novosibirsk: Institute of Archaeology and Ethnography, Siberian Branch of the Russian Academy of Sciences, 2017.

Chernykh, E. N., L. I. Avilova, and L. B. Orlovskaya. *Metallurgical Provinces and Radiocarbon Chronology.* Moscow: Institute of Archaeology, Russian Academy of Sciences, 2000.

Christian, D. *The Great History: How Everything Began and What Comes Next.* Translated by A. D. Gromova. Moscow: Kolibri, 2019.

——— *Origin Story: A Big History of Everything.* New York: Little, Brown and Company, 2018.

Chumakov, A. N. "On the Subject and Boundaries of Global Studies." *The Age of Globalization* 1 (2008): 7–16.

Clement of Alexandria. *Stromata.* Translated by E. V. Afanasiev. St. Petersburg: Oleg Abyshko, 2003.

Dao De Jing. Translated by Yan Hsin-Shun. In *Ancient Chinese Philosophy: A Collection of Texts,* vol. 1, 114–138. Moscow: Mysl, 1972.

Darwin, Charles. *On the Origin of Species by Means of Natural Selection.* In *Darwin: Works,* Vol. 3. Moscow: Academy of Sciences of the USSR, 1939.

Denisov, A. O. "The Rife Mountains in Western European Medieval Cartography." In *T. N. Jackson, I. G. Konovalova, A. V. Podosinov, and A. A. Frolov, Northern Eurasia in Ancient and Medieval Cartography,* 98–155. Moscow: Aquilon, 2017.

Deniz, R. *Amerigo Vespucci, Martin Waldseemüller—The Secret Deal.* Baku: Palmarium Academic Publishing, 2020.

Derevyanko, A. P. *New Archaeological Discoveries in Altai and the Problem of Homo Sapiens Formation.* Novosibirsk: Institute of Archaeology and Ethnography, Siberian Branch of the Russian Academy of Sciences, 2012.

——— *The Earliest Human Migrations in Eurasia in the Early Paleolithic.* Novosibirsk: Institute of Archaeology and Ethnography, Siberian Branch of the Russian Academy of Sciences, 2009.

——— *Three Global Human Migrations in Eurasia, vol. 2: The Initial Settlement of Northern, Central, and Middle Asia.* Novosibirsk: Institute of Archaeology and Ethnography, Siberian Branch of the Russian Academy of Sciences, 2017.

Detlova, E. V., and S. V. Kuzminykh. "Hero von Mergart and His 'Memoirs of Soviet Russia.'" In *G. von. Mergart, Memoirs of Soviet Russia: Letters,* 4–9. Krasnoyarsk: Krasnoyarsk Regional Local History Museum, 2019.

Diogenes Laertius. *On the Lives, Teachings, and Sayings of Famous Philosophers.* Translated by M. L. Gasparov. Moscow: Mysl, 1986.

Dobrovolsky, G. V., L. O. Karpachevsky, and E. A. Kriksunov. *Geosphere and Pedosphere.* Moscow: GEOS, 2010.

Eliade, Mircea. *Histoire des Croyances et des Idées Religieuses. Vol. I: De l'Âge de la Pierre aux Mystères d'Eleusis*. Paris: Payot, 1976.

——— *History of Belief and Religious Ideas*. Vol. 1, *From the Stone Age to the Eleusinian Mysteries*. Translated by N. N. Kulakova, V. R. Rokityansky, and Y. N. Stefanov. Moscow: Criterion, 2002.

Ferguson, Niall. *Empire: How Britain Made the Modern World*. Moscow: AST-Corpus, 2013.

Fernandez-Armesto, Felipe. *Amerigo: The Man Who Gave His Name to America*. Moscow: A. Yelkov, 2019.

Fiedel, Stuart J. "Initial Human Colonization of the Americas: An Overview of the Issues and the Evidence." *Radiocarbon* 44, no. 2 (2002): 407–436.

Freeman, David. "Introduction." In *Man and His Symbols* by Carl G. Jung. Moscow: Silver Threads; St. Petersburg: AST, 1997.

Freud, Sigmund. "Zur Einführung des Narzißmus." In *Jahrbuch der Psychoanalyse*, vol. 6., edited by S. Freud, 1–24. Leipzig; Wien, 1914.

——— *Introduction to Psychoanalysis: Lectures*. Moscow: Nauka, 1989.

Fromm, Erich. *The Greatness and Limitations of Freud's Theory*. Moscow: AST, 2000.

——— *The Heart of Man: Its Genius for Good and Evil*. Moscow: Respublika, 1992.

Gamboa Hinestrosa, Pablo. *El Tesoro de los Quimbayas: Historia, Identidad y Patrimonio*. Bogotá: Planeta Colombiana, 2002.

Gerasimov, I. P., and A. A. Velichko, eds. *Paleogeography of Europe for the Last Hundred Thousand Years*. Moscow: Nauka, 1982.

Gobri, I. *Luther*. Moscow: Young Guard, 2000.

Godlevsky, S. F. *E. Renan: His Life and Scientific-Literary Activity*. St. Petersburg: I. G. Gershun's Printing House, 1895.

Gorelov, N. S. "The Founder of Fantastic Indology." In *Letters from the Imaginary Realm, 185–189*. St. Petersburg: Azbuka-Classics, 2004.

Gruber, D. *Theologian of the Holocaust: Shamash—Messianic Theological Messenger*. Rovno: Sphinx, 2004.

Grusse, R. *Genghis Khan: Conqueror of the Universe*. Translated by E. A. Sokolov. Moscow: Young Guard, 2007.

Hain, V. E. "On the Basic Principles of Constructing a Truly Global Model of Earth Dynamics." *Geology and Geophysics* 51 (2010): 753–760.

——— "The Alpine Folded Geosynclinal Area." In *Mountain Encyclopedia*. Vol. 1. Moscow: Soviet Encyclopedia, 1984.

——— "The East African Rift System." In *Mountain Encyclopedia*. Vol. 1. Moscow: Soviet Encyclopedia, 1984.

——— "The Ural-Mongolian Folded Belt." In *Mountain Encyclopedia*. Vol. 5. Moscow: Soviet Encyclopedia, 1991.

——— *Tectonics of Continents and Oceans*. Moscow: Nauchny Mir, 2001.

Harmand, S., Jason E. Lewis, J. E. Feibel, C. S. Lepre, S. Prat, A. Lenoble, X. Boes, R. L. Quinn, M. Brenet, A. Arroyo, N. Taylor, S. Clement, G. Dover, J.-Ph. Brugal, L. Leakey, R. A.

Mortlock, J. D. Wright, S. Lokorodi, Ch. Kirwa, D. V. Kent, and H. Roche. "3.3-Million-Year-Old Stone Tools from Lomekwi 3, West Turkana, Kenya." *Nature* 521 (2015): 310–315.

Hazen, Robert. *The Story of Earth: The First 4.5 Billion Years, from Stardust to Living Planet.* 4th ed. Moscow: Alpina Non-Fiction, 2018.

Hellman, X. *Great Confrontations in Science: Ten Most Exciting Disputes.* Moscow: I. D. Williams, 2007.

Herberstein, Sigismund von. *Notes on Muscovy.* Moscow: Moscow State University, 1988.

Herodotus. *History in Nine Books.* Translated and annotated by G. A. Stratanovsky. Leningrad: Nauka, 1972.

Hesiod. *Complete Collection of Texts: Theogony, Works and Days, The Shield of Heracles, Fragments.* Moscow: Labyrinth, 2001.

Hodgson, L., ed. The Second World Conference on Faith and Order (Edinburgh, August 3–18, 1937). New York: The Macmillan Company, 1938.

Hogarth, D. G. *The Nearer East.* New York: D. Appleton and Company, 1902.

Homer. *The Iliad, The Odyssey.* Moscow: Artistic Literature, 1967.

Hublin, Jean-Jacques, A. Ben-Ncer, S. E. Bailey, S. E. Freidline, S. Neubauer, M. M. Skinner, I. Bergmann, A. de Cabec, S. Benazzi, K. Harvati, and Ph. Gunz. "New Fossils from Jebel Irhoud, Morocco and the Pan-African Origin of Homo Sapiens." *Nature* 546 (2017): 310–315.

Ibn Hisham. *The Life of the Prophet Muhammad, As Told by al-Bakkai, from Ibn Ishaq al-Muttalib (Early 8th Century).* Translated by N. A. Gainullin. Moscow: Umma, 2007.

Ibn Ishaq—Ibn Hisham. *The Life of the Prophet: The Great Battle of Badr.* Translated by A. B. Kudelin and D. V. Frolov. Moscow: Institute of Europe, Russian Academy of Sciences; Russian Souvenir, 2009.

Ivanov, A. N. "Vasily Nikitich Tatishchev." In *People of Russian Science: Essays on Outstanding Figures in Natural Science and Engineering, Geology, Geography,* 306–316. Moscow: Fizmatlit, 1962.

Jackson, T. N., I. G. Konovalova, A. V. Podosinov, and A. A. Frolov. *Northern Eurasia in Ancient and Medieval Cartography.* Moscow: Aquilon, 2017.

Jordanes. *On the Origin and Deeds of the Getae (Getica).* Moscow: Aletheia, 2001.

Jung, Carl G. *Man and His Symbols.* Moscow: Silver Threads; St. Petersburg: AST, 1997.

Kahn, Charles. *Anaximander and the Origins of Greek Cosmology.* New York: Columbia University Press, 1960.

Kandel, Eric R. *In Search of Memory: The Emergence of a New Science of Mind.* New York: W. W. Norton and Company, 2006.

Kaplin, P. A., and A. O. Selivanov. *Changes in Sea Levels in Russia and the Development of Coasts: Past, Present, Future.* Moscow: GEOS, 1999.

Karamzin, N. M. *History of the Russian State.* Vol. 8, edited by A. N. Sakharov. Moscow: Nauka, 1993.

Kazakh-Russian Relations in the 16th–18th Centuries: A Collection of Documents and Materials. Alma-Ata: Academy of Sciences of the Kazakh SSR, 1961.

Kefeli, I. F., and A. P. Mozelov. "The Idea of Globalism and the Idea of Communism: For and Against." *Scientific and Practical Journal of the Northwestern Academy of Public Service* 2 (2006): 74–80.

Kerzhentsev (Tarasov), B. Yu. *Serfdom Russia: A History of Popular Slavery.* Moscow: Veche, 2020.

——— *The Accursed Time: Russia in the 17th–18th Centuries.* Moscow: Veche, 2018.

Klyashtorny, S. G. *Monuments of Ancient Turkic Writing and the Ethno-Cultural History of Central Asia.* Moscow: Nauka, 2006.

Klyashtorny, S. G., and D. G. Savinov. *Steppe Empires of Ancient Eurasia.* St. Petersburg: Faculty of Philology, St. Petersburg State University, 2005.

Kobe, David H. "Copernicus and Martin Luther." *American Journal of Physics* 66, no. 3 (1998): 190–196.

Komissarov, S. A., and M. Yu. Ulyanov. "Linzzi." In *Great Russian Encyclopedia*, vol. 17, 529–530. Moscow, 2010.

Korochkova, O. N., V. I. Stefanov, and I. A. Spiridonov. *Sanctuary of the First Metallurgists of the Middle Urals.* Yekaterinburg: Ural University, 2020.

Kovalevskaya, V. B. *The Caucasus—Scythians, Sarmatians, Alans of the 1st Millennium BC to the 1st Millennium AD.* Pushchino: ONTI PNTS RAS, 2005.

Kudelin, A. B. *Historiographical and Literary Aspects of the "Life of the Prophet" by Ibn Ishaq—Ibn Hisham.* Moscow: Institute of Oriental Studies, Russian Academy of Sciences, 2017.

Kustine, A. de. *Russia in 1839.* Vol. I. Translated by V. A. Milchina and I. K. Staff. Moscow: Sabashnikov Publishers, 1996.

Las Casas, B. *History of the Indies.* Moscow: Nauka, 2007.

Lebedeva, E. Yu. "Archaeobotany and the Study of Bronze Age Agriculture in Eastern Europe." In *OPUS: Interdisciplinary Research in Archaeology*, no. 4, edited by M. V. Dobrovolskaya, 50–68. Moscow: Institute of Archaeology, Russian Academy of Sciences, 2005.

Lenel-Lavastin, A. *Forgotten Fascism: Ionesco, Eliade, Cioran.* Moscow: Progress-Tradition, 2007.

Lindsey, B. *Globalization: Repeating the Past. The Uncertain Future of Global Capitalism.* Moscow: Irisen, 2006.

Linschoten, Jan H. *His Discours of Voyages into ye Easte and West Indies.* London: John Wolfe, 1598.

Losev, A. F. *Diogenes Laertius—Historian of Ancient Philosophy.* Moscow: Nauka, 1981.

Losleben, Elisabeth. *The Bedouin of the Middle East.* Minneapolis: Lerner Publications, 2003.

Lovell, J. *The Great Wall of China.* Moscow: AST, 2008.

Lu, Liancheng. "The Eastern Zhou and the Growth of Regionalism." In *The Formation of Chinese Civilization: An Archaeological Perspective*, edited by Susan Allan, 200–247. New Haven: Yale University and New World Press, 2005.

Lubsan Danzan. *Altan Tobchi (Golden Chronicle)*. Translated by N. P. Shastina. Moscow: Nauka, 1973.

Lurie, S. Ya. *Democritus: Texts, Translations, Studies*. Leningrad: Nauka, 1970.

Magidovich, I. P. *Introduction*. In *The Book of Marco Polo*. Translated by I. P. Minaev; edited by I. P. Magidovich. Moscow: Geografizdat, 1956.

McNeill, William H. "The Rise of the West after Twenty-Five Years." *Journal of World History* 1, no. 1 (1990): 1–21.

Men, A. *History of Religion: In Search of the Way, Truth, and Life*. Vol. 1: Origins of Religion. Moscow: Slovo, 1991.

Merezhkovsky, D. S. "Which of You?" *Judaism and Christianity. New Ship* 4 (1928): 14–21.

——— *Reformers: Luther, Calvin, Pascal*. Edited by Temiry Pakhmus. Tomsk: Vodoley, 1999.

Merhart, G. von. *Memoirs of Soviet Russia: Letters*. Edited by E. V. Detlova and S. V. Kuzminykh. Krasnoyarsk: Krasnoyarsk Regional Local History Museum, 2019.

——— "The Palaeolithic Period in Siberia: Contribution to the Prehistory of the Yenisei Region." *American Anthropologist* 25, no. 1 (1923): 23–55.

——— *Bronzezeit am Jenissei: Ein Beitrag zur Urgeschichte Sibiriens*. Wien: Abtob Schrol and Co., 1926.

——— *Daljoko: Bilder aus Sibirischen Arbeitstagen*. Edited by Hans Parzinger. Wien: Böhlau-Verlag, 2008.

Miller, K. *Mappae Mundi: Die Ältesten Weltkarten*. Heft 6: Rekonstruierte Karten. Stuttgart: Roth, 1898.

Monin, A. S., and Yu. A. Shishkov. *History of Climate*. Leningrad: Gidrometeoizdat, 1979.

Müller, A. *History of Islam: From Pre-Islamic Arab History to the Fall of the Abbasid Dynasty*. Moscow: Astrel, ACT, 2004.

Näcke, Paul. "Die Sexuellen Perversitäten in der Irrenanstalt." *Wiener Klinische Rundschau* 27–30 (1899): 445–454.

Napol'skikh, V. V. "Twice Forgotten (D. G. Messerschmidt—The First Researcher of Udmurt Language and Culture)." *Art (Lad)* 4 (1998): 46–56.

Nesje, Atle, S. O. Dahl, and J. Bakke. "Were Abrupt Lateglacial and Early-Holocene Climatic Changes in Northwest Europe Linked to Freshwater Outbursts to the North Atlantic and Arctic Oceans?" *Holocene* 14, no. 2 (2004): 299–310.

Newton, R. R. *The Crime of Claudius Ptolemy*. Baltimore: Johns Hopkins University Press, 1977.

——— *The Crime of Claudius Ptolemy*. Translated by N. B. Malycheva. Moscow: Nauka, 1985.

Obruchev, S. V. "Creative Collaboration: Correspondence of V. A. Obruchev with E. Suess." *Priroda* 10 (1963): 51–53.

Obruchev, V. A., and M. I. Zotina. *Eduard Suess*. Moscow: Journal and Newspaper Union, 1937.

Parzinger, Hans. "Archäologisches in Daljoko." In *Gero von Merhart: Ein Deutscher Archäologe in Sibirien 1914–1921: Deutsch-Russisches Symposium 4.–7. Juni 2009, Marburg*, edited by Andreas Müller-Karpe, Christiane Dobiat, Stefan Hansen, and Hans Parzinger, 49–61. Marburg, 2010.

Patton, Harry, Andrew Hubbard, Knut Andreassen, Alexandre Auriac, Peter L. Whitehouse, Anders P. Stroeven, Chris Shackleton, Morten Winsborrow, Jörgen Heyman, and Andrew M. Hall. "Deglaciation of the Eurasian Ice Sheet Complex." *Quaternary Science Reviews* 169 (2017): 148–172.

Pearson, C., S. Leavitt, B. Kromer, S. Solanki, and I. Usoskin. "Dendrochronology and Radiocarbon Dating." *Radiocarbon* 64, no. 3 (2022): 569–588.

Perea, Antonio, A. V. Casanova, and A. G. Usillos, eds. *El Tesoro Quimbaya*. Madrid: CSIC; MECD, 2016.

Pitulko, V. V. "The Earliest Evidence of Human Settlement in the Arctic." *Russian Polar Research* 1, no. 23 (2016): 17–20.

Pitulko, V. V., A. N. Tikhonov, E. Y. Pavlova, P. A. Nikolskiy, K. E. Kuper, and R. N. Polozov. "Early Human Presence in the Arctic: Evidence from 45,000-Year-Old Mammoth Remains." *Science* 351, no. 6270 (2016): 260–263.

Plutarch. *Parallel Lives*. Vol. 1. Translated by A. H. Clough. Moscow: Nauka, 1961.

Podosinov, A. V. "The Hyperboreans in the Romance of Hecataeus of Abdera (Interpretative Issues)." *Vestnik Drevnei Istorii* 1 (2013): 116–129.

——— "The Riphean Mountains in Ancient Tradition." In *Northern Eurasia in Ancient and Medieval Cartography*, edited by T. N. Jackson, I. G. Konovalova, A. V. Podosinov, and A. A. Frolov, 35–97. Moscow: Aquilon, 2017.

Polo, Marco. *Il Milione*. Edited by Luigi Ferdinando Benedetto. Firenze: L. S. Olschki, 1928.

Polushin, A. "Orthodoxy—The Ship of Salvation." *Orthodox Newspaper* 46 (655), December 5, 2011. Ekaterinburg.

Porshnev, B. F. *Social Psychology and History*. 2nd ed., revised and expanded. Moscow: Nauka, 1979.

Procopius of Caesarea. *The Gothic Wars*. Translated by S. P. Kondratiev. Moscow: Nauka, 1950.

Prokofyeva, N. I., and L. M. Alekhina, eds. *Notes of Russian Travelers of the 16th–17th Centuries*. Moscow: Soviet Russia, 1988.

Ptolemaeus, Claudius. *Cosmographia*. Translated by J. Angelus; edited by Ph. Beroaldus et al. Bologna: Dominicus de Lapis, 1462.

Ptolemy, Claudius. *Geographia*. Ulm: Lienhart Holle, 1482.

Ratzinger, Joseph. *Pope Benedict XVI: Jesus of Nazareth*. St. Petersburg: Azbuka-Klassika, 2009.

Renan, Ernest. *Histoire Générale et Système Comparé des Langues Sémitiques*. Paris: Imprimerie Impériale, 1855.

Rerikh, Y. N. *The Beast Style of the Nomads of Northern Tibet*. Moscow: International Center of the Roerichs, 1992.

Revunenko, E. V., and A. M. Reshetov. "Sergey Mikhailovich Shirokogorov." *Ethnographic Review* 3 (2003): 100–119.

Rosenberg, Arthur. *Der Mythus des 20. Jahrhunderts: Eine Wertung der Seelisch-Geistigen Gestaltenkämpfe unserer Zeit*. Hoheneichen; München, 1930.

Rovinsky, D. A. *Russian Folk Pictures*. Vol. I: Atlas: 1–135. St. Petersburg: Expedition for the Preparation of Government Papers, 1881.

Rubruck, William of. *Journey to Eastern Lands*. In *Travels to Eastern Lands by Plano Carpini and Rubruck*, edited by N. P. Shastina, 87–194. Moscow: Geografizdat, 1957.

Saltykov-Shchedrin, M. E. *The History of a Town*. In *The Collected Works of M. E. Saltykov-Shchedrin*, vol. 8, 265–433. Moscow: Artistic Literature, 1969.

Schaff, Philip. *History of the Christian Church*. Vol. 7. New York: Charles Scribner's Sons, 1910.

Selivanov, A. O. *Sea Level Changes in the Pleistocene-Holocene and Coastal Evolution*. Moscow: Institute of Water Problems, RAS, 1996.

Semikhatov, M. A. "Riphean." In *Mountain Encyclopedia*, vol. 4, 343–344. Moscow: Soviet Encyclopedia, 1989.

Shandong Provincial Institute of Cultural Relics and Archaeology. *Linzi Qi Tombs. Vol. 1*. Beijing: Shandong Provincial Institute of Cultural Relics and Archaeology, 2007.

Shastina, N. P., ed. *Travels to Eastern Lands by Plano Carpini and Rubruck*. Moscow: Geografizdat, 1957.

Shirokogorov, S. M. *Ethnical Unit and Milieu: A Summary of the Ethnos*. Shanghai: Edward Evans and Sons, Ltd., 1924.

——— *Ethnos: Study of the Fundamental Principles of Changes in Ethnic and Ethnographic Phenomena. Reports of the Oriental Faculty of the State Far Eastern University*, 67. Shanghai: Shanghai Commercial Press, 1923.

——— *The Place of Ethnography Among the Sciences and the Classification of Ethnic Groups: An Introduction to the Course on Ethnography of the Far East, Given in 1921–1922 at the Far Eastern State University*. Vladivostok: Svobodnaya Rossiya, 1922.

Sidikhmenov, V. Ya. *China: Pages of the Past*. Moscow: Rusich, 2000.

Slovtsov, P. A. *Historical Review of Siberia*. Vols. 1 and 2. 2nd ed. St. Petersburg: I. N. Skorokhodov Printing House, 1886.

——— *History of Siberia: From Ermak to Catherine II*. Vol. 1. Moscow: Veche, 2006.

Sobolevsky, S. I. *The Greek Language of Biblical Texts: Koine*. Moscow: Moscow Podvorie of the Holy Trinity Sergius Lavra, 2013.

Sozomen, Hermias. *Ecclesiastical History*. St. Petersburg: Fisher Printing House, 1851.

Spengler, R. N. "Agriculture in the Central Asian Bronze Age." *Journal of World Prehistory* 28, no. 3 (2015): 215–253.

Spinoza, B. *Theological-Political Treatise*. Kharkov: Folio, 2000.

——— *Theological-Political Treatise*. Vol. 2 of *Selected Works*. Moscow: State Political Publishing House, 1957.

Stöllner, Thomas, and Lasha Cambashidze. "Gold in Georgia II: The Oldest Gold Mine in the World." In *Anatolian Metal V*, edited by Deutsches Bergbau-Museum, 187–200. Bochum: Deutsches Bergbau-Museum, 2011.

Strabo. *Geography in 17 Books*. Translated by G. A. Stratanovsky. Moscow: Nauka, 1964.

Suess, Eduard. *Die Entstehung der Alpen*. Wien: Wilhelm Braumüller, 1875.

Suleimenov, B. S., and V. Ya. Basin. *Kazakhstan within the Russian Empire in the 18th and Early 20th Century*. Alma-Ata: Nauka, 1981.

Svet, Ya. *Columbus.* Moscow: Molodaya Gvardiya, 1973.

Tacitus, Cornelius Publius. *Works: Vol. 1: Annals; Minor Works.* Translated by A. S. Bobovich. Moscow: Ladomir, 1993.

Tatishchev, V. N. "Rus or, as It Is Now Called, Russia." In *V. N. Tatishchev, Selected Works on the Geography of Russia,* 107–138. Moscow: Geografgiz, 1950.

Theophanes the Confessor. *Chronography.* In *The Chronicle of the Byzantine Theophanes from Diocletian to the Reigns of Michael and His Son Theophylact,* translated by V. I. Obolensky and F. A. Ternovsky. Moscow: Imperial Russian Academy of Sciences, 1887.

Tishkov, V. A. *Requiem for an Ethnos: Studies in Socio-Cultural Anthropology.* Moscow: Nauka, 2003.

Trifonov, V. G., T. P. Ivanova, and D. M. Bachmanov. "Evolution of the Central Part of the Alpine-Himalayan Belt in the Late Cenozoic." *Geology and Geophysics* 53, no. 3 (2012): 289–304.

Tsintsov, Z. "The Long-Hidden 'Noble' Secrets of Alluvial Sediments: River Gold in Present-Day Bulgaria." In *Gold and Bronze: Metals, Technologies, and Interregional Contacts in the Eastern Balkans During the Bronze Age,* edited by St. Aleksandrov et al., 59–69. Sofia: National Archaeological Institute with Museum, 2018.

Tyumenov, A. M. *Jews in Antiquity and the Middle Ages.* Moscow: Kraft, 2003.

U, Weihua. "Archaeological Observations on Funerary Customs of the Qi State of the Eastern Zhou Dynasty." *Cultural Relics and Archaeology/Tube Journal* 3 (2018): 92–95.

Vasilyev, L. S. *History of Eastern Religions.* Moscow: University, 2000.

Veller, M. *Heretic.* Moscow: AST, 2019.

Vernadsky, V. I. *Scientific Thought as a Planetary Phenomenon.* Moscow: Nauka, 1991.

——— *The Biosphere and the Noosphere.* Moscow: Nauka, 1989.

Vinokurov, E. Y. *Theory of Enclaves.* Kaliningrad: Terra Baltica, 2007.

Vits, B. B. *Democritus.* Moscow: Mysl, 1979.

Volkov, G. *By the Cradle of Science.* Moscow: Young Guard, 1971.

Vozgrin, V. E., E. N. Bespyatykh, and E. A. Savelieva, eds. *Notes by Captain Philipp Johann von Strahlenberg on the History and Geography of the Russian Empire of Peter the Great: Northern and Eastern Europe and Asia.* Vol. 1. Moscow and Leningrad: Institute of History, USSR Academy of Sciences, 1985.

Walker, M., S. Johnsen, S. O. Rasmussen, T. Popp, J.-P. Steffensen, P. Gibbard, W. Hoek, J. Lowe, J. Andrews, S. Björck, L. Cwynar, K. Hughen, P. Kershaw, B. Kromer, T. Litt, D. J. Lowe, T. Nakagawa, R. Newnham, and J. Schwander. "Formal Definition and Dating of the GSSP (Global Stratotype Section and Point) for the Base of the Holocene Using the Greenland NGRIP Ice Core, and Selected Auxiliary Records." *Journal of Quaternary Science* 24, no. 1 (2009): 3–17.

Wegener, A. *The Origin of Continents and Oceans.* Leningrad: Nauka, 1984.

——— *Die Entstehung der Kontinente und Ozeane.* 4th ed. Braunschweig: F. Vieweg and Sohn, 1929.

Werner, E. *Myths and Legends of China.* Translated by S. Fedorov. Moscow: Tsentrpolygraph, 2007.

Wood, Frances. *Did Marco Polo Go to China?* London: Secker and Warburg, 1995.

Yakupov, R. I. *Introduction to General Ethnology: A Brief Course of Lectures.* Ufa: Publishing House of Bashkir State Pedagogical University, 2016.

Yelnitsky, L. A. *Ancient Knowledge of Northern Lands.* Moscow: Geografgiz, 1961.

Yukhimenko, E. M. *Pokrovsky Cathedral (St. Basil's Cathedral) on Red Square.* Moscow: Interbook-Business, 2011.

Zakhvatkin, A. Z. "D. S. Merezhkovsky and N. A. Berdyaev: Reflections on the Church." *Herald of the Russian Christian Humanitarian Academy* 13, no. 2 (2012): 124–131.

Zelenkov, A. I. "The Glory and Misery of Globalization." *Philosophy and Social Sciences* 3/4 (2013): 4–12.

Zweig, Stefan. *Amerigo: Geschichte eines historischen Irrtums.* Stockholm: Insel Verlag, 1944.